A THEORY OF /CLOUD/

Cultural Memory
in
the
Present

Mieke Bal and Hent de Vries, Editors

A THEORY OF /CLOUD/

Toward a History of Painting

Hubert Damisch

Translated by Janet Lloyd

STANFORD UNIVERSITY PRESS

STANFORD, CALIFORNIA 2002

A Theory of /Cloud/ was originally published in French in 1972 under the title *Théorie du /nuage/*, © Editions du Seuil, 1972.

Stanford University Press
Stanford, California
English translation © 2002 by the Board of Trustees of the
Leland Stanford Junior University

This book has been published with the assistance of the
French Ministry of Culture — National Center for the Book.

Printed in the United States of America on acid-free,
archival-quality paper

Library of Congress Cataloging-in-Publication Data

Damisch, Hubert.
  [Thâeorie du nuage. English]
  A theory of cloud : toward a history of painting /
Hubert Damisch ; translated by Janet Lloyd.
      p. cm. — (Cultural memory in the present)
    ISBN 0-8047-3439-9 (cloth) — ISBN 0-8047-3440-2 (paper)
1. Painting—Philosophy.   2. Clouds in art.   I. Title.
II. Series.
    ND1140 .D2813   2002
    759.94—dc21   2002001376

Original Printing 2002

Last figure below indicates year of this printing:
11   10   09   08

Typeset by Tseng Information Systems, Inc. in 11/13.5 Adobe Garamond

Contents

Illustrations

Painting consists of material hellishly woven, ephemeral and of little worth, because if the superficial coating is removed, nobody any longer pays any attention to it.

—JACOPO DA PONTORMO, LETTER TO BENEDETTO VARCHI,

18 FEBRUARY 1548

Sign and Symbol

Cupolas

Correggio's frescos in the cupolas of Parma:

The Ascension of Christ, or rather, *The Vision of Saint John on Patmos* (church of San Giovanni Evangelista, 1520–24) (Plate 1a)
The Assumption of the Virgin (Parma cathedral, 1526–30)[1]

The Principle

Correggio's predecessors and (for a long time) many of his successors in the art of painting ceilings strove, when offered such surfaces, to affirm and emphasize the ceilings' architectonic nature, often in an illusionist fashion — when, that is, they did not treat cupolas, vaults, and ceilings simply as so many walls on which to arrange and paint figures and panels conceived according to the norms of a painting set on an easel. In contrast, the solution associated with Correggio amounts to a negation: of the building itself, and even of the fact that it is a closed space. This effect is created on a key part of its overarching cover, by a decor conceived in such a way as to seem to "pierce" the stone fabric and create a fake opening onto a sky that is itself painted in trompe l'oeil. (It was a solution already adopted as early as 1472–74, by Mantegna, for the ceiling of the Spouses' Chamber in Mantua [Plate 1b], to which we shall be returning.)

TROMPE L'OEIL

On the inner surface of the cupola, the concave form of which is skillfully exploited by this decor, a circular, or rather hemispherical, composition is displayed. It is devoid of any architectonic reference or indication and appears, in its entirety, to be organized solely according to the requirements of aerial perspective and foreshortened figures. In San Giovanni Evangelista, the cupola rests upon a very low, almost blind drum that is treated as an entablature and decorated with a frieze in the ancient style. The more or less slumped circle of the Apostles rests upon a mass of grayish, solid-looking clouds with a scattering of putti (cherubs). In the center of the ring, the figure of Christ is to be seen, floating against a background of golden glory. The whole "machine," as Annibale Carracci was to call it, seems to have no visible connection with the building, which appears to be cut off at the level of the base of the vaulting, giving way to a celestial vision. In the cupola of the cathedral, in contrast, the figures of the Apostles, set between the oculi in the drum, are solidly implanted along the lower ledge, in front of a balustrade bearing candlesticks and a throng of adolescents who effect a transition to the celestial level above. The trompe l'oeil here is more assured, and there is a more marked opposition between those who, for all they are reaching heavenward, nevertheless keep their feet firmly planted on the ground, and the distant, misty figure of the Virgin positioned in the middle of the apse, floating away on a gleaming, fluid spiral composed of an endless swirl of angels, the elect, and clouds. (In the cupola of San Giovanni Evangelista, however, there is one, almost hidden figure—visible only from the end of the apse—who is also positioned on the ledge: it is Saint John himself, kneeling at prayer.)[2]

DESCRIPTION (BURCKHARDT, 'THE CICERONE')

"Between 1520 and 1524, Correggio painted in San Giovanni, and probably the first thing was the beautiful and severe form of the inspired Evangelist in a lunette over the door of the left transept. Afterwards came the dome. . . . It is the first dome devoted to a great general composition: *Christ in Glory*, surrounded by the Apostles sitting on clouds, all introduced as *The Vision of Saint John*, seated on the edge below. The Apostles are genuine Lombards of the noble type, of a grandiose physical form. . . . The view from below, completely carried out, of which this is the earliest preserved instance, and certainly the earliest so thoroughly carried through [Die völ-lig durchgeführte Untensicht, von welcher dieses Beispiel das frühste erhal-tene und jedenfalls das frühste so ganz durchgeführte ist], appeared to con-temporaries and followers a triumph of all painting. They forgot what parts of the human body were most prominent in a view from below, while the subject of this and most later dome paintings, the glory of heaven, would only bear what had most spiritual life. They did not perceive that for such a subject the realization of the locality is undignified and that only ideal architectonic composition can awaken a feeling at all in harmony with this [Man empfand nicht mehr, das für diesen Gegenstand die Raumwirklich-keit eine Entwürdigung ist und dass überhaupt nur die ideale, architekto-nische Composition ein Gefühl erwecken kann, welches demselben irgend-wie gemäss ist]. Now here the chief figure, Christ, is foreshortened in a truly frog-like manner, and with some of the Apostles the knees reach quite up to their necks. Clouds, which Correggio treats as solid round bodies of defi-nite volume [Als Raumverdeutlichung, Stütze und Sitz, malerisch auch als Mittel der Abstufund und Unterbrechung dienen die Wolken, welche Cor-reggio als consistent geballte Körper von bestimmten Volumen behandelt], are employed to define the locality, also as means of support and as seats, and pictorially as means of gradation and variety. Even on the pendentives of the cupola are seated figures, very beautiful in themselves, but exagger-atedly foreshortened; an Evangelist and a Father of the Church on clouds, where Michelangelo in a similar place would have given his prophets and sibyls solid thrones [Während noch Michel Angelo seinen Propheten und Sibyllen an ähnlicher Stelle feste Throne gegeben hatte].

"At last, Correggio, in 1526–30, painted the dome of the Cathedral, and therein gave himself up altogether, without any limit, to his special conception of the supernatural. He makes everything external, and dese-crates it. In the centre, now very much injured, Christ precipitates himself towards the Virgin, who is surrounded with a rushing crowd of angels and a mass of clouds. The impression is certainly overpowering; the confused group of numberless angels, who are here rushing towards each other with the greatest passion and embracing, is without example in art [Der Knäul zahloser Engel, welche hier mit höchster Leidenschaft einander entgegen-stürzen um sich umschlingen, ist ohne Beispiel in der Kunst]. Whether this is the noblest consecration of the events represented is another question. If so, then the confusion of arms and legs, which has been described in the well-known witticism of *un guazzetto di rane*,[3] was not to be avoided; for if the scene were real, it must have been something like this. Further below, between the windows, stand the Apostles, gazing after the Virgin; behind them, on the parapet, are Genii busy with candelabra and censers. In the Apostles, Correggio is not logical: no one so excited as they are could stand still in his corner; even their supposed grandeur has something un-real about it. But some of the Genii are quite wonderfully beautiful; also many of the angels in the paintings of the cupola itself, and especially those which hover round the four patron saints of Parma, on the pendentives. It is difficult to analyse exactly the sort of intoxication with which these figures fill the senses [Es ist ein eigenthümliche Art der Berauschung, womit diese Gestalten den Sinn erfüllen]. (Best light for ascending the cupola, towards noon.)"[4]

SPACE/LIGHT: THE DESTRUCTION
OF THE PERSPECTIVE CUBE

From the point of view of the concept of a "Baroque" or "pictorial" or "painterly" (*malerisch*) style, whose most striking feature is possibly its "antipathy to any form with a clear contour,"[5] historically speaking the cupolas painted by Correggio could not fail to mark, as it were, a new de-parture: his work, remarkably precocious and confident, constitutes one of the first manifestations of an art in which the propositions of the perspec-tive cube and the orthogonal and closed tectonic space of the Quattrocento (fifteenth century) seem to fall apart.

"The church interior, its [the Baroque style's] greatest achievement, revealed a completely new *conception of space directed towards infinity*: form is dissolved in favour of the *magic spell of light*—the highest manifesta-tion of the painterly. No longer was the aim one of fixed spatial propor-tions and self-contained spaces with their satisfying relationships between height, breadth, and depth. The painterly style thought first of the effects of light: the unfathomableness of a dark depth, the magic of light stream-ing down from the invisible height of a dome, the transition from dark to light and lighter still are the elements with which it worked.

"The space of the interior, evenly lit in the Renaissance and conceived as a structurally closed entity, seemed in the Baroque to go on indefinitely. The enclosing shell of the building hardly counted: in all directions one's gaze is drawn into infinity. The end of the choir disappears in the gold and glimmer of the towering high altar, in the gleam of the '*splendori celesti*' [celestial splendours], while the dark chapels of the nave are hardly recognisable; above, instead of the flat ceiling which had calmly closed off the space, loomed a huge barrel-vault. It too seems open: clouds stream down with clouds of angels and all the glory of heaven; our eyes and minds are lost in immeasureable space.

"True, these effects of an all-consuming decoration are the product of a later period of Baroque, but the determining feature of this art, the uncontrolled revelling in space and light, was strongly felt from the first."[6]

Prestige

But it was not only stylistic analysis that enabled historians such as Wölfflin and Riegl (and already Burckhardt) to see in the cupolas of Parma the point of departure for a whole series of cupolas that they were among the first to describe as "Baroque"—cupolas where clouds, great banks of clouds, invade the entire composition or even overflow its frame onto adjacent parts of the architecture. The line of descent of such compositions is firmly established historically, as is abundantly attested.

OPERATION

"I could not resist immediately going to see the great cupola that you have so often praised to me and I am still stupefied by the sight of such a great machine, so well understood in all its parts, so well observed all the way up from the bottom to the top, and with such rigour, yet always with such judgement, with such grace, and with colouring that is truly that of flesh [una cosi gran macchina, cosi ben intesa ogni cosa, cosi ben veduta di sotto in sù con si gran rigore, ma sempre con tanto giudizio, e con tanta grazia, con un colorito che è di vera carne]! Upon my word, Tibaldi, Nicolino [Nicolo dell' Abbate], even—dare I say it?—Raphael are no longer of any account."[7]

These lines, sent from Parma by the young Annibale Carracci to his cousin Luigi, on April 10, 1580, certainly convey the considerable—and *reasoned*—prestige that Correggio's painted frescoes, above all his great cupola,[8] possessed in the eyes of the future masters of the Bologna school. *Even Raphael is no longer of any account.* . . . The fact is that it is *taste* that wins out here, taste, which was to be the major concern of eclecticism, and which a writer such as Mengs, two centuries later, would still recognize to be the distinctive mark of Correggio's art. Following Agostino Carracci, Mengs was to associate that taste with the *expression* (drawing) of Raphael and the *truth* (colors) of Titian.

But what was taste for a man of the eighteenth century if not, in the same fashion as a physical taste (the flavor of some foodstuff), an action exerted upon the optic nerve by an object intended to be agreeable to the eye, an *operation* (to use the vocabulary of Mengs, but also of Wölfflin) that could be regarded as "a kind of style or manner"?[9] We cannot be certain that this was also the Carraccis' understanding of style, even if, for academicians, a certain kind of Platonism, according to which physical beauty constitutes the visible appearance of perfection, has always fitted in well with sensuality. Where the cupolas of Parma are concerned, what is important is that Annibale Carracci saw fit to distinguish between the arrangement of the parts, the perspective organization of the "machine," and the judgment, the grace and the quality of the color, all of which Mengs was to call "style" or "manner."[10] The comparison prompts a number of questions as to the mechanisms of the operation and the relations between "style" (is style just a matter of execution?) and the structure of the work ("a machine so well understood").

MODEL/SERIES

If the search for "sources" has a meaning (and in the present context it certainly has, but perhaps not the most obvious one), it would at least appear that we have here the model, explicitly recognized, acknowledged, appreciated as such, of a mode of illusionist decor, both spectacular and filled with dynamic movement. The historians were to label this "Baroque," and it was to predominate in Rome a whole century, at least, after Correggio: cupolas, vaults, and ceilings where a Virgin, a saint, a family, or even a whole religious order can be seen ascending in "glory" or "triumph" into a sky swirling with clouds. Key examples are provided by the cupola painted by Lanfranco in Sant' Andrea della Valle (*The Assumption*, 1625–28), the ceiling by Pietro di Cortona in the Barberini Palace (*The Glory of the Family of Urban VIII*, 1633–39), the vaulted ceiling of the nave of the Gesù (church of Jesus), painted by Baciccia (*The Triumph of the Name of Jesus*, 1672–83), and the ceiling of Sant' Ignazio, by Father [Andrea] Pozzo (*The Triumph of the Jesuits*, 1691–94).[11]

A BLOCKED DEVELOPMENT

As we have noted, the Carraccis took a great interest in Correggio's "machines." Yet the decoration of the vault of the Farnese gallery (1597–1604) constitutes, in its very principle, a contradiction to Correggio's *continuous* composition, which adopts a perspective that passes from the bottom upward (*dal sotto in su*), as in the cupolas of Parma. Correggio, as early as 1520, appears to have been prompted by the vertical relation between the spectator and the scene presented before his eyes to establish the vanishing point of the composition at its zenith. But the Carraccis, at the beginning of the following century (and also Pietro di Cortona, around the middle of that century, in the nave of the Santa Maria in Vallicella church) continued to apply the Venetian schema of the *quadro riportato*. In such a schema, a painting, or a cycle of paintings, was conceived in accordance with the norms of an easel painting and a horizontal perspective, but was arranged on the vault or ceiling in such a way that the floor level swung through ninety degrees in relation to the wall, and any visitor who wanted to study the collection of pictures was obliged to perform a feat of veritable optical gymnastics. (As Wölfflin perceived, such a feat was inevitably necessary, given the way in which the composition was organized: "Michelangelo's clear, architectonic arrangement has been sacrificed in order to achieve a richness beyond one's full comprehension. The principles of the design are difficult to recognise and in the face of this intangibility, the eye remains perpetually in a state of unrest. Image overlaps image, and it seems as if removing one will only reveal another; the corners open out into unending vistas.")[12]

The solution adopted by Correggio may at first sight seem more "natural," or at any rate more homogeneous and coherent, and less contradictory given that it is founded in all its parts, not just the "corners," on a unifying principle. Nevertheless, as the result of a historical blockage (the reasons for which—among other things—the present study aims to discover), more than a century was to pass before that solution came to predominate in Rome, and from there to spread throughout the whole of Europe. (In the eighteenth century, though—this kind of quarrel was nothing new—it was claimed that Correggio himself had got the idea of his cupola from an earlier master, one of whose works was in Santi Apostoli, in Rome.)[13] We know from Ludovico Cigoli, who helped to introduce this type of decor in Rome,[14] that that earlier Florentine painter—who was, as we shall see, a friend of Galileo's, who had settled in Rome—had made the journey to Parma. When in 1625, following a series of dramas and rivalries that need not concern us here, the Theatine monks of Sant' Andrea della Valle wanted a cupola of their own, it was to an artist from Parma that they appealed, namely Giovanni Lanfranco. Bellori relates that, while on a visit to Correggio's hometown, Lanfranco had made a model of the cupola of the cathedral.[15] He was later to win acclaim for this kind of painting in Naples (the church of Gesù Nuova, 1635–37, etc.). As for Giovanni Battista Gaulli, known as Baciccia, he was to be celebrated as one of the masters of ceiling painting on the strength of his *Triumph of the Name of Jesus*, a fresco painting executed on the vaulted ceiling of the Gesù. And a few years earlier, when working on the pendentives of Sant' Agnese in Agone, he likewise had not failed to make the trip to Parma, to study Correggio's techniques.[16]

A Grid

So it is that, however prestigious Correggio's cupolas may have been right from the start, they nevertheless occupy an enigmatic position in the history of art, since over one hundred years separate them from the first Roman works for which they provided the explicit model. Leaving that problem aside for the moment, it is somewhat astonishing to find that at the turn of the last century, at the time of the Vienna Sezession (and also that of Cézanne, Seurat, and the German Impressionists), Aloïs Riegl, the most famous representative of the Vienna school and the first of the *theorists* of art, in the sense in which the present text understands the term, saw fit to agree with Burckhardt in recognizing this provincial artist as *the most modern of all the painters* of the Italian Renaissance.[17] It is important to see how the works of Correggio find their place, as both a point of reference and an illustration, as the linchpin in a theoretical opposition that even today, whether we like it or not, functions as a grid for a reading and a framework around which every interpretation proposed for these works must be articulated. (But it is up to the reader to pass through that screen and substitute his own grid, his own text, for the original grid and the received text.)

CONCEPTS

Formed as he was by eighteenth-century taste (Meng's taste), Stendhal judged Correggio's frescoes, not surprisingly, to be "sublime" and was "moved to tears" by them.[18] But Riegl? If the latter allotted the painter a fine role in his system, that was not solely because he regarded some of his works as the point of departure, if not the model for a form of art (the "Baroque" — or even, as Burckhardt was already calling it, the "Rococo"),[19] the concept of which historians were then endeavoring to elaborate. Correggio's productions seemed to Riegl to occupy a decisive position in a central issue on which hinged the entire development of Western art, from Greco-Oriental antiquity to the days of Impressionism. This concerned both the axis or spectrum that joins the objective pole of art (where beings and things are "rendered" as they are first presented to the sense of touch in the immediate proximity of their surface) to the subjective pole (where they are such as they are revealed to sight, at a distance, in three-dimensional space) and, accordingly, also the distance that separates the tactile and the optical modes of representation. If the Renaissance was regarded as the moment of compromise between flat composition, in the ancient manner, and modern composition, with the emphasis on spatial depth, it was Correggio's works that for the first time, with no restrictions, revealed a determination to construct the picture from the point of view of the *subject*— the subject defined in Kantian terms — where space was a given, constitutive feature of awareness.

How should this be understood, from a systematic rather than a historical point of view, if it is true that Riegl[20] was the first to try to analyze works of art on the basis of their formal conditions of possibility,[21] and to replace an iconographical enquiry, concerned solely with the images' content, that is with what they *signified,* by a study of their formative principles and the constitutive features of the appearance of things as painting presents them to the eye? Riegl was interested in the relations between the figures and the background from which they stand out, between shapes and the place where they are disposed, outlines, relief, shadows cast, colors, chiaroscuro, foreshortening, atmospheric perspective, and so on. As we shall see, those principles, now treated systematically by Riegl, were some of the very aspects and rubrics that Mengs selected in his day in order to discuss Correggio's taste and his contribution to the art of painting.

What we must do here is return to the subject of the *taste* that the academicians ascribed to Correggio, in order to pinpoint the concept behind it and to distinguish the ideological field within which, since Riegl, the theory of art has operated. If taste was the prerogative of Correggio, does that mean that Raphael and Titian lacked it in the areas of art in which they excelled—drawing (*expression*, grandeur) for Raphael, color (*truth*) for Titian? Given that academic doctrine held that no artist of the past had ever managed to prove his excellence in every part of art at once, Correggio was bound to delight the eclectic spirits who praised him for having possessed all the various excellencies to the full, and also having added to grandeur and truth "a certain elegance, which generally bears the name of taste, and which signifies the proper and determinate character of things, excluding all the indifferent parts, such as the insipid and useless."[22] But the important point here is that Mengs saw fit to declare that harmony, the "intermediate style"—at which this painter excelled—corresponded to his determination "to delight the sight and the souls of spectators." According to Mengs, *giving pleasure to the eyes* was the operation (the "manner or style") that boiled down to taste, and Correggio was the first to achieve that end by exploiting every part of his art and using his talent solely "for the satisfaction of his heart, *and according to his own sensations*,"[23] since, after all, "painting becomes attractive by means of the eyes."[24]

Whatever its sentimental connotation ("Correggio possessed very delicate sensibility, a tender and affectionate heart . . ."),[25] the essentially visual and subjective quality that Mengs perceived in Correggio's art thus tallied well enough with Riegl's assertions[26] and the position that the latter assigned to this painter at the inception of modern art—an art that, in the last analysis, was interested only in sensations or optical impressions. But there is more that needs to be said: for the reason why, for Riegl, the opposition between the two poles, the tactile (objective) and the visual (subjective), takes on the value of a theory and allows him to classify works of art along an axis that is no longer determined by history and succession, regarded as an explanatory principle, is that it serves to introduce a protocol of formal analysis. An art devoted to the apprehension of bodies considered as materially impenetrable, solid entities will resort to outlines that isolate the figures from the background against which they stand out, and will endeavor to eliminate relief and foreshortening, all effects of overlapping or distancing, and all expressive features that might interrupt the objective continuity of the flat surface or disrupt the order of the figures on it. The means of such an art will be essentially *linear*, quite the opposite of the *pictorial* means that characterize an art that takes as its object (or referent) space itself, and as its subject (not to be confused with any "fable" or "story" that the image illustrates or, in its own particular way, recounts) the representation of singular forms as they appear and reveal themselves within a free and unlimited depth, considered as a luminous and aerial substance.

When Wölfflin later listed the distinctive features of the "linear" and the "pictorial" styles (that is to say the opposed pairs of limited/limitless, flat presentation/presentation in depth, open forms/closed forms, plurality/unity, etc.), he was simply readopting the *theoretical* system defined by Riegl a few years earlier,[27] and doing so for purposes that the history of academic art was soon to reduce to those of a strictly descriptive classification. The art of Correggio certainly seems to provide a good illustration of the "pictorial" style, if it is true, as Mengs had already suggested, that this painter rejected the straight and simple lines by which external shape is defined and always tried to interrupt his outlines by chiaroscuro, at the same time remaining unrivaled in the art of aerial perspective, thanks to having noticed that nature never presents several objects all with the same degree of force ("for whatever is in the forefront of the picture is assumed not to be veiled by the illuminated corpuscles that circulate in the air; whereas the background, which is set further away, must be regarded as being covered by several layers of those corpuscles").[28] If, as Riegl was to argue, Correggio was the painter closest to the moderns and was also the first of the "Baroque" artists, it was because "the contours of his predecessors, which were too meagre and too cramped, could not accommodate his celestial spirit."[29] However, we should beware: "By increasing the masses of light and of shade, like a river overflowing its banks, he swept his spectators out into the vast sea of all that is graceful, and from there, seduced by the song of its sirens, many were led into a land of errors."[30]

Index

The scientific spirit demands that the complexity presented to it should be analysed in such a way as to enable it to extract one single feature and use it as a key to the whole.
—Louis Hjelmslev, *Prolegomènes à une théorie du langage*

Process: Texture

Let us for the moment pass over the pathological connotation—"the land of errors"—that Mengs chose to attach to some of the *historical* consequences of Correggio's pictorial practice: for he was speaking not so much as a theorist, but rather as the guardian of an order and its *limits*. But the consequences and resonances of Correggio's oeuvre were not solely plastic; it also provided food for thought for generations of souls of varying degrees of sensitivity, and there is no denying that commentaries such as Stendhal's conferred upon it a somewhat confused prestige. Even within the field of interpretations expressed in linguistic terms (as opposed to "pictorial" interpretations produced by other painters), a division becomes perceptible. The musings of Mengs and, to a lesser degree, the analyses of Riegl focus solely on the making of the paintings, to the exclusion of any semantic elaboration. They nevertheless gave rise to an interpretation that assigns Correggio's oeuvre a place of its own in the ideological field known as "history of art." But the meaning that is conferred upon the oeuvre from the point of view of a history of "styles" remains transcendental, in the strictest sense of the term: it is imposed from outside by an interpretation that sees Correggio's productions only from a historical and doctrinal point of view and on the basis of a strictly technical, formal description that pays no attention to the ongoing signifying process, the trajectory of which these productions mark.

The fact is that there is no reason to believe that the set of material features listed under the rubrics of outline, color, and so on is systematically articulated to produce a *meaning* (perhaps all that they combine to produce is an *effect*), let alone, to borrow Louis Hjelmslev's terms, that these *figures* are subservient to (iconic) *signs* that logically determine their existence. And even if one set out to see them hypothetically as elements in a signifying order, it seems unlikely that they would be amenable a priori to a *finite* inventory, or that there would be any grounds for attempting to reconstruct a signifying system in the way that linguists set out to do for elements—phonemes and distinctive features—that correspond to the lower levels of linguistic analysis. And yet, if there is any foundation for the idea of a semiology of art, a description at the level of truly pictorial elements could surely not be limited to registering an indefinite number of conventional tricks, studio recipes, and characteristic features. It ought to be possible, in one way or another, to mesh it in with an analysis that operates at a "higher" level,[31] and that corresponds to a truly signifying articulation (it would be easy enough to show that any "effect" is bound to be of a symbolic order). Even in the context of strictly representational art, the pictorial process is never geared solely to depiction, solely to the denotation of beings and objects, by means of color and/or drawing. Wherever painting is concerned, iconic signs possess a *texture* of their own, a texture that is frequently a predominant component in the operation of "taste" — so much so, indeed, that it may be thought that style owes it part of its prestige. Thus the by now classic opposition between the "linear" and the "pictorial" (or "painterly"), in so far as it is relevant, refers one back to the material texture of the pictorial process, making one doubt that its importance is purely decorative. Is not the "treatment" of the image, rather, a second-order signifier, where the connoted message corresponds to the "style" of the representation and is somehow superimposed upon the denoted message? The latter, for its part, is based upon imitation, upon an analogy between the signifier and the signified that, however, is not assumed to have the force of a code.[32]

STYLE/SYSTEM

Now, the important point with respect to the notion of a pictorial process is precisely that one can and should distinguish between "imitation" and "stylization," between the figurative presentation or "denotation" as such and the expressive connotations that accompany it. In the field of the history of art the notion of *style* is so ill-defined and the extension of the term so loose that that question is, quite literally, never *addressed*. (Wölfflin, for instance, uses both the word *style* and the word *language* indifferently, to designate the linear and the pictorial modes of presentation: any form of presentation, however linked to "a certain idea of beauty," and even if it possesses "particular virtues of its own," for all that constitutes a language "capable of expressing anything.")[33] There is no a priori justification for drawing a distinction between, on the one hand, the style, individual or collective, that confers upon the processes to be analyzed their quality and their specific "taste" and, on the other, the "system" that underpins them. And there is even less justification for postulating the existence, over and above the diversity of works of art, of an *aisthesis*, a network of structural constraints and formal and expressive possibilities that might bestow upon the pictorial products of any given period their historical coherence. No justification at all, apart from a few signs or indications to which the context lends a certain emphasis, a particular *charge*—signs around which it seems possible to construct the project of an analysis that might, by means of a kind of symptomology, afford access to the deep structures of the images of paintings, seized upon within the unity of the semiotic process of which they are the object.

Correggio's /Cloud/

We are here to attempt a semiological analysis that does not set out by acknowledging its dependence upon the linguistic (phonetic) model, but instead aims to define the specific semiotic function that constitutes the mainspring of pictorial production. Such an analysis cannot possibly proceed simply by a functional division of the painted surface into its constitutive parts, and then by breaking down those parts, in their turn, into the elements of which they are composed. On the contrary, it needs to circumvent the flat surface upon which the image is depicted in order to target the image's texture and its depth as a *painting*, striving to recover the levels, or rather the registers, where superposition (or intermeshing) and regulated interplay—if not entanglement—define the pictorial process in its signifying materiality. However, this should be done without presupposing their relative coherence or the possibility of drawing up a more or less exhaustive list of terms and functions that belong to the same class. In the absence of any explicit theory, the method to be followed must perforce be inductive. It would amount to inferring from an analysis of the pictorial process itself some concept of what we have called its various levels or registers, and then *describing* these, in the physical sense of the term, and revealing their relative organizational role.

Now, there is one motif that seems to provide a good index to Correggio's manner and that, through its sensible appearance and the functions that it assumes, seems to combine, as in a bundle, the principle features by which the "art" of this painter is generally characterized. This motif, which it should be possible to use as a guiding thread for an in-depth analysis of Correggio's work, this theme or *index* is /cloud/, the pictorial graph denoted as *cloud* at the level of description.[34] The word will appear between two forward slashes (as in the title to the present work) every time that the analysis requires it to be identified as a signifier; it will appear in italics only when used in a strictly denoting capacity, and between quotation marks when it refers to that which is signified. /Cloud/ is used for a variety of purposes and in a variety of figurative guises in many of Correggio's compositions. (We have already noted the various innovative purposes for which Correggio used it, in such great profusion, in his decorative compositions in Parma: he even went so far as to separate off a number of heaps of cloud on the pendentives of his cupolas, to serve as seats for the figures of the Evangelists and the Church Fathers.) Here and there (in Florence, Dresden, etc.) we find a Madonna supported by a bank of cloud, or enthroned against a background formed by a glorious assembly of seraphims, with an edging of clouds; elsewhere (*Night* in Dresden, *The Gypsy Girl* in Naples, *The Madonna of the Bowl* in Parma), little angels play and tumble about on balls of cloud floating in the top part of the composition. And sometimes we find clouds that are less chaste and less Christian: a luminous, sunny one from which a stream of gold pours down upon the *Danae* of the Borghese Gallery; and another (in the Vienna museum) upon which, against a background of indistinct mists, the figure of Jupiter can be seen embracing Io.

 Cloud is a key term in the figurative vocabulary of Correggio, and—as we shall see—is also one of his favorite themes. But it is a theme that, thanks to the textural effects to which it lends itself, contradicts the very idea of outline and delineation and through its relative insubstantiality constitutes a negation of the solidity, permanence, and identity that define *shape*, in the classic sense of the term (although it must be said that in some compositions the banks of cloud are forcefully drawn and take on an altogether solid, material appearance). According to Mengs, Correggio rejected rectilinear outlines and straight, simple shapes and persistently endeavored to break them up, preferring flowing, internal forms and curving lines.[35] If that is so, he was bound to be attracted to nebulous structures, both on account of their plasticity and because they provided him with the means to position, split up, and confuse the figures that he set among them, just as he pleased. Bodies entwined in clouds defy the laws of gravity and likewise the principles of linear perspective, and they lend themselves to the most arbitrary of positions, to foreshortenings, deformations, divisions, magnifications, and fanciful nonsense. As Burckhardt pointed out, in the church of San Giovanni Evangelista, the knees of some of the Apostles seem to be joined to their necks. And as for the cupola of the cathedral, all that can be made out there are fragmentary, split-up figures drowning in a mass of equally patchy, indefinite cloud. You have to agree with Mengs: such a composition is more like "*a product of the imagination or a dream than a fine ideal.*" [36] (And, to repeat Burckhardt's comment, "the confused group of numberless angels who are here rushing towards each other with the greatest passion, and embracing, is without example in art. . . . The impression given by these figures is almost one of a kind of intoxication.")

Correggio's "celestial" art and his "vaporous" manner thus seem to be achieved and to manifest themselves through one particular element, a theme whose value and functions Burckhardt was the first to recognize. For although /cloud/, considered as a pictorial graph, may appear to contradict the assumptions and principles of an art founded upon a strict delineation of forms and upon geometric perspective, and although it furthermore makes it possible to free figures from the physical laws that govern bodies, and sanctions many aerial effects, raptures, and paradoxical ruptures and juxtapositions, it is not necessarily the case that its occurrence, its manipulation, and its very treatment stem solely from whim or from "mannerism," in the pejorative sense of that term. The "intoxication" proffered by Correggio is precisely calculated, and *cloud* plays its part in that calculation as a free vector that lends itself to operations the nature of which, as Burckhardt correctly recognized, is semiotic, both in a signaling sense and syntactically. /Cloud/ is not just an instrument adopted by a style; it is the very material of a *construction*. Even if Correggio did provide the scenes of Glory in the beyond with "a measurable cubic space that he filled with figures in powerful flight," that space is not defined according to the architectonic norms of linear perspective. But neither is it defined according to the atmospheric and luminous norms of aerial perspective. For the skies of Correggio are never without *clouds*, the ambiguous appearance of which was well perceived by Burckhardt, who noticed the way in which they made it possible to combine two kinds of perspective: "the designation of the space, the painting's substratum and the surface upon which the picture is painted, and all the gradations and nuances of the painting are expressed by clouds, which Correggio treats as resistent, compact masses with a determinate volume."

SIGNS AND FIGURES

Cloud is thus an iconic element regularly used in Correggio's paintings for precise and limited purposes, and always in combination with other motifs. However, in some instances, it swamps the entire figurative field and underpins an entirely original mode of organizing, articulating, and defining (*Verdeutlichung*) or designating the space, the substratum, the surface upon which the picture is painted. *Designating*: the choice of word is important, for by its very nature, the idea of a sign is surely always linked with that of designation.[37] Contrary to the phenomenological theory according to which an object cannot be given simultaneously both as an image and as a concept,[38] analysis associates a given pictorial graph with a linguistic sign (its *interpretant*, as Peirce puts it): it designates that pictorial graph as *cloud*, at the same stroke establishing it as a *representamen*, an iconic sign. But though the graph functions as a sign, it is not a sign only through language, description, and the separation that the latter establishes between the level of the signifier and that of the signified. The /cloud/ graph does not play a solely pictorial or decorative role; it also serves to designate a space. Its iteration and proliferation particularly deserve our attention given that the sensible appearance of such a "sign" is apparently expected to satisfy the demands of a particular taste (if not style), which Annibale Carracci saw as a necessary corrective to the rigor of the machinery that used it as its material. But this is the point at which the shoe pinches and we perceive the equivocality and the limits of the problematic of a sign in the context of the analysis of artistic compositions. The very same units operate at several levels, playing several roles at once, now as the signified, now as signifiers. Although they may be identical from the point of view of an analysis of the pictorial process in its iconic constituents, the objects are different with regard to the functions that they assume and the operations that define them. The unit denoted as *cloud* may, at the level of the making of the painting, play the role of a *figure*, in the Hjelmslevian sense (that is, a nonsign that, as part of a sign, belongs to a system of signs:[39] that is a purely operational definition, but one that does not exclude ambiguity when "style" is also taken into account, along with its *figures*, which correspond to a "higher" level than that of the sign, in the strict sense of the term). The same unit seems to shift, without modification, from one level to another when it functions as a sign that serves to designate a space, a situation, or a spatial relationship. Without modification? The fact that the pictorial element /cloud/ can, as Burckhardt pointed out, take on the appearance of a solidly delineated body of a determinate volume (thereby contradicting the vaporous, evanescent connotations that are attached in the semantic field to a signified "cloud") indicates clearly enough that the relative organization of the various pictorial levels or registers needs to be conceived in a dialectical fashion, rather than it being assumed that elements from lower levels are mechanically integrated with those at higher levels. This will be seen even more clearly when we consider the symbolic (iconographic, rhetorical, thematic, etc.) functions that this element assumes within the corpus of works to which Correggio's name is attached (although, it must be said, the very notiŏn of an "element" now seems questionable).

The Machine and the Dream

Themes

MOVEMENT

"In Correggio, everything is in movement."[40] That is obviously a metaphorical formula. For there are no mobile elements or parts in these paintings, these machines that are animated by no mechanism or motor and that, unlike many present-day works of art, do not aim to create any optical cinematic effects. The "movement" claimed to characterize Correggio's art is in no sense *real*. Nor, however, does it depend in any way upon the perceptive mechanisms of illusion. Yet if the assertion carries any theoretical weight, the metaphor must not be solely a piece of critical jargon and the works themselves must present a *figure* of movement that can be defined—an image the mainspring of which can be apprehended in rhetorical, if not thematic, terms. But the question is whether it is possible to conduct an analysis solely on a formal level, when (as Mengs realized) the image also seems to operate on registers other than that of perception: on the level of the imagination, for instance, and perhaps on that of dreaming.

In this respect, too, *cloud* seems to be a kind of guiding thread, or—to borrow Gaston Bachelard's expression—an inductive sign that can serve as a point of departure for an argument that is both logical and analytical. If, as is claimed by the thesis upon which so-called thematic criticism is based, an image's value lies not so much in its configuration as in its mobility, its internal dynamism, and the scope of the imaginary variations to which it lends itself, then cloud certainly does provide dreaming (and any analysis of it) with incomparable material. And if it is true that the psychology of the imagination neither can nor should work upon static figures—if it can only learn from images that are in the process of deformation,[41] it will be agreed that this most amorphous of objects must be one of the most valued oneiric themes. For does it not give access to a world of shapes in movement and deformed by movement, and lend itself to constructions whose constant mutations bring the formal powers of dreaming fully into play?[42] But once the door is opened to *material imagination* (the kind which, according to Bachelard, is defined by its links with the four elements: air in the case of Correggio) and to *dynamic imagination* (which, for its part, is said to operate at the level of elementary impulses: a desire for verticality, dreams of flying, etc.), then, surely, the deciphering of images bit by bit should give way to interpretation that pays attention to what they change into, to their continuous linkages. Such an interpretation should be founded, again according to Bachelard, upon the revelation of the element of choice that they involve and on a direct engagement in their specific dynamism, or even the fundamental impulse to which they are geared.[43]

However alien Bachelard's speculations on the *imagination of movement* are to the resolutely antipsychologistic aims of the present study, they nevertheless provide a good reason to reflect upon the manner in which the sensible texture of a "style" affects the themes that it serves. In the case of the works of Correggio—first and foremost among them the cupolas of Parma—a seamless passage of continuity appears to lead from pictorial matter to imaginary substance, from motif to theme, and from a strategic element, identifiable as such, to the very principles that dictate the ordering of the images. If the pictorial graph of cloud seems to constitute one of the key terms of Correggio's vocabulary, it is not so much because of the signaling and possibly syntactical uses that are imparted to it—uses whose legitimacy and regularity are attested by their recurrence—but rather on account of the remarkable scope of the functions that it assumes in the large fresco paintings. It was cloud that made it possible to introduce the sign of a divine presence (in the form of a few putti or little angels) into a composition organized according to the principles of perspective, or to lift an enthroned Virgin or an amorous nymph out of ordinary space. In the major fresco decors that very same element—or motif—takes on the dimensions of large swathes of cloud that define the place in which cohorts of angels and the elect move about or come to rest and, together with them, serves to designate (to borrow Burckhardt's expression) a particular space (*Raumverdeutlichung*). In other words, the development of the valences assignable to it is not effected in a linear fashion, solely in accordance with the dimensions of a *sign*. In a context such as this, the graph, on the contrary, operates so as to link together the various levels and agencies of the signifying process, and functions simultaneously both as a constructive element and as a schematic way of introducing a thematic development.

So is it really necessary to appeal to psychology, along with those who practice thematic criticism, and to take into account a hypothetical "projection of being" that is supposed to manifest itself, in its constitutive unity, at every level in the work of expression?[44] It is perfectly possible, without doing so, to see that Correggio's "style" seems to *signal* itself through the artist's favorite element, which is at once matter, a form and possibly a fetish, a lexical term, a syntactical tool, and a thematic emblem the repetition and even proliferation of which is prompted by a logic that is manifestly not simply imaginative, but plastic, in the strict sense of the term. From the motif (of cloud denoted by a signifier made "in its image") one moves, again with no break in continuity, to the *theme* (the miraculous vision, the opening up to divine space). What happens is that, in this case, in contrast to the mechanisms that iconographical analysis reveals, the motif does more than simply illustrate a text elaborated in advance, capturing it by iconic means—just as in phonetics writing can capture speech. To the extent that it appeals to the senses, the material object seems to anticipate the theme, for the fundamental experience that is said to give Correggio's style its own peculiar tonality can, in the unanimous opinion of art critics, be summed up as a kind of *ecstasy, ravishment,* or *effusion,* a participation in the substance of the heavens. But by the same token, the justification for the semiological project itself seems to be placed in doubt, for the symbolic process that produced these paintings seems to be at once infra- and suprasemiotic: infrasemiotic in that its source seems to lie deeper than the source of forms (signs) and their manifest articulation; and suprasemiotic to the extent that Correggio's themes seem to imply the negation of any discontinuity between the level of the material and the level of the image, and also implies an immediate transition between the two, with no reference at all to a signifying order and solely through the mediation of the imagination. The "style" of Correggio seems to be so much in tune with his sensible and his thematic material that it would be pointless, if not wrong, to seek to analyze it in an objective fashion in order to reveal its formal bases. Far better to "live" it through a kind of subjective participation guided by its movement, and eventually to recognize these images for what they are: *operators of elevation* (the expression Bachelard used to describe the poetic images of Shelley, who, incidentally, was the author of a poem entitled *The Cloud*).

THE AIR, THE HEAVENS

Mengs, who respected the established tradition of interpretation, regarded Raphael as *divine* and Correggio as *celestial*. The nuance is not without importance. The reason why Mengs declared that the cupola of Parma cathedral resembled a product of the imagination or even a dream was not solely that Correggio was indifferent to the "fine ideal": is not the *intoxication* encouraged by his painting of a dreamlike nature, linked to the dynamics of an ascensional imagination, and connected, first and foremost, to a *dream of flying*; and does not Correggio seem to be the prototype of a "vertical" painter, altogether devoted to the aerial element? First there are the almost too perfect examples that the Parma cupolas provide of a surge polarized both by height and by light (and furthermore punctuated by characteristic ambiguous and elusive signs of the aerial imagination, such as wings and clouds). In addition, many of his other most famous compositions also manifest a predominantly vertical axis, the axis by which communication between earth and heaven operates. One is the ceiling of the Camera di San Paolo, with its fake tunnel of greenery pierced by openings to the sky (a theme introduced earlier by Mantegna, in *The Virgin of Victory*, now in the Louvre), affording glimpses of putti in the ancient manner.[45] Another is the Dresden *Night*, in which a bank of cloud laden with cherubs is associated not only with a pillar (an obligatory accessory for any Nativity), but also with a staircase, the top of which is truncated. And then there are images of Madonnas, in which the double register (earth/heaven) is underlined by the addition of a cloud of putti to the earthbound group of figures. In one of these (*The Gypsy Girl*, in Naples) it is the putti who push aside a screen of foliage; in another (*The Madonna of the Bowl*, in Parma) that task is left to Saint Joseph. Elsewhere, the painter tries out a variety of substitutes: for instance, in *The Virgin of Saint Jerome* (Parma), a large curtain suspended from the branches of a tree takes the place of clouds, and at the same time adds the sense of an unveiling to the upward surge of the painting.

Meanwhile, in another series of paintings by Correggio, the ascensional, aerial element, along with clouds, is treated in an unusual way. The four mythological panels commissioned by Federigo Gonzaga, the second duke of Mantua, were said by Vasari to have been intended for Emperor Charles V, but it has now been shown that they were originally destined for a room in the Palazzo del Te, known as the Hall of Ovid.[46] These paintings, now dispersed, were designed both iconographically and thematically as a single unit. Each of the four panels in this cycle—the first in the history of modern painting to be devoted to the loves of Jupiter—illustrates a tale of the god's union with a mortal creature. The compositions take a variety of forms, but all are aerial or belong in some way to "dreams of Ouranos."[47] In one (*Io*) Jupiter appears to the object of his desire as a cloud; in another (*Danae*) as a shower of gold pouring from the clouds; in another (*Leda*) as a swan with outspread wings; in the fourth (*Ganymede*), transformed into an eagle, he is bearing the loved one aloft. The allegorical (and Christian) interpretation of these mythological morality tales is perfectly in keeping with the thematic interpretation. The abduction of Ganymede (traditionally regarded as a prefiguration and substitute for John the Evangelist, the same Saint John whose vision Correggio also painted) is supposed to symbolize the intellect, liberated from all terrestrial desires, soaring into the heaven of contemplation.[48] In the *Danae* panel, the presence at the foot of the bed of two little cupids, one with wings, the other without, is supposed to represent the opposition between celestial Venus (sacred love) and terrestrial Venus (profane love), while a more mature Cupid, with a larger pair of wings (which Cesare Ripa's *Iconologia* regards as the essential hallmark of *desiderio verso Iddio*, desire the object of which is God), stands beside the girl in the posture of an intercessor.[49] The composition of the *Leda* panel poses an interesting problem of interpretation. It is supposed to be an allegory founded on the opposition between the Apollonian and the Dionysiac principles of music.[50] But that does not rule out detecting in the group of bathing women in the right side of the picture other elements of the story, revolving around the central figure of Leda in the embrace of the swan, elements that constitute a complete narrative sequence, from the appearance of the divine bird through to its final departure as it soars up into the sky, conferring upon the image its vertical and aerial dimension.

Fantasies

'io'

But the Vienna *Io* remains the most remarkable painting of the series, on a number of counts: its title (*Io, eo, I,* the mark of subjectivity), the fable that it illustrates, the novelty of the version that it proposes, and the function that it assigns to the cloud, which is here treated as a fantastical object of desire. Cloud, in the ever changing variety of its forms, may be considered the basis, if not the model, of all metamorphoses. But in the ancient sources, first and foremost in Ovid, and likewise in the modern commentaries, it is represented simply as the instrument chosen by Jupiter to further his designs: the god wished to deny Inachos's daughter any chance of escape, or perhaps his purpose was to hide their lovemaking from prying eyes. "Coperto di nebia a dimostrare che nel viso humano son le cose divine occulte [covered by cloud, to show that divine things are hidden from the human face]": the purpose of the allegory on the page facing the fable of Io in the 1522 edition of the *Metamorphoses* was clearly apologetic, even if Jupiter may have been intent upon concealing himself from Juno, rather than from people in general. That, at any rate, was the interpretation chosen by Paris Bordone in his version of *Jupiter and Io*, in which the god, in human form, is depicted sitting with the girl, surrounded by a cloud that occupies the rest of the composition.[51] Correggio's version is far more erotic and equivocal; and the fact that a fallow deer drinking in the background may be seen as a traditional emblem that confirms the painting's allegorical meaning (that of a *desiderio verso Iddio*)[52] is not particularly important. The ecstasy and enjoyment conveyed here by the means of painting are that of a "subject" whose arms find nothing to embrace but a cloud, the fetishistic aspect of which, in the Freudian sense, as a fantastical substitute for the real object desired, is made abundantly clear.[53] /Cloud/ may be one of the signs used to indicate mystical desire for union with God, but in this picture it is presented as a hallucinatory object, which, as such, is connected with a fundamental but by no means spiritual "ascensional" experience: the experience of an erotic tension and its resolution through a wet dream.

FUNCTIONS

At this point both the allegorical interpretation and the thematic one seem to have found their limits: the first, the allegorical interpretation, because it fails to explain the relation between the claimed "morality" of the images and their power of erotic suggestion,[54] and the relation between the emblematic meaning of this or that iconographic feature or schema and the representational function of a decorative composition apparently designed to illustrate the power that Eros holds over even the duke of Mantua. The latter, in an *impresa* (publication) issued by the Gonzaga family and still displayed on the mantelpiece of the Hall of Ovid, is explicitly assimilated to Jupiter, the master of Olympus.[55] The second, the thematic interpretation, the psychological nature of which has been discussed above and that sets out to move beyond the painted figures and reconnect with the *lived* movement from which the "hieroglyphs" of art emerge,[56] reaches its limit because, as a result of the very decision to serve the fantastical powers of painting, it is always in danger of allowing itself to be diverted into the path to which Stendhal steered it when he wrote that, in order to understand Correggio, it was necessary sometimes to have played the fool in the service of amorous passion. Both interpretations proceed from a logic that is alien to the pictorial order: from the logic that is the root of humanistic rhetoric, in the case of the allegorical interpretation; from the logic that dictates the configurations of desire, in the case of the psychological interpretation, the psychoanalytical veneer of which should not mislead us. Even if the pictorial order does borrow a certain amount from the impulses of fantasy and the prestige of discourse, surely that does not mean that it possesses no necessity, no efficacy, no legitimacy of its own.

The fact that Correggio's oeuvre finally, or almost finally, leads to the figure of Io (the subject) grappling with a lure, may itself be regarded as emblematic, if not enigmatic: it reveals the ambivalence of an element (cloud) the nature of which is hard to define. Does it belong to an autonomous signifying order or does it fulfill symbolic functions other than those assigned to it by the cultural code or the rhetoric of the subconscious? To set out to seize upon Correggio's oeuvre in its unity at once perceptible and thematic is no doubt to seek to encompass too much, at the risk of clutching only at forms and structures that are insubstantial. Every signifying process implies discontinuities and articulation. It is only at the level of what is signified and the process of subjective "comprehension" that the passage from one level or register to another may seem to be an unbroken sequence. But no more does the allegorical interpretation, which refers the image to texts alien to its own order, give access to the pictorial process in all its specificity. If the pictorial graph designated as *cloud* in many cases certainly does not have the vaporous, unstable appearance that would make this element the material connotation of an effusion and the dynamic connotation of an ascent, then one must perforce recognize that the iconic process is not necessarily founded upon a natural analogy, a bi-univocal correspondence between the sign and the real object that it is supposed to represent (denote). Both signs and figures are determined, even in their perceptible appearance, by the functions that they assume at different levels and their relations with units of the same type with which they may be associated or to which they are opposed. Any endeavor to evolve a theory of art should begin, it would seem, by assessing those functions and relations.

Deviance and the Norm

The Land of Errors

Mengs declared that, by attacking the outline and the network of limits within which his predecessors had confined their forms, Correggio led many of his followers into "a land of errors." His declaration makes it quite clear that it was not because the painter presented his public with dreamlike images and themes that he played a determining and possibly—to some extent—harmful role in the history of art. The crux of the matter was form, in the first instance drawing, which theory assimilated to *expression*. We need to understand why calling outline into question was considered such a mistake in the Neoclassical period. Over and above the superficial tremors in the history of taste or of "the life of forms" that it caused, this reaction manifested the *resistance* of a representational structure that was founded upon the supreme excellence of the graphic functions of drawing and that, apparently, "Baroque" experiments never fundamentally undermined.

At this point it is tempting to appeal to Wölfflin's ideas: the dissolution of outlines may have seemed to undermine the order of figurative representation in that its corollary was an exacerbation of *pictoriality* that

excluded any overstrict delineation of forms. But even if Correggio's productions are historically clearly dated and localized, they were also part of
a much more general situation: that of the painted image considered in
its specific difference from other images (namely that it is an image that is
painted and, as such, is irreducible to any other kind of image). That situation is regularly manifested (usually in association with particular named
works) by the appearance, in the plastic field, of features that are soon condemned as aberrant, or even pathological. From the point of view of accepted norms, any innovation has the air of a deviation. But among those
deviations there are some that, far from simply introducing a stylistic variation or a more or less profound change in the rules governing figurative representation, appear to undermine the whole hierarchy of pictorial functions
and, by the same token, pose the question of the status of the painted image
in altogether new terms. A pictorial image cannot be reduced to an icon
whose texture—as we have called it—is a matter of indifference: as a painting, it carries a certain weight, an ongoing power to appeal to the senses
that bears no comparison to the insubstantiality of a color film printed on
a piece of sensitive paper that is in itself neutral and of little value (but that
is nevertheless an integral part of the finished product). It is this texture
that a particular tradition of thinking—a cultural system whose persistent
influence is manifested by the principally iconographical trend of studies
on art—strives systematically to obliterate or repress, pretending that all
that images offer is the information that they convey—information that is
measurable, analyzable, and, as such, something that can be *exchanged*.

'UT PICTURA POESIS'

In this respect, the same applies to painting as to poetry—*ut pictura
poesis*—albeit in a very different sense from that understood by humanist
theory.[57] In the ordinary conditions of linguistic exchange, the aim of the
message in itself, the presentation of the spoken or written signifier, in its
sensible materiality, appears to be an extra function that is subordinated
to other expressive functions. But poetic language, on the contrary, seems
to be characterized by the predominant role that is conferred, within the
overall economy, on the "palpable" aspect of signs, the deliberate and regulated manifestation of which determines the organization of the message.[58]
As the Russian Formalists pointed out, the sounds of poetry are not simply
elements in an external harmony, nor do they merely constitute an accom-

paniment (or setting) for the meaning. They possess a meaning, a verbal function of their own, and they serve an autonomous poetic purpose.[59] On the other hand, we should not forget that among certain authors, some of whom were to rank among the founders of structural linguistics, those theses owed much to the influence, not only of so-called "transrational" poetry (which used words without meaning) and music that accommodated "noise" of various kinds, but also of a kind of painting that was then endeavoring to produce, both in theory and in practice, what Mondrian was soon to call fundamental "plastic means": surface, line, dots, and also—in a more secret and transgressive fashion—color, the "truth" so long subservient to "expression" (i.e., to the sign), which was now regarded as an irreducible pictorial function, an autonomous formal instrument, complete in itself.[60]

The reception that common awareness (common sense) reserved for the paintings of the "Fauves," when they first appeared, may have been prompted by critical pretensions, but it also testified to the pathological ("senseless") character that the pictorial culture of the West is inclined to detect in any deviance from tradition that leads to allowing the sensible (material) components of painted images to prevail over its truly iconic components (in this case allowing color to prevail over drawing). The order in which Alberti, in *Della pittura* (On painting), presents the "parts of painting"—(a) circumscription (or delineation), (b) composition, (c) the distribution of light and colors—is one that establishes a cultural system that claims to be transhistorical and whose dominance for several centuries no experiment, however apparently radical, managed to threaten. "Truth" (color) must not predominate over "expression" (drawing). As Rousseau (and later Kant, in virtually identical terms) was to write,

the feelings that a painting excites in us are not at all due to colours. . . . Beautiful, subtly shaded colours are a pleasing sight; but this is purely a pleasure of the sense. It is the drawing, the imitation which gives life and spirit to those colours. The passions they express are what stir ours; the objects they represent are what affect us. Colours entail no interest of feeling at all. The strokes of a touching picture affect us even in a print. Without those strokes in the picture, the colours would do nothing more.[61]

As Jacques Derrida notes in his commentary on this text, aesthetics is here mediated by a *semiology* and made dependent upon the *sign*. In aesthetic experience, as defined by Rousseau, the subject is affected not by things (sensations), but by signs, by imitation that depends upon lines drawn, delineation, outlines. Now, the same prohibitions that ensure the coherence of the system (and of practice founded on signs and imitation) by, for example, making color depend *technically* upon the outline, which, by delimiting color, provides the reason for it[62]—those same prohibitions also operate at the level of delineation itself: *limits* must not be subjected to any undermining of the rules,[63] nor should they encompass too much *body*. To be sure, as Alberti remarks, one cannot speak of painting in the manner of mathematicians, who consider the form of things purely intellectually, with no regard at all for anything material: whoever seeks to make things more visible may have to paint "a fatter Minerva."[64] But within the system of linearity itself, outlines should be, if not invisible, at any rate as unobtrusive as possible, for otherwise they would not "represent" (show) the outline of forms and objects but would appear on the canvas or the wall as fissures (*fessura*),[65] cracks similar to those that separate the various pieces in a marquetry design—where (as is proved by the example of Gauguin's technique) the overaccentuation of the graphic means seems to transgress the rules governing a representational structure that is founded upon the strict subordination of pictorial means, both graphic and colored, to the iconic functions of the painted image.

What needs to be resolved is the question of the extent to which the problematic of the sign, to which the semiological approach to the pictorial phenomenon may seem in principle to be subordinated, is linked to a predetermined cultural system, the historical and geographical limits of which coincide with its own limits of validity. What happens when the problematic of the sign is applied to nonfigurative works, creations in which the very notion of representation, and with it that of the iconic sign, is challenged? The present work does not set out to resolve that question directly. Rather, it aims first to show how the question of the sign and its special links with "drawing" arises within the representational system itself and, historically, has never ceased to be at work there. But such a way of being at work (which is to be understood in the sense of humidity being at work within wood) is one of the fundamental features of what we call the "history of art." This suggests an analogy between poetic practice and pictorial practice. Both seem to threaten the hierarchy of functions that constitutes the basis for the interchange of signs. Indeed, when linguistic formalism was first being elaborated, those who promoted it thought that the only way to define poetic language was to oppose it to everyday language. They suggested that in poetic language, in contrast to what seemed to them to be the rule in prosaic language, the formal linguistic constituents (sounds, morphological elements, etc.) acquired a value of their own, and the practical goal of communication (the transmission of information) took second place.[66] Formalism defined itself, on the strength of being a science of literature, as a science of "literarity" (*Litteraturnost*), the task of which was to define the distinctive features that characterize a literary object as such. Art theory, on the other hand, was never particularly concerned to pick out the specific difference of painted images, that is to say the features that distinguish painted images from every other kind of image. A painted image cannot be reduced to whatever it is that it depicts and/or represents (its "subject"), nor to its contents; and as for plastic formal "constituents" (if one can accept such an idea), they cannot be regarded simply as the instrument of communication whose success depends upon negating the matter there to be perceived. The phenomenological theory of the image has it that sensible elements that function as an analogous representative of the object targeted by the image must be "neutralized" (Husserl) or even "annihilated" (Sartre) in order for the imagizing synthesis to operate. But this theory fails to explain the paradox of a form that is not a signifier but does signify *itself*,[67] a form that is its own contents, matter which, right from the start, already functions as form, something that is signified that cannot under any circumstances be separated from its signifier. That is a paradox well suited to reveal the constitutive deviance of a practice in which means of expression are put to work in ways that are alien to the ordinary circuits of communication.

But the fact is that the conception of poetic language as a *deviation* from normal language tends to harden, as Julia Kristeva has pointed out,[68] into a commonplace that makes it impossible to study the morphology peculiar to poetry. In particular, one cannot limit oneself to affirming the autonomy of poetic means and suggesting that this corresponds to but the first stage, soon to be superseded, in the formal theory. And the above remarks are equally valid for pictorial "language," which never presents itself in the guise of a particular code that must be referred to a more general (iconic) code that incorporates it and whose rules it violates. The question of deviance, and the norm and the prohibitions upon which the norm is founded and by which it is precisely circumscribed, must be formulated in plastic terms, in terms *of* plasticity, for every innovation—even when it stems from causes of a social or ideological order—in the last analysis operates within the pictorial series, where it finds expression in a new distribution of levels, means, and functions. There is no a priori justification for reducing artistic products to signs or systems of signs, or for reducing specifically plastic effects to communicative effects. But if the sensible quality of the *analogon*—its grain, color, texture—is to be anything more than a mere accompaniment to the information that it conveys, it remains necessary to seize upon the relation that exists between the pictorial order and the order of signs, and the complementary relationship that unites the system of figurative rules—which confer legibility upon the image—with the set of formal variations that constitute it as a *painted* image.

The Normal and the Pathological

In this respect, Correggio's oeuvre provides a good historical point of departure. The contradiction between, on the one hand, the manifest interest of the decoration of the cupolas of Parma in the eyes of his contemporaries and his successors and, on the other, the fact that more than a century passed before this decorative solution that appears to have found its definitive form right from the start came to triumph—that contradiction suggests that some implicit prohibition was indeed transgressed. It is important to determine the nature of that prohibition and assess its impact, for it may have operated either at the level of plastic means or else solely at the level of what Correggio's paintings signified. In the domain of art, all innovations appear deviant, not only in relation to what has gone before, which has come to be seen and respected as a norm, but also in relation to practice, to custom, to functions that are considered to be regular. But there is deviance and deviance: sometimes it can be tolerated by the norm, which seems to accommodate it as being possible to integrate within the "system"; at other times it seems to upset the balance and the way that the system functions, and in that case society, through its institutions, may work to reject it and sweep it aside.[69] But again, such deviance only reveals its full character and precise impact once it is restored to its context, that of a practice the norm of which is indissociable from the transgression that produced it and that establishes it to be a norm.

If the sensible texture of the icon appears, in certain conditions, to be the principal organizer of the "message," does that mean that a gap appears between the figurative (iconic) functions and the pictorial ones, because the emphasis placed on the material treatment and components of the image seems to hamper the synthesis that produces the image and the legibility of its figures? If we took Formalism, as originally expressed, literally, we would be tempted to extend to communication what René Leriche said about the "normal" aspects of life, namely that what is "healthy" is "life [communication] amid the silence of the organs":[70] in "normal" conditions of communication very little attention is paid to the sensible substance of expression, for to do so would introduce more or less extensive disorder into the transmission of messages. But what appears pathological with regard to the ordinary practice of expression appears, on the contrary, to be the norm in the realm of art. Literature is an exercise of writing that restores a "tongue" (in the sense of "language") to the reader, just as a tongue in the physical sense, the organ of the body, conveys to the subject the message that that body belongs to him or her. (But should this body speak for itself, either in extreme pain or in extreme pleasure, then the "subject" has to recognize that he or she is possessed by it, as he or she also is possessed, from birth to death, by language).[71] As for painting, its evolution since Cézanne shows well enough that there can be no vision—as Merleau-Ponty observed—unless it is connected with a *body* (a social construct, just as language is) and is passed on by that body to other bodies and to nature itself.[72] But if literature and painting have these powers, it is only because they are intimately linked with the very substance of expression, with its very material. And, once again, such an alliance inevitably seems paradoxical, or even pathological, from the point of view of ordinary forms of expression.

STAINS

The functions assigned to the element "cloud" in the works of Correggio manifest the determining role that is played by the sensible component—paint—in pictorial expression. At the same time they convey the fantastical connotation that arises from the disproportion between the thickness of the layer of paint deposited on the canvas and the vast effects that it aims to produce. On a conceptual level, a "cloud" is an unstable formation with no definite outline or color and yet that possesses the powers of a material in which any kind of figure may appear and then vanish. It is a substance with neither form nor consistency, onto which Correggio imprints the emblems of his desires, just as Leonardo, before him, imprinted his onto the stains on a wall. That comparison is no more arbitrary or anachronistic than the pathological connotation ascribed in sixteenth-century painting to an element—a lure?—that was alien to the order of expression (drawing), if not to that of "truth" (color). This is borne out by Vasari's life of Piero di Cosimo, which, it is worth noting, follows on immediately after that of Correggio, which, for its part, is separated from that of Leonardo only by the biographical portrait of Giorgione. Vasari's life of Piero di Cosimo ends with a paragraph in which Vasari tries to demonstrate the strangeness of the painter's character and behavior, the bizarre nature of his mind, and the attraction that he felt toward difficult and unusual things (*la stranezza del suo cervello ed il cercare che egli faceva delle cose difficili*). All those things were attested by the attention that he paid to cloud formations and also to the marks left on walls by the accumulation of "spit left by sick people," out of which he conjured up all kinds of fantastic battles and cities and the most grandiose landscapes.[73] One hardly needs the testimony provided by Vasari's text to see that such exercises are regarded as *symptoms*, symptoms of a sickness, the stigmata of which are not borne solely by the art of Piero di Cosimo. However, a characteristic that in the case of one particular individual may be regarded as neurotic takes on a quite different significance when it appears to be a response to some objective historical cause,[74] as is here the case: the very same interest in stains and cloud formations that Vasari regarded as a symptom of a morbid disposition in the case of Piero di Cosimo, in Leonardo's text appears in the guise of a *method* to promote inventiveness, a method long followed in China (albeit in different forms and in a different context) and that was to encounter equivocal fortunes in Western painting.

Clouds, Painting

THE WORK OF CHANCE

"I shall not fail to include among these precepts a new discovery, an aid to reflection (*una nuova inventione di speculatione*), which, although it seems a small thing and almost laughable, nevertheless is very useful in stimulating the mind to various discoveries (*a destare l'ingegnio a varie inventioni*). This is: look at walls splashed with a number of stains (*muri umbrattati di varie machie*) or stones of various mixed colours. If you have to invent some scene (*se havai a inventare qualche sito*), you can see there re-semblances to a number of landscapes (*potrai li veder similitudini di diversi paesi*) adorned in various ways with mountains, rivers, rocks, trees, great plains, valleys, and hills. Moreover, you can see various battles, and rapid actions of figures (*atti pronti di figure*), strange expressions on faces, cos-tumes, and an infinite number of things, which you can reduce to good, integrated form (*le quali tu potrai ridure in integra e bona forma*). This hap-pens thus on walls and varicoloured stones, as in the sound of bells, in whose pealing you can find every name and word you can imagine (*come del sono delle campane che ne loro cocchi vi troverai ogni nome e vocabolo che tu oti imaginerai*).

"Do not despise my opinion when I remind you that it should not be hard for you to stop sometimes and look into the stains on walls, or the ashes of a fire, or clouds, or mud, or like things (*nelle machie de muri, o nella cenere del fuoco, o nuvoli, o fanghi, o altri simili lochi*), in which, if you consider them well, you will find really marvellous ideas. The mind of the painter is stimulated to new discoveries, the composition of battles of ani-mals and men, various compositions of landscapes and monstrous things (*e di cose mostruose*) and similar creations, which may bring you honour, be-cause the mind is stimulated to new inventions by obscure things (*perche nelle cose confuse l'ingegnio si desta a nuovi inventioni*)."[75]

Even in antiquity attention was paid to images the formation of which was due solely to natural mechanisms, figures presented as a result of chance or the work of Fortune: configurations in the sky,[76] "imagis-tic" stones,[77] *mirabilia* of various kinds, not forgetting that, according to Pliny, "variations of colour and shape are seen in the clouds in proportion as the fire mingled with them gains the upper hand or is defeated."[78] But it was only in the Hellenistic period that—whether as a rhetorical figure of speech or as a theory—a connection was made between such phenomena and artistic products.[79] Pliny himself considered the phenomena primarily as an extra argument that demonstrated the extent of the powers of For-tune, the same Fortune that was not above sometimes coming to the aid of particularly ambitious or demanding artists, enabling them to produce the desired illusion by means that were nothing if not automatic. One example is provided by Protogenus who, despairing of reproducing the appearance of the froth issuing from a dog's mouth, was said to have hurled a sponge at his painting, thereby, without intending to, achieving his goal: a repre-

sentation the truth of which owed nothing to art (art which, when manifest, contradicts the truth of imitation).[80] It was an anecdote frequently
pressed into service: Leonardo himself reused it, changing the protagonist
to Botticelli, who, he related, decided that it was not worth bothering with
a particular landscape since all he had to do was squeeze out a sponge on
the required spot in order to produce a most satisfying representation:[81]
Fecitque in pictura fortuna naturam, Fortune/Chance produces the effect of
nature in painting. Pliny likewise believed such haphazard inventions to
be so many "lucky strokes of fortune" reserved for consummate artists. It
was Philostratus who, putting the words into the mouth of Apollonius of
Tyana, first posed the following question: "And the things that are seen in
heaven whenever the clouds are torn away from one another, I mean the
centaurs and stag-antelopes . . . and the horses, what have you got to say
about them?"[82] Are they works of imitation, in which case it would have to
be recognized that God is a painter who, leaping from his chariot, delights
in drawing pictures in the sand, like a child? The hypothesis is untenable,
for the figures that appear in the sky make no sense (*ta t'auto men asēma*),
they are purely chance effects. It is man who, being naturally inclined to
imitation, confers a meaning upon them as well as a relative permanence,
by associating them with the idea of the creatures that they evoke. Only
a person who has been trained in drawing can produce imitations (*apo-
mimeisthai*) that deserve to be called paintings (*graphikai*). Painting, for its
part, is purely a matter of art.[83]

As H. W. Janson noticed, the above is a key text an allusion to which
may perhaps be detected, learned though it may be, in the curious "imagistic" (or better still "informed") clouds that are to be seen in two paintings
by Mantegna. In his *Saint Sebastian*, now in Vienna (reproduced on the
cover of this book), there is a cloud in which can be glimpsed the figure
of a horseman, which corresponds to the sculpted fragments—a head, a
sandaled foot, a number of busts—with which the ground is strewn. It is
as if, through the effect of the contrasted interplay of *similarities*, which,
according to Michel Foucault (to whom we shall be returning), dictated
the configurations of knowledge in the Renaissance, these archaeological
items were rivaling the inspiration of the Nature that this humanist painter
claimed to take as his master—Nature, the figures of which were hence-
forth taken as *signs*, on a par with those that ancient texts preserve.[84] In
the same way, in *The Triumph of Virtue* (now in the Louvre), the gigantic

face that appears in the clouds, close to a crumbling section of mountain, seems, to borrow Lucretius's words,[85] to echo the anthropomorphic shrub (to be found in so many ancient fables, beginning with that of Daphne) that occupies the left side of the scene. The painting, having first opened up a space of analogies, now opens up a space of metamorphoses, metamorphoses from which art has always drawn inspiration, as from a spring. Painting is bound to gain from an "invention" that enables the painter to find inspiration in the ever changing shapes of the clouds,[86] provided that what he sees in them are images that demand to be interpreted, deciphered, and eventually produced by painterly means, just as the sound of bells may conjure up a whole collection of names or words for anyone silently trying to interpret them (although in this case, of course, the reference to *phone* is clear).

MIRRORS

It is precisely at this point that the exercise, if not the method, becomes questionable. For although cloud may offer a particularly inspiring prop for daydreaming and flights of the imagination, this seems to be thanks not to its outline, but on the contrary to whatever it is about it that defies the regime of delineation and pertains to its material nature, its "matter" aspiring to "form." It is on that account that it seems to epitomize the "vaporous" style that Correggio, rightly or wrongly, is said to have initiated. If that is the case, one can see how it was that the ancient authors were not given to regarding clouds as a source of inspiration for painters, and even less to seeing in their ever changing forms a paradigm of pictorial composition. If painting was first and foremost drawing (one and the same Greek term, *graphein*, being used to designate the acts of both writing and drawing *or* painting), so that painting started with outlines,[87] and color played no more than a secondary role, as a kind of *extra*, the attractiveness of which rendered it all the more disturbing and suspect, then the connotation of cloud was right from the start that of an element outside the norm (and, as we shall see, it is doubtful whether the painters of antiquity allowed any room for it in their representations [see the first section of Chapter 4]). Proof—proof that concerns painting—that cloud (in the physical, meteorological sense of the term) is linked with color rather than outline is to be found in book 3 of Aristotle's *Meteorologica*. This discusses the apparitions that can be seen forming in the sky on a clear night ("chasms, trenches, and blood-red colours"),[88] or when clouds are close to the sun and are reflected by some liquid surface: the cloud, in itself colorless, then seems to be full of "rods."[89] Aristotle offers the same explanation for all these phenomena, namely *reflection*. But in the cases of solar rods, halos, and mock suns, what is reflected is not shape but only color. Cloud, constituted as it is of many more or less dense and aqueous components of minute dimensions, functions in its mass as mirrors that, unlike those that reflect shapes, show only colors:

Colours only are reflected in mirrors that are small and incapable of subdivision by our sense of sight. In these shape cannot be reflected. If it could be, it would be capable of subdivision, as all form has the characteristics both of shape and of divisibility. Since, then, something must necessarily be reflected, but shape cannot be, the only remaining possibility is that colour should be.[90]

We may leave aside the assertion—the theoretical importance of which needs no underlining—that the notion of shape is linked to that of *divisibility*, whereas color, in each of these microscopic mirrors, appears unrelated to shape "and, as a point, incapable of subdivision."[91] But what does need to be retained from this attempt at a scientific explanation is the distinction that is introduced between the refraction of shapes and that of colors. Cloud interests Aristotle from the point of view of optics, insofar as the effects of reflection that it occasionally produces elude the order of shapes (but not, it must be emphasized, that of *specularity*). And Aristotle indeed paid no attention to the shapes that cloud formations take on, being interested only in the phenomena of color refraction that they dramatically display. But what can be the relation of those phenomena to painting, if it is true that—as Philostratus was to declare—painting should only resort to colors for the purposes of imitation? He believed that if imitation was not its purpose, painting would constitute nothing but a vain play with coloring materials, a kind of cosmetology.[92]

THE TWO DIMENSIONS TO THEORY

It was necessary to pose the question in the above terms in order to show what is at stake behind certain occurrences of cloud in painting: namely, the fate of a particular *order*. However that may be, for Aristotle cloud and in general meteorological phenomena do represent a kind of deviation, or at least signal a change. They constitute the *sign* or *indication* of a variation, if not a perturbation, of which they are the harbingers. Does that mean that /cloud/ has the same significance in the pictorial field, and that its massive occurrence in painting from the sixteenth century on can also be understood as a harbinger, a symptom? The moment that landscape art became officially acceptable, despite the qualms indicated by the anecdote about the "sponge throwing," which Leonardo associates with Botticelli, Pontormo was to recognize /cloud/ to be one of the most difficult parts of painting, one that, in his opinion, the painter should practice assiduously—even going to the lengths proposed by Leonardo, namely studying the stains on a wall as a model.[93] It is as if Pontormo meant to associate this sign, or emblem, with a definition of painting that he had produced. He saw painting as a canvas woven in hell, ephemeral, and of little worth. If its surface finish or superficial coating (*quello riciolino*) were removed, nobody would pay any further attention to it.[94] The implication is that a theory of /cloud/ might give one access to the most secret impulses of a practice readily classified as *infernal*, the moment it began to borrow from things that were "confused." "Theory" should here be understood to have a double meaning: first, to justify carrying out a few historical investigations, and then, by dint of further probing and extending the analytical field, as a means of introducing a reasoned reconstruction capable of explaining the equivocal, evolving role played by *cloud* in Western painting.

2

Sign and Representation

> The object of representation can be nothing but a
> representation of which the first representation is
> the interpretant.
> —Charles Sanders Peirce, *Principles of Philosophy*

'Iconomystica'

Use Values

A sign—in the problematical and deliberately ambiguous sense in which the word is used in the present text, so as to assess the limits of the validity of the concept within the field of pictorial analysis—a sign has a use value that defines it. A rapid survey of the occurrences of *cloud* in the works of Correggio and his immediate contemporaries seems to lead one to distinguish, through comparison, between the most conventional valences of the sign and other, quite exceptional and possibly transgressive combinations of such signs. The latter, essentially, are represented in the "celestial" cupolas of Parma: here, the cloud motif is used as a constructive element that serves to designate—once again to borrow Burckhardt's expression—a space.

The figures of Michelangelo on the ceiling of the Sistine Chapel had no need of clouds in order to move from one spot to another through the air. And, except in the "sky" of *The Last Judgment*, in which angels carrying the Cross and the pillar of the Passion are flying about, cloud, as a constructive element, has no part to play in the composition. Michelangelo's organization of the scene does not imply a negation of the edifice. On the contrary, it introduces into a constructed volume devoid of any visible structures a fake network of cornices, pilasters, and grisaille arches, into which are inserted a series of scenes conceived as so many easel paintings.[1] As for the figures of the prophets, the sibyls, and their train of slaves, disposed on the lower parts of the vaulted ceiling, these are positioned in trompe l'oeil architectural features, and not, as Burckhardt points out, on banks of cloud, as Correggio's Evangelists were to be in the pendentives of the Parma cupolas. Conversely, leaving aside the matter of the fantastical connotation that may be attached to them, the use that Correggio makes of cloud in some of his panels is prompted by the conventional practice that made it possible to distinguish between, or rather to associate within the same composition, two levels that appear to be mutually exclusive, namely the "terrestrial" and the "celestial." A number of other examples also spring to mind: among the "divine" Raphael's works, *The Virgin of Foligno*, which is often regarded as a prototype (even though it was not without precedents), the apparitions of Yahweh to Isaac, Jacob, and Moses, in the Vatican Loggia, and above all the famous *Dispute over the Holy Sacrament*, in which the arrangement of the two levels, one above the other, is effected on a monumental scale (while on the ceiling of the very same Stanza della Segnatura, oculi in the trompe l'oeil setting display allegorical figures positioned on clouds). Then there are the enigmatic *Virgin of Saint Sixtus* (now in the museum in Dresden), who seems to be advancing toward the spectator across a carpet of clouds; the paintings in the Hall of Psyche, in the Villa Farnesina, which is always evoked in connection with Correggio's Camera di San Paolo and his *Loves of Jupiter*, in which the mythological figures are positioned among arcades of foliage, in little gaps in the leaves, where a few wisps of cloud have collected, while on the ceiling the gods make their way, in procession, toward a cloud-swathed Olympus. Yet another example is the small panel of *The Vision of Ezekiel* (Florence, Pitti Palace): here, the landscape in which the tiny figure of the prophet can be glimpsed is reduced to a narrow band of land and sea relegated to the lower part of the composition and represented on quite a different scale from the silvery bank of cloud upon which can be seen a Creator who has a look of Jupiter about him, in a burst of golden glory and surrounded by angels and symbols of the Evangelists.

ZURBARÁN

Raphael's work was devoted to conciliation, interaction, and *composition* (the area of painting in which, according to the Academy, he excelled). But although his works are said to be so eminently "legible," they nevertheless pose great problems of interpretation, given that their novelty borrows so subtly from a superficial conventionality. So they do not really provide a good term of comparison. Let us, rather, consider a corpus of works that is later (but contemporary with the triumph of the great ceiling art of Rome), and that, despite being alien to the aristocratic circles of humanist culture, testifies to many of the most common impulses of iconic communication as understood by the classical period. The great religious cycles painted by Francisco di Zurbarán in Seville between 1629 and 1640 are linked, through the processes by which they were commissioned, to the activities of the Spanish monastic orders, at the crossroads of the Old and the New worlds; and from the point of view of iconography (whatever their prestige and pictorial splendor), they belong to the traditional horizon of pious imagery. Whereas Raphael won an international audience thanks to the diffusion of his works through their engraved reproductions, Zurbarán, in contrast, like so many of his contemporaries, was happy to seek models for his compositions in prints, most of them either German or Flemish, and often enough he even exaggerated their most traditional aspects.[2] From a semiological point of view, there is probably no notion more equivocal than that of *archaism*. As Jakobson remarked, "when the temporal factor comes into play in a system of symbolic values, it becomes a symbol itself and may be used as a stylistic means."[3] But whether deliberate or not, the archaism, which critics are generally in agreement in regarding as one of the predominant features of Zurbarán's art, does offer an extra guarantee that in his case we are presented with a conventional repertoire and accepted norms. It remains to be seen what use is made of cloud in such a context, for purposes and in ways that are, quite clearly, not solely those of "painting," as museums understand the term.

Images

ECSTASY AND RAPTURE

Zurbarán's *Vision of the Blessed Alonso Rodriguez* (Plate 2) was originally designed to adorn the altar of the sacristy of the church of the Jesuits in Seville. The lower part shows Father Rodriguez, the janitor of the College of Jesuits in Majorca and the author of famous mystical works, kneeling at prayer in a dark passage that opens on to an architectural vista. A copy of Thomas à Kempis's *Contentus mundi* has just slipped from his hand to the floor, which is paved with black and white tiles rendered in accordance with strict perspective. At his side stands an interceding angel. Both are gazing at the upper part of the composition where, framed by clouds filled with a mass of cherubs' heads, Christ and the Virgin are seated, each holding a heart-shaped purse from which pour rays of light that are directed at the blessed Rodriguez's breast. On a cloud positioned slightly lower and to the right stands a group of angel-musicians. The desire for realism — insofar as the word is meaningful in such a context — seems to compete with the expression of a somewhat equivocal mysticism, but essentially the signifying process depends upon the opposition between the interplay of shadow and light in the constructed space and the dazzling brightness diffused from the cloud. The division of the picture into two levels, the one terrestrial and ordered according to the rules of linear perspective, the other celestial and of indefinable depth, is, however, by no means exceptional in the work of Zurbarán. Let us consider just a few of the more telling images in this respect. *The Portioncule* (The little cell of Saint Francis), in the museum of Cadiz — probably contemporary with the painting mentioned above (1630) — shows Saint Francis on the flagstoned floor of his cell, in ecstasy before the apparition of Christ and a mediating Virgin, amid cloud and surrounded by angels. In *The Pentecost* (also in Cadiz), a divine cloud is spread above the Virgin and the Apostles grouped around her. In a later work, *The Mass of Father Cabanuelas*, part of the Hieronymite cycle of Guadalupe (1638–39), it is likewise amid cloud, but one of rather modest dimensions, that the symbol of the Eucharist appears to the priest kneeling before the altar. In the *Saint Jerome* that is part of the same cycle, the saint is lifted on a cloud intermingled with putti up to the heaven to which, in yet another painting in the same museum, *Saint Joseph Crowned by Christ* has already acceded.

It does not seem particularly difficult to understand these images. The figurative schemata adopted by Zurbarán are perfectly conventional, even when not directly borrowed from pious prints, as in the case of *Saint Bonaventure at Prayer*, now in the Dresden art museum, painted for the Franciscans of the College of Saint Bonaventure. The composition of this picture is very close to that of *The Mass of Father Cabanuelas*, and is imitated from an engraving produced in Antwerp in 1605.[4] The constructional ploy of a cloud to introduce a divine group or symbol into a perspective composition is extremely common in religious paintings of the sixteenth and the seventeenth centuries. The interest of such compositions, most of which depict an apparition that the image presents to the spectator both as a "real" scene and as a miraculous vision, does not chiefly lie in the spatial dichotomy that they impose. Rather, it lies in the modalities of communication established between the terrestrial and the celestial levels and, within each of these, between the figures that inhabit them. In some, the elect figure remains firmly set upon the ground, upon the earth (satisfied with this world, as he is?), but seems to welcome the vision by which he is honored and also to display it to the spectator, who accedes to it thanks to his mediation (*The Vision of Blessed Alonso Rodriguez, The Portioncule*). In others, the saintly figure is swept up to heaven or temporarily lifted out of ordinary space. In yet others (dramatic compositions at which Zurbarán excels), the vision is granted to one figure alone, while all those surrounding him are excluded from it; in this way, the spectator is privileged, thanks to the image, to take part in two modes of perception simultaneously: Saint Bonaventure's companions watch him respectfully as he is absorbed in his meditation; but they have no inkling of the vision, the advent of which tears through the network of sensible certainties.

THE MYSTICAL CLOUD

It is tempting to draw a number of comparisons between these images and some of the characteristic themes of Spanish mysticism. The writings of Saint Teresa are full of descriptions of the ascents to heaven of men "greatly committed to prayer," resplendent in glory, and in a few cases (as an extraordinary favor) accompanied by angels or even by Christ himself. In her *Libro de su vida* (Book of her life), the saint also describes "ravishments," "ascents," "spiritual flights," even "transports," as follows: "The Lord takes the soul and raises it utterly from the ground, just as clouds or the sun attract vapor, or so I have heard it said. The divine cloud rises into the heavens, carrying the soul with it, and begins to disclose to it the splendors of the kingdom that is prepared for it." It is worth noting the mobilizing and initiatory force that Saint Teresa attributes to the divine cloud, and the power of attraction that it is believed to exert upon the soul that it carries along with it (the separation between the two levels being thus clearly underlined). Zurbarán's saintly figures may not leave the ground (in conformity with the mechanism of *union* in which, unlike in ravishment, "we remain on our own terrain"). But that was not the case for Saint Teresa, whose body, like that of Saint Diego, in Murillo's *Angels' Kitchen*, did sometimes lose all its weight: "My soul was uplifted and ordinarily my head followed, it being impossible to restrain it. Sometimes even my whole body was also uplifted and no longer touched the ground." This phenomenon usually occurred without witnesses; but when it did happen in public, the saint's followers were obliged to hold it down. "However, that very seldom occurred."[5]

In the writings of Saint Teresa, cloud and banks of cloud thus play a role comparable to that imparted to them in the paintings of Zurbarán, and also in the works of Correggio. But there is one difference between the two painters: in his paintings in the cupolas of Parma, Correggio allows the greatest possible room to an event that not only contradicts the norms of common experience but also interrupts the regularity of the constructive order. Disposed all around the tambour, the Apostles remain "inside" the cathedral, on *this side* of the balustrade, as does the saint huddled in a corner of the cornice in the church of San Giovanni Evangelista. That is a simple effect of proportion between the part of the composition devoted to the manifestation of the divine and that reserved for the world below; not that the opposition between the two levels seems to operate entirely to the benefit of the *other world.*[6] Not only does the *cloud* liberate those whom it supports from the laws of gravity, but at the same time it shows how profane space may open onto *another* space, which imbues the former with its truth. Where spiritual flights, ravishments, and miraculous visions are concerned, from Giotto's or Zurbarán's *Saint Francis* across the board to Madame Bovary,[7] and including Bernini's Saint Teresa, cloud is the obligatory accompaniment — if not the motor — of ecstasy and all other forms of ascent or rapture. More generally, it is regularly associated with an irruption of *otherness* or of the *sacred*. Sometimes cloud parts to allow the elect person to perceive the object of his or her adoration—a gratification that is denied to the followers of Saint Bonaventure and the companion of Father Cabanuelas in Zurbarán's paintings, for divine realities can only manifest themselves through rents in the screen that conceals them from common awareness. At other times cloud appears as an immediate manifestation of the sacred in the guise of a divine expanse of cloud that descends to share in the exile of human beings, as did once the pillar of cloud that served as a guide to the children of Israel when they departed from the land of Egypt.[8]

The Hierophanic Code

SIGNS/SYMBOLS

If the theme of rising into heaven and the various forms of ascent or "crossing over" constitute a perhaps necessary component of any vision or ecstasy, and if cloud and banks of cloud seem directly involved in celestial transcendence, that may be because, even before it acquired a religious valorization, the sky seemed, from a phenomenological point of view, to be a place of transcendence, at once the origin of all material things and also of law, the seat of power, and sovereignty.[9] Does that mean that, by studying the canvases of a seventeenth-century provincial painter, we can accede to an order of data that, because of their essential nature, elude history, or even to an a priori form of religious consciousness? To borrow from the vocabulary of the history of religions, cloud seems to have a hierophanic significance; in other words, it is an object that manifests that which is sacred, or contributes to its manifestation. However, the identification of an isolated hierophanic element is not, in itself, particularly interesting. It is of no theoretical significance so long as one is content to regard it as an archetypal emblem stemming from a symbolism claimed to be eternal, an emblem that underlies a whole variety of religious forms without ever becoming exhausted by any of them.[10] The distinction between the sacred and the profane presupposes an ever-renewed division between objects and beings, and the same applies to signs of the sacred as applies to any other sign: they belong to complexes, historically constituted systems, in which the lateral link between one sign and another is more important than the direct, vertical link between signifier and signified by which a symbol is currently defined.

SEQUENCES

Even a very superficial examination of the work of Zurbarán suggests that it might provide prime material for research of a comparative order. Many of the elements, the "figurative objects"[11] that are noticeably present in Zurbarán's painting—a limited number of elements that reappear remarkably frequently—may be described as "hierophanic." Significantly enough, many of them are elements that a phenomenologist would consider to belong to a symbolism closely related to celestial symbolism. The association of cloud with a massive pillar, generally positioned at the axis of the composition, clearly stems from the "symbolism of the centre," studied by Mircea Eliade.[12] However, that reference is, in itself, of no operative value. The same pillar around which the terrestrial scene is arranged and the top of which is lost in the clouds, thereby establishing communication between earth and heaven, regularly reappears in Zurbarán's paintings, from his *Annunciation* and two *Adorations* in the Grenoble museum, across the board to *The Mass of Father Cabanuelas*, and including *The Apotheosis of Saint Thomas Aquinas*, in the museum of Seville. In *The Conversation Between Pope Urban II and Saint Bruno*,[13] painted for the charterhouse of Santa Maria de las Cuevas, a massive pillar occupies the center of the image, with the two principal figures positioned symmetrically, one on each side of this axis. In this case, the top of the pillar is not lost in the clouds but is concealed by both the canopy above the papal throne and a curtain draped above the saint. It is a sequence that is also to be found in the portrait of *Gonzalo d'Illescas, Bishop of Cordova*, in the Guadalupe cycle of paintings: the pillar stands in the center of the composition, while a billowing curtain "resembling a cloud" seems to signify that the prelate is writing at God's dictation. Those two elements also serve to introduce into the composition a *distance* between the bishop, working in his den, and an outdoor scene, set in the middle distance, showing monks dispensing charity at the entrance to the monastery. But the fact that one object may be substituted for another in this way within the economy of a hierophanic text (a fact already noted in connection with Correggio's *Virgins*) is instructive: /cloud/ has no particular meaning in itself; its only meaning is that which stems from the relations of consecutiveness, opposition, and substitution that link it to other elements in the system.

The Functions of Representation

EXPERIENCE AND ITS FIGURES

What of the status of an image, a figurative representation, when the painter seems to have borrowed his data from an order other than the order of painting, so that in the last analysis his role seems reduced to that of a mere illustrator trying, by means of his art and for the purposes of edification, to record the memory of unnamable experiences described by mystical writers?[14] The relations between the reality of mystical experience and its literary or plastic expression are, indeed, extremely ambiguous. After describing, in the terms quoted above, the ravishment that she experienced, Teresa adds the following remark (one that is extremely valuable on account of what it implies about the functions of metaphor [or of "images"] in the economy of the phenomenon of mysticism): "I do not know if the comparison is exact. But at any rate, that is what truly happens."[15] But if the representation, even if erroneous, of a physical process—in this case that of evaporation—can be accepted as a *form* of the experience, then that experience can no longer be described as what comes first, and it is impossible, strictly speaking, to distinguish between the "reality" of the ecstasy and its "expression": *como se vede en la pintura* ("as is seen in the painting"): in the last analysis, the experience is identified—as is dazzlingly clear in Saint John of the Cross[16]—with its poetic expression. But if it is possible for the experience to borrow its form from an image, then surely the painter is not reduced simply to noting and recording a vision in the elaboration of which he played no part. At his own level, and using his own means, he works to establish a connection between man and God, and to define the means and the configuration of that communication.[17]

And the fact is that, even if there does come a point in mystical experience when one's powers and the imagination itself are abolished, nevertheless neither Saint John of the Cross nor Teresa of Ávila condemned the use of painted or sculpted images. On the contrary, they pitied heretics who, through their error, had deprived themselves of such a source of consolation and—even more—of the help that the soul may derive from these when the Lord is absent or leaves it "in aridity."[18] Saint Teresa reckoned that the contemplation of an image of the Savior could facilitate dialogue with him,[19] and Saint John of the Cross, for his part, did not hesitate to blame people who treated such images with scant respect, along with the craftsmen "who sculpt them so badly that, instead of encouraging devotion, they remove it," and who ought to be banned from exercising an art that they practice in such an inept and clumsy fashion.[20] To be sure, images should not be valued for their own sake or for their sensual attractions; but clearly their functions are not solely decorative, and the contemplation of works of art can sometimes encourage and sustain prayer.

ON THE RIGHT USE OF IMAGES

That was a view that was in perfect conformity with the position of the Council of Trent on *the legitimate use of images*,[21] and also with the teachings of Saint Ignatius. The latter approved resorting to both great and minor means in order to establish the "colloquium" or dialogue that was the consummation of prayer: after the recollection of a story or theme that was the subject of the meditation, each of the necessary "spiritual exercises" involved a *composition*, which, in its turn, depended on "seeing with the eyes of the imagination" the material place in which the biblical scene took place, but also, figuratively speaking, the doctrinal point or article of faith that the person in retreat wished "to contemplate" (the cave of the Nativity, the house of Anne or Caiphas, the "vale" in which human beings, compounds of soul and body, were exiled, among the unreasoning animals).[22] Such a "colloquium" climaxed in a form of total representation (an "application of the senses") that was supposed to harvest the fruits of the day-long meditation.[23]

When, in the seventeenth century, the Jesuits came to regard images as they did poetry or the theater, that is to say as some of the most effective weapons of a propaganda totally founded upon *representation*, they were, in this respect, altogether faithful to Saint Ignatius and the spirit of the Council. But whereas the mystic writers aimed above all to open up ways of meditation to those in retreat, in the solitude of their oratories, the clerics were to pursue far more extensive ends designed to get images to serve both the training of Church leaders and the edification of the masses. *Iconomystica*,[24] the science of images that "profitably, energetically, and deliciously" teach the mysteries of faith, can on occasion take on the appearance of an esoteric technique. But it is one that only makes sense when organized to produce effects that, far from being contradictory, are in fact complementary, or at any rate presuppose the existence of a common language, a language that is comprehensible to every individual at his or her particular level and perspective, among both the common people and those who claim to be its "guides."

"There is nothing more delectable or that more easily gets a thing to slip into the soul than painting, nothing that etches it more deeply on the memory, or that more effectively affects the will, giving impetus to it and forcefully moving it."[25] The list of functions imputed to painting in the *cursus studiorum* suffices to show that images were no longer created for a closed caste, as those of Raphael sometimes were. The very same means were used to enrich the experience of a handful of solitary figures as were used to indoctrinate the masses, without this being considered contradictory. The same period that produced the great mystics simultaneously saw the Church triumphantly establishing itself in the contemporary age; and it was thanks to the explicit presence of this debate in his painting that Zurbarán appeared as one of the rare artists who had drawn the best of his inspiration from the teaching of the Church, unlike Murillo, who was to find himself obliged to fall in line with the norms of sentimental religiosity. *The Conversation Between Urban II and Saint Bruno*, in which the pope is seen endeavoring to persuade his confessor to renounce his retreat from the world and take a more active part in the work of the papacy, suggests well enough the mainsprings of a propaganda that was only fully effective when it presented the masses with images of holy figures whose interventions in the ordinary world were all the more impressive because they had shown themselves capable of totally withdrawing from its seductions. In truth, the Church had long since realized that the teaching of a figure such as Loyola provided it with one of the most appropriate means of ensuring for itself the contingent of saints that it required from each generation:[26] the mystical current and the worldly current appeared as the two complementary sides of a single policy that aimed to affect the greatest possible number of individuals through the intermediary of a handful of exceptional figures (or *images?*). If it was to edify the masses, the mystical experience had to acquire a form in which it could become the object of a *representation* that was at once legible and efficacious. The paradox of Zurbarán's art is that of painting of monastic inspiration in which, in the last analysis, the demands of representation and symbolic efficacy override all other considerations. Hence the ambiguous status assigned, in the pictorial order, to the elements of an iconography that was at once intellectual and figurative and that functioned on several levels at the same time.

The Two Modes of Representation

Writing and Representation

PRESENTATION/REPRESENTATION

The late fifteenth and early sixteenth centuries produced many religious compositions in which a saint, a prophet, a martyr, or a Church dignitary presents the spectator with a group (generally a Virgin and Child) that takes on the properties of an icon. In Correggio's cupolas and—in a quite different context—the paintings of Zurbarán, it is a matter no longer of the presentation of an image, but rather the representation of a vision. The point—and also the reason why the Parma paintings and Zurbarán's pictures elude the norms of pious imagery—is that the spectator only accedes to the miraculous vision thanks to the intervention or intermediary of an intercessor, himself in many cases assisted by further intermediaries, some angelic, others not. The point is worth repeating: it is thanks to the painting that the spectator gets to take in at a glance both the representation of an individual at prayer and that of the apparition that is supported by the latter's presence (as is clearly denoted by the attitudes of Zurbarán's Alonso Rodriguez and Saint Francis and that of the Apostles in the cathedral of Parma, who seem to be quite literally supporting the celestial scene on their wide-flung arms). It is a support, in the form of an intermediary, that may in some cases be indispensable: although a certain application may be needed to spot him, a *witness* is present, even if he is no more than a tiny figure (as in Raphael's *Vision of Ezekiel*) or is virtually invisible, as in the case of Correggio's Evangelist, so well concealed by the jutting cornice that for a long time the painting was regarded as an Ascension rather than the vision of Christ that appeared to Saint John on the island of Patmos, as described in the Apocalypse.

Yet Correggio did not follow the text of Scripture to the letter: "Behold, he cometh with clouds; and every eye shall see him, and they also which pierced him."[27] The figure of Christ, far from being set on a swathe of cloud (as are the figures of the Apostles who encircle it) is suspended in an indefinite space, against a golden background, and his posture shocked many of the painter's contemporaries, as indeed it did Burckhardt.[28] To be sure, the adoption of foreshortened perspective (*dal sotto in su*) created difficulties for a representation of the descent of Christ on a carpet of clouds, following the principle adopted by Raphael in *The Virgin of Saint Sixtus*. But even before the sixteenth century and in the seventeenth still, the staging in a painting (as in a theater) of raptures, ascensions, and apparitions of Christ seems to have posed problems of a figurative as well as a theological nature. Far from following the teaching of the biblical texts, however explicit, it seems that, where both the Ascension[29] and an Apparition[30] were concerned, many were reluctant, not to recognize Christ's powers of levitation, but to pretend that those powers depended upon some kind of apparatus, an accessory with sometimes equivocal associations.[31] They could accept, at a pinch, that the elect might need to be carried by clouds up to heaven,[32] and even that the Virgin might. But Christ? Titian, who had set the Virgin of the Frari church (in Venice) on a solid-looking cloud, painted his Urbino Christ rising well above the ground without the help of any support. There is no shortage of examples of figurative representations, many of them contradictory, that reveal the existence, in the sixteenth century, of a problem of representation that is also made manifest by the fact that because Correggio did not see fit to follow the Gospel text to the letter, and dispensed his Christ from relying upon the form of support that the Apocalypse ascribed to him, the *Vision of Saint John* was for a long time regarded as an Ascension. It is as if the divine essence of the Son of Man would have been uncomfortable with such support, even when clouds seem to constitute the lining and edging of his glorious golden background, as is the case in the Parma cupola, the Frari *Assumption*, and many other compositions that either have a celestial setting or that incorporate both an earthly and a heavenly level.

The use of cloud or of banks of clouds in the depiction of an Ascension, an Assumption, or even a celestial gathering, a scene of triumphant glory or of a "Paradise," implied an implicit choice with regard to the nature and aims of representation.[33] In the seventeenth century, when Piero Testa planned to paint a "glory in Paradise" in the apse of San Martino ai Monti, in Rome, he declared that he would break with the principles then honored and would "represent it without clouds," saying that it was a very grave error to surround the Throne of Light and the Dwelling of the Elect with clouds. "For such places were havens of peace and everlasting serenity, and clouds only made them turbulent and dark."[34] But what can be said of a painter who, not content to use clouds as accessories to what must be called a dramatic "setting," makes cloud an attribute, if not the actual clothing of some of the figures in his drama?

In Dürer's *Apocalypse* (1498), the Evangelist is seen consuming the book proffered to him by an angel whose feet seem to be pillars but whose body is concealed both by clouds and by the rays that shine forth from his countenance. It is an astonishing image, but one that in fact follows the Gospel text to the letter.[35] Scripture abounds in texts that justify the use of cloud for purposes at once theatrical and figurative. It was in the form of a pillar of cloud that Yahweh led his people out of Egypt, according to Raphael's representation of the scene in the Vatican Loggia—or rather a pillar in which cloud alternated with fire, either because Yahweh wished to illuminate the night or because he desired to thicken its shadows (Exod. 13:21–22, 14:19–20). But that alternation does not suffice to account for the hierophanic functions of cloud and its relationship with the fire that it sometimes contained: far from cloud representing an immediate manifestation of the God of Israel, it appears primarily as a *screen* that he spread above his people in the daytime, as if to shade them (Ps. 105:38; Wisd. of Sol. 10:17–18), a screen that, when it covered the Tabernacle where he was to meet them and that was filled with his glory, prevented the children of Israel from entering (Exod. 40:34–38). It is written that Yahweh descended in that cloud in order to speak with Moses. But that does not mean that Moses, even if he spoke with him "face to face as a man speaketh unto his friend," was allowed to contemplate his countenance:[36] the glory of Yahweh is a consuming fire that can only appear *inside a cloud*.[37] One cannot look upon the face of Yahweh and live, so it must remain hidden, like his name, as the pharaoh's men were to learn to their cost. Moses himself only sees God's *back*. The glory of Yahweh, present in the cloud that both sanctions and cloaks its manifestation, only presents itself to the eye in a concealed guise, in the same way that his name, known to the people only as an unsayable tetragrammaton, can be pronounced by the priests only in a whisper, indistinctly, when prayers are recited.[38]

Sign and Representation

"Moreover, brethren, I would not that ye should be ignorant how that all our fathers were under the cloud, and all passed through the sea;

and were all baptised unto Moses in the cloud and in the sea. . . . Now all these were our examples . . ."[39] These lines from the First Epistle to the Corinthians not only manifest the importance of the cloud that announces divine glory in the mental iconography of Christianity; they also offer a kind of justification, in advance of the humanists, soon followed by artists and clerics and led by the Jesuits, for the setting up of a descriptive science of allegorical images, "devices," "emblems," and other symbolic illustrations that had been capturing people's imaginations ever since the discovery of Horapollon's *Hieroglyphica* in 1419. There, too, cloud that reveals even as it veils, that shows even as it conceals, appears as a key term, even an index to a discipline that is linked with the written word in its most fundamental endeavor. For that discipline aimed to discover a modern equivalent of hieroglyphs that were then — as can still be seen in Father Kircher — considered to be a purely figurative form of writing the signs of which call for a metaphorical interpretation.[40] *These things were figures of what concern us*: for the humanists, that applied as much to Egyptian inscriptions as to Bible stories, oracles from the prophets, the parables of Christ, and the fables of the ancients. The French translator of Cesare Ripa's famous *Iconologia* was to write that the inventors of such figures generally "covered them with thick cloud, so that the ignorant and the educated might understand them in different ways, and not penetrate equally into the secrets of Nature."[41]

Cesare Ripa refers to "le immagini fatte per significare una diversa cosa da quella che si vede con l'occhio,"[42] images designed to signify something other than that which is presented to the eye, images that are those of painters, that is of people who know how, by means of colors or any other medium, to represent something that is different from that *analogon* but that *resembles* the thing that is signified.[43] Images such as these, which (to borrow the terminology of Paolo Giovo)[44] constitute the signifying body of the figure, must arouse in the mind a desire to discover what is signified, the concept or idea that is its soul. But that "soul" is not something that each and every one is able to perceive, for it is concealed by veils and clouds of metaphor and allegory, even by extremely subtle ones that cast an aura around certain representations that are articulated, as Ripa's proem puts it, in the manner of a definition.[45] The philosophical connotations of that theory,[46] both Neoplatonic and Aristotelian, suggest that the same can be said of emblematic collections as of the Council of Trent's all too famous decree on the legitimate use of images, which, historically, falls in between the appearance of the first example in the genre, the *Emblemata* by Alciato (Lyons, 1530) and Ripa's *Iconologia* (1st ed., Rome, 1593), which set out a systematic theory relating to that use: far from constituting the point of departure of a period that is sometimes confidently described as emblematic,[47] this literature in many respects marks the end of over a century of humanist speculation. It is worth noting, simply, that although that literature is today enjoying a remarkable vogue among scholars, very few of them are concerned to come to terms with not only its iconographical but also its quite literally *iconological* implications, or to bring to light the principles, both semantic and syntactical, that govern the veritable figurative statements that are described and explained in many of these collections. Yet it is at this level, and only at this level, that there is a chance to recover a development that appears to have been of an essentially *theoretical* nature. Far from being contemporary with the appearance of the earliest collections, which simply confirmed the triumph of an allegorical vein to which Raphael had already paid tribute, this development was to make its mark only at the end of the sixteenth century, at which point Ripa endeavored to define, though still in equivocal terms, the secrets of an imagistic language that the eighteenth century would not hesitate to regard as *universal*.[48]

IMAGE AND DEFINITION: THE IMAGE OF BEAUTY

A single example borrowed from Ripa's *Iconologia* will suffice to indicate what seems to have happened, on the theoretical level, at the turn of the classical period. When it came to depicting "Beauty" (*se dovendo dipingere la Bellezza*), that is to say depicting it by means of an image articulated in the same way as a proposition, an image that conveyed the measure of the thing signified (*la cosa significata nell'immagine*), just as definition is the measure of what is defined (*come la definitione è misura del definito*), the painter could not follow the example of "moderns" who claimed to show the essential quality through its contingent effects. Beauty is not a predicate, but a proportion; but that is not to say that a beautiful and well-proportioned image expresses it adequately. For that would simply be a tautology (*un dichiarare idem per idem*), if not an attempt to represent something unknown by something equally unknown, and to claim, for example, to represent the sun distinctly by means of a candle. Such a figure would have nothing more than an accidental link with the thing signified: the "resemblance" (which in itself implies a difference) would be lacking,

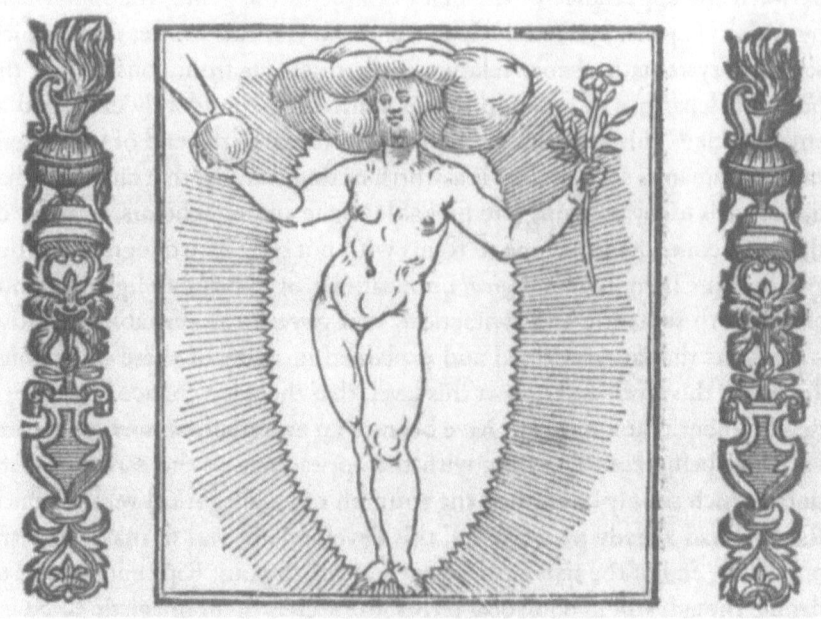

FIGURE 1. *Beauty.* A figure from Cesare Ripa's *Iconologia* (Rome, 1593).

and it is resemblance that constitutes the soul of an image (*non haverebbe la similitudine, che è l'anima*).

So how should a painter proceed when, using the means of his art, he wished to signify something other than what he "presented to the eye"? He should endeavor to produce a figure the parts of which correspond, term for term, to the parts of the thing that is signified, while being disposed in an order in conformity with that of the elements of representation,[49] in such a way that they constitute a whole so coherent (*che tutte queste parti facciano insieme un'armonia talmente concorde*) that it is hard to tell what is most admirable: the exact proportion that exists between the two things, or the good judgment of the person who has managed to organize the parts so well that the spectator need recognize only one thing, but one that is perfect and delectable. He will represent Beauty in the guise of a naked woman, holding in her left hand a lily and in her right a sphere and a compass, a woman whose head is lost in the clouds, while the rest of her body is hard to distinguish, presented—as it is—in the midst of dazzling *glory*.

As Jean Baudoin put it in his translation, the fact is that "it is no less difficult to paint her than to look at her, without being dazzled by the rays that surround her. And although she is unwilling to listen to the praises heaped upon her by Fame, who cannot speak in worthy fashion of her, nevertheless the heads of both of them are enveloped in cloud." [50] Here, glory and cloud are again, significantly, paired together. But it is worth noting the terms in which Cesare Ripa himself justifies the use of such signals:

Beauty must be painted with her head lost in the clouds; for there is nothing about which it is more difficult to speak in a mortal language, or that can be less easily understood by human intelligence. For Beauty, of all created things, is nothing other than a splendour that stems from the light of God's face, as the Platonists would say, primary beauty being something that is associated with Him, but the Idea of which was then communicated, thanks to his goodness, to his creatures, and seems to be the cause that is discovered in those creatures that received it as their lot. And just as those who look at their faces in a mirror then forget them immediately, as Saint James says in his canonical epistle, just so, when we behold beauty in mortal things, we can hardly raise ourselves high enough to see that pure and simple light in which all forms of light originate. [51]

THE ELEMENT OF REPRESENTATION

We cannot contemplate the face of Beauty any more than the face of God, in which it participates. We can only glimpse it through a cloud, as one can also recognize its reflection in sensible beings. The image of Beauty proposed by Ripa cannot therefore be reduced to an allegory (if one sets aside the attributes with which it is endowed, which designate not its essential quality but only its accessory features). That image is constructed to seem like a definition, and appears as a substitute for one—a definition that follows the natural articulations of the idea, which a good dialectician should be able to analyze, in all its elements, without destroying any part of it. [52] It would thus appear (but, as we shall see, the appearance is misleading) that we are here close to the element that Michel Foucault has defined as that of representation, in the logical, discursive sense of that term: the relation between the signifier and that which is signified seems to be one of *standing for it*, of substitution, reciprocity, and a strictly binary organization in which the sign appears as a representation that duplicates and reduplicates itself. As *The Logic of Port-Royal* was to declare,

The sign encloses two ideas, one of the thing representing, the other of the thing represented; and its nature consists in exciting the first by means of the second. . . . When one looks at a certain object only in so far as it represents another, the idea one has of it is the idea of a sign, and that first object is called a sign.[53]

So much for the theory of signs to which the classical age subscribed. But the above also suffices to show that it is not possible to come to any absolute conclusion about the relations between images and language: those relations have a history, and each period seems to decide about them in its own way, which is perhaps in part what confers upon it its coherence, its unity. Let us turn, once again to Foucault, and borrow from him the following formula, which sets Ripa's endeavor in perfect focus: from the classical age on, "the sign is the *representativity* of the representation in so far as it is representable."[54] There can be no sign for Beauty except insofar as Beauty presents itself as a representation, which, as such, can in its turn be represented by a statement—either an image or a definition—that can be substituted for it and that is absolutely transparent in relation to it. And even the sign of opacity, namely *cloud*, in which the limit of representation, of what is representable, is revealed, must tally exactly with the thing that is signified, which it denotes. Beyond the clouds what opens up is the realm of whatever cannot be the object of a representation because it cannot be given a name—the infinite or indefinite space the silence of which was to bother classical minds until such time as they discovered a language that made it possible to interrogate it, analyze it, and make it *speak*.

SIGN AND SIMILARITY

In the cases of God or Beauty, it is only their faces that are hidden from sight. They can at least be named and, through knowledge of the names, one can accede to knowledge of the thing signified. And if the fundamental task of classical "discourse" is indeed to "ascribe a *name* to things and in that name to name their being,"[55] understandably enough speech, the science of images (but they come to the same thing if a science is quite simply a well constructed language), or iconology, is perfectly prepared to call upon names for help. According to Ripa, there are some images—common, trivial ones—that anyone can recognize at a glance, and others that, on the contrary, are bound to remain enigmatic. But the curiosity aroused by an image conceived in accordance with the rules, the desire that the "body" of the image must excite in the mind to discover the meaning that is its "soul," that desire is bound to increase in the visible presence of the name of the thing that is signified ("Questa curiosità viene ancora accresciuta dal veder i nomi"). As *The Logic of Port-Royal* also held, it is in the nature of a sign to excite the idea of the thing that represents through the idea of the thing that is represented. Given the recognized primacy of the thing signified, and the perfect transparency between the sign and the idea, and between the idea and the sign, and given the representation's privilege, in its discursivity, to represent itself, semiology and hermeneutics overlap and fuse: to analyze signs and to discover their meanings are one and the same, since the signifier is only valid as such to the extent that it is perfectly in accordance with the thing that is signified.[56] To know a thing is firstly to be able to name it ("*per che,*" as Ripa writes, "*senza la cognitione del nome non si puo penetrare alla cognitione della cosa significata* [because without knowledge of the name, it is not possible to accede to knowledge of the thing signified]"). The possession of the name, if not its visible presence, authorizes both the reading of the image and its interpretation, that is say—according to the meaning that the classical age gave to the word—knowledge of the thing that is signified. But even if the image thus seems to constitute a mode of knowing that gives access to the thing that it represents, it is not so much on account of resemblance of its visible parts to the beings and objects of the external world (however much "imitation" continues to provide the objective mainspring of depiction). Rather, it is on account of the resemblance that exists between the image and the thing that is signified, a similarity founded upon order and measurement, which turns the image thus conceived into a kind of definition (*per che questa sorte d'Imagini si riduce facilmente alla similitudine della definitione*).

The image must be the measure of the thing signified, just as a definition is the measure of the thing defined. But we should not be misled by the terms used above: *measurement* is here less important than comparison, an analogy that defines the poetic space of the image. The theory of metaphor, of expression that operates both to veil and to clothe, of images that both reveal and conceal—that theory, in truth, is not original on a philosophical level but is derived from the twofold Neoplatonic and Aristotelian traditions. If even comparison proceeds from the order of thought, following the articulations of an idea, it is still not free from an element of resemblance (*similarity*). Now, as we have seen above, Foucault has shown that similarity played a fundamental role in Western knowledge down to the end of the sixteenth century, organizing forms of knowledge and orienting the exegesis or understanding of natural phenomena and even the art of representation: similarity founded upon contiguity, sympathy, analogy, the relations of like to like, similarity that confers upon signs the value of metaphors.[57] Ripa criticizes the *dichiarare idem per idem*, metaphor that is reduced to tautology; but all the same he does not spurn the prestige possessed by figures of similarity, the kind that Descartes, following Bacon, was to condemn in the name of order and measurement.[58] Classical logic was to make signs dependent upon a binary organization, founded upon the reciprocity between the thing representing and the thing represented. But for Ripa the two kinds of resemblance that he considered to constitute the nerves and force of any well-formed image (*e sono queste due sorti di similitudine il nervo, e la forza della imagine ben formata*) still implied a relationship that involved three terms, as in classical rhetorical theory: the first kind of similarity consisted in the equal proportion in which two things distinct from each other related to a third (*nell' egual proportione che hanno due cose distinte fra se stesse ad una sola diversa da ambdue*: to resemble strength, one might paint a pillar holding up a building just as, in a human being, it is strength that enables him or her to resist adversity). The second kind of similarity came into play when two distinct things met in another that was different from both of them (*quando due cose distinte convengono in una sola differente da esse*): to denote magnanimity one might take a lion. This way of proceeding, which already owes less to analogy but still does associate two ideas in a single sign, was, in Ripa's opinion, less praiseworthy, albeit more common.

However ambiguous or even contradictory Ripa's terms may be, his text nevertheless serves a perfectly explicit intention: his aim is to establish a universal language (or science) of images. It was a project that inevitably chimed confusedly with the plan for a general grammar that the classical age was to pursue. But Ripa's *Iconologia*, at least with respect to its theoretical and encyclopedic plan, is a work that was in many ways exceptional.[59] And there is no reason to believe that the art of his day borrowed from emblematic literature anything other than iconographical motifs, themes, and programs, or that the theory of figurative expression that it obeyed can account for contemporary developments in painting. From a historical point of view, this literature, which appeared in the second half of the sixteenth century, occupied a paradoxical position. According to the traditional view, Quattrocento art stood for conditions and rules of a type of "objective" representation that was liberated from the norms of medieval symbolism. But it seems that in the following century the vogue enjoyed by the works of emblematists and "iconologists" reflected, on the contrary, the resurgence of a taste for allegory, a passion for metaphor that scholars believe to be echoed in the figurative arts. It is as if it was expected of painters—just as of poets: *ut pictura poesis*—to assume an allegorical role and to speak the language of resemblance, of similarity, in an age that on an intellectual level wished to recognize only identities and differences upon which *mathesis*, the science of order and measurement, could operate. And yet, allegory, as conceived by Ripa, as it was visibly used, was part and parcel of the pictorial order established in the Quattrocento to which classical representation still subscribed (as is shown, in the seventeenth century, by the quarrel between Abraham Bosse and the Academy, over the status of perspective) (see Chapter 3, pp. 116 ff.).

INTARSIA I

That is confirmed, indirectly, by the systematic attack launched against allegorical poetry, in particular that of Tasso, by one of the creators of modern science. In his *Considerazioni al Tasso*, Galileo declares that allegory constrains the reader to interpret every element as an abstruse allusion to something else. It forces natural narrative, "originally perfectly visible and designed to be seen face on," to adapt to "a meaning aimed at obliquely and no more than implied,"[60] in the manner of paintings that offer to the uneducated eye nothing but a chaos of lines and colors, or some amorphous landscape in which, however, by adopting the appropriate point of view, it is possible to discern portraits, forms, even entire scenes that are hidden and implied there. In principle, anamorphosis is part and parcel of perspective construction, of which it appears to be an aberrant form of development.[61] But the analogy with painting does not stop there. Galileo goes on to say that Tasso's story resembles a work of marquetry more than an oil painting. Its outlines are hard, emphatic, its figures dry, without relief or roundness; and it is too crammed with elements set one alongside another, as if the poet had sought systematically to fill up every empty space. As Panofsky rightly perceived,[62] Galileo's critique testifies to his taste for a less graphic, more "pictorial" style, but his comments were also timely: they drew attention to the link that existed between the procedures of allegory and a type of craftsmanship that was associated right from the start with the institution of a code of perspective; and by the same token, they confirmed the ambiguity of the iconological project. For even if the theory of signs developed by Ripa in his proem is in some respects in accord with the theory to which the classical age was to rally, notwithstanding, through its way of focusing upon and manipulating various elements in the image, it also belonged to a long-standing figurative tradition.

THE CLOUD OF REPRESENTATION

Ripa's *Iconologia* tends to assign to *cloud* and to the functions that it assumes in figurative representations, not only the potentialities that it can have in the visual domain in the representation of rapture or miraculous apparitions, but also other, more purely discursive potentialities. This element thus comes into play at the junction of two apparently distinct realms. Is it possible to establish a link of implication (or succession) between the two modes of representation, the visual and the discursive? Or should we, on the contrary, recognize them to be mutually alien, and wrongly associated by language, which has the same term apply to both? The question is of theoretical interest, since it concerns the general essence of representation, understood as the element in which the sign finds its privileged place. But it is also of historical interest, insofar as it may be claimed that the era of discursive representation — founded upon substitution — succeeded an age in which representation was, on the contrary, regarded as repetition, a theater of life, a mirror of the world. But the problem would then be to place the art of the Quattrocento in its own epistemological context: in an age that seems devoted to the mirror-play of similarity, was not art, far from *imitating* space[63] — as Foucault puts it — already working in opposition to the forms of knowledge and endeavoring to define the objective conditions of representation in both the visual and the discursive sense of the term? Was it not trying to construct, by the means of painting, a theater in which classical tragedy would bring into play the mainsprings of *mimesis*, to open up the field of substitutions using the means of repetition, and to produce a *stage* for it? If so, representation in the classical sense of the term would, on the aesthetic level at least, be part of a larger framework, with a far more extensive background.

Here again, cloud can serve as a theoretical guide. For the classical definition of a sign seems to constitute but one figure among other possible ones in the representational field itself, one that does not necessarily possess the purity and transparency with which it is credited in *The Logic of Port-Royal.* With respect to the idea that it produces (its *interpretant*), a sign, or *representamen*, to borrow the by now classic definition of Peirce, stands in the stead and in place of something (its object). But

the object of representation can be nothing but a representation of which the first representation is the interpretant. . . . The meaning of a representation can be nothing but a representation. In fact, it is nothing but the representation itself conceived as stripped of irrelevant clothing. But this clothing can never be completely stripped off; it is only changed for something more diaphanous.[64]

These declarations not only echo Ripa's observations to the very letter; they moreover confer an unexpected resonance upon both the visual and the discursive functions that cloud assumes in the representational order. *Cloud* (and at this point we should remember the importance that Wölfflin attached to the motif of the veil in Baroque imagery)[65] reveals only as it conceals: in every respect, it appears to be one of the most favored signs of representation, and manifests both the limits and the infinite regress upon which representation is founded.

Painting as Theater

The Object of Representation

ART AND SPECTACLE

If representation, in the discursive sense of the term, can be reduced in the context of classical epistemology to an interplay of substitutions founded upon the reciprocity of signifier and signified, of the thing that represents (as *The Logic of Port-Royal* puts it) and the thing represented, do not the ambiguous relations established between art and theatrical spectacle long before the seventeenth century indicate a secret kinship between the two modes of representation? What is the position of painting if it *presents itself* as a representation and as the equivalent or substitute of a theatrical spectacle, some of whose means it may borrow but which, in its turn, also claims to imitate life, or even to depict the passions? And what is the position of representation itself if the discursive and the visual (or spectacular), far from being mutually exclusive, may become linked and complementary on a theater stage and also in an oil painting?

It is at this point worth repeating that in the seventeenth century imagery, as well as the theater, was considered as among the most effective weapons of religious propaganda, first and foremost that of the Jesuits. The latter managed to combine their activities as producers of *imprese*—printed images accompanied by a saying or motto—with those of *impresarios*, organizers of spectacles of every kind: theatrical representations, operas, ballets, and parades and processions of various types.[66] Artists, too, from Brunelleschi to Rubens and from Leonardo to David, were likewise quite prepared to act as theater directors, designers, gadget makers, or even—why not?—makers of rhymes and playlets. At the level of the men involved and their activities, links are thus attested between the various sectors of representation, among which poetry should also be included if it is true that, as d'Alembert believed, "representations imagined on canvas are possibly the touchstone by which to assess the beauty of poetic images."[67] Does that mean that the borrowing and exchanging of motifs and procedures, and even the transposition of representations from one series to another, was the rule at a time that seems to have been particularly sensitive and attentive to the prestige of both spectacle and imagery, whether poetic or pictorial? And does it mean that we are reduced to trying to find the sources of any particular given image?

THE DUPLICATION OF REPRESENTATION

There were means by which it was possible to introduce into a painting, or even onto the theater stage, a distinction, both discursive and visual, between an earthly register where the laws of weight obtained and a celestial register in which attraction seemed to operate in contradiction to the norms of this world, *dal sotto in su* (from below to above). And those means cannot be dissociated from those that either operate as signals or dramatically, by which the various figures were characterized: a particular figure—that of Christ, the Virgin, a hero, or perhaps an allegorical figure of a Virtue—would be recognizable, decipherable, interpretable from certain differential features (costume, attributes, conformity to a conventional type, etc.) and also from being inserted into a particular scene or network of dramatic relations (a particular "apparatus" or element in the decor, the figures' engagement in their respective characteristic actions, etc.). But the division of the composition into two registers, one positioned above the other, the one terrestrial and the other celestial, also affects the impact of the duplication that constitutes the mainspring of any representation. It may be—as we have seen in, for example, Zurbarán's *Alonso Rodriguez* or his *Portioncule*, that the canvas simultaneously shows two scenes or two scenic elements that are complementary but whose coexistence within the unity of a single vision is problematical. In contrast, where the actors do not participate in the grace bestowed upon the saint or the blessed protagonist (whose companions were often reduced, as was customary in most accounts of levitation, to observing their transports and ravishment through a keyhole), the spectator is admitted, albeit as a supernumerary, into the circle of an exclusive vision. Just as in the theater when, within a representation another representation is enacted, which some of the actors are invited to watch— a play within a play, the model of which remains the play that Hamlet presented to those close to him—it may appear that by embedding one scene within another[68] the representation engineers its own destruction, or at least shows *itself* up to be simply a representation: what is presented to the eyes of the public is no longer a representation, but the representation of a representation, which ought to be sufficient in itself if representation, in its essence, were nothing but the production of a spectacle.

But such is the paradoxical nature of representation that it is never more assured, as such, than when it openly presents itself as a representation, or even as *the representation of a representation*. This is something that people all too often forget when looking at many paintings, ranging from Giotto to David (and beyond). Yet it is something to which particular attention should be paid when it comes to elucidating not only the relations between art and reality, but also the relations that are formed on occasion between the pictorial series of representations and the various forms of spectacle (or theater). There can be no doubt that many of the discussions on the so-called "realism" of the art of the Renaissance, or even of the late Middle Ages, are irrelevant where the representations of paintings refer not only to the natural order of sensible appearances but also, and often primarily, to the instituted order of spectacle that culture sets up for itself, in many different guises. It is today accepted that throughout the Middle Ages the development of plastic depiction ran parallel to that of a paraliturgy designed to provide dramatic illustration of the episodes of the Christian legend, with each of the two systems of presentation, each of those two

series reacting upon the other and borrowing a measure of each other's prestige.[69] This interaction between painting and the theater, between dramatic
spectacle and figurative representation, was to continue well beyond the
Renaissance and the Neoclassical (or Baroque) age. Francastel has shown
how the Renaissance changed the terms and conditions of a dialogue in
which the interlocutors were often interchangeable, by introducing a new
repertory of themes and accessories for both the theater and figurative representation, and by replacing medieval forms of representation by modern
forms that were elaborated in the theater and in painting simultaneously.
And as for the "Baroque" theater, whose *machines* were certainly not alien
to those of painting, George Kernodle has tried to show the connection between, on the one hand, the development of perspective and new styles of
plasticity and, on the other, the construction of the Italian stage. Initially,
this was based on linear norms, but it later developed an ever more pictorial illusionism, with stage architecture progressively giving way to painted
scenery.[70]

To limit ourselves to the signaling elements used by Correggio and
by Zurbarán, there can be no doubt that, in the "machines" of the former
and the two-tiered compositions of the latter, *cloud* is not used for purely
illusionist ends and that its status remains ambiguous within the figurative
network. /Cloud/ is not necessarily depicted in such a way that it resembles
a "real"—or rather, a "natural"—cloud. Its volume, its consistency, its very
solidity declare that the functions assigned to it are not solely descriptive.
As we have seen, it provides the material, if not the means, for a construction while at the same time ensuring communication between the terrestrial
and the celestial levels and sanctioning figurative and plastic effects that
could not be justified by reference to the order of natural reality or even
to the order of supernatural vision. The illusion—if that is what it is—is
played out on another level: that of a representation, in the theatrical sense
of the term, produced by means of painting, painting that is—to repeat—
organized in such a way as to be a representation of a representation.

Substitution

CULT VALUE AND EXHIBITION VALUE

A painting's value as a representation is often wiped out as a result
of the work being removed from its original context and presented in con-

ditions that impart new functions to it. A work of art is not necessarily designed to be contemplated. Without invoking the example of entire traditions — Egyptian or Etruscan painting — that were designed to be hidden from view, Walter Benjamin has shown that the assignation of a ritual function to images implies that their presence is more important than their visibility, or even contradicts it: only the priest has access to the *cella* in which the statue of the god is kept; and countless Virgins, both painted and sculpted, were supposed to remain veiled; quantities of Christian relics and treasures were kept locked away, except on the rare occasions when the faithful were permitted to glimpse them; and in normal conditions, a large proportion of the decor of a sacred edifice was more or less invisible.[71] Once a work was liberated from its ritual functions and acquired its own material autonomy, there were more occasions on which it could be exhibited. Panels and canvases are easier to move from one place to another than mosaics, frescoes, or stained-glass windows. Once placed on view in a museum, their function as exhibits outweighs every other, even if it does not become any easier to get close to them. Long before Malraux, and in a far more critical and perceptive fashion, Walter Benjamin assessed the extent of the metamorphosis introduced by museums. Far from it being the case that in a museum a work of art no longer has any function other than that of being a work of art, the absolute preponderance of its value as an exhibit — and, it should be added, the gathering together of productions from many distinct, if not heterogeneous cultural fields, and their consequent reduction to a common order — ensure that the object is assigned totally new functions, among which its specifically artistic function, in extreme cases, may even appear somehow superfluous. Once it is relegated to the status of an object in a collection, if not to that of an exchange commodity, the work provides material for an essentially nominal and scholarly knowledge, and makes no real theoretical or critical impact, subjected as it is to the imperatives of property and speculation.

This metamorphosis, far from being an effect of the institution of museums, on the contrary manifests a mutation in the cultural order, of which the museum itself is simply a consequence. A work exhibited on a picture rail is removed from the space of representation in its concrete sense (and the word space is not used at all metaphorically here). The example that Walter Benjamin borrows from Herman Grimm, that of Raphael's *Virgin of Saint Sixtus*, illustrates that point well enough, for the puzzle posed by this painting is resolved as soon as an attentive examination of it reveals its paradoxical status. The painting is now in the Dresden art museum, but tradition has it that it was originally painted for the high altar of the charterhouse church in Piacenza, where it was still to be found in the eighteenth century, when the elector of Saxony acquired it. In the painting, two green curtains, hanging from a very visible curtain rod, are drawn aside to reveal the Virgin, carrying the Child, and stepping forward on a carpet of clouds. Kneeling at her feet are Saint Sixtus and Saint Cecilia, the former looking up at the Virgin and pointing his finger in the direction of the spectator, the latter with her eyes cast down toward the lower level of the composition where lies — as will be seen, the word is not out of place — the crux of the problem: a kind of wooden lintel upon which is placed a miter and upon which two cherubs are leaning. "It is truly an extremely rare and singular thing,"[72] was Vasari's comment on this painting that was generally recognized to be a "masterpiece" but whose presence in a distant province and also the fact that, exceptionally for a work by Raphael, it was painted on canvas, not on wood, greatly intrigued critics. Some of them thought that the work might have served as a standard carried in processions.[73] Returning to the enquiry and adopting a functional point of view, Grimm meditated upon the significance of the painted wooden lintel that marks the bottom limit of both the composition and the cloud upon which the Virgin and the saints are positioned. He then produced the hypothesis that the painting had been commissioned from Raphael on the occasion of the solemn laying to rest of Pope Julius II, and that it had been designed to be placed at the end of the funerary chapel, so that the Virgin would appear to be emerging from a kind of niche, closed off by curtains, and advancing in the direction of the coffin at which Saint Sixtus is pointing and toward which Saint Cecilia is looking, the coffin upon which the aforesaid cherubs are leaning. Roman Catholic ritual ruled that images exhibited in the course of funerals could not subsequently be exhibited upon high altars, so the painting was then sent far away, to prevent the story of how it had been used becoming known.[74]

Was Walter Benjamin right when he said that *Virgin of Saint Sixtus* was primarily valuable as an *exhibit*? If we accept Grimm's explanation — which does have the merit of accounting for some of the more unusual features of the painting — we are bound to admit that the function originally assigned to Raphael's canvas is somewhat at odds with the religious role that the painting was to take on subsequently, once it was placed above the high altar of the charterhouse of Piacenza, and also with its present function in a museum. This work was not intended to be revered; but nor was it intended to be a permanent exhibit in the context of a highly acclaimed collection, where its value as a *representation* is blurred and—to borrow the words of Eugène Müntz—the scene takes on the aspect of an intimate and mystical drama unrelated to the concrete universe, taking place, without pomp or ceremony (as can be seen from the humble metal curtain rod to which the curtains are fixed), in a place of light and poetry where neither the notion of time nor that of space applies.[75] Wölfflin's intuition of the work's significance as a spectacle may seem to have been closer to the mark: he thought that the painting required to be hung quite high, in order to produce its full effect and to convey that the Virgin was descending from heaven and was appearing to the faithful in what was only a fleeting vision.[76] But Grimm's enquiry has shown that the image is not so much a representation of a vision, a double vision (since the Virgin appears to the two saints at the same time as to the spectator), but rather a substitute for a *tableau vivant* designed to be positioned within the framework of a solemn funeral, on a specific date and in a specific place, and in the context of material surroundings in which the curtain rod, the curtains, and the wooden lintel depicted in the painting fitted in perfectly.[77]

POSSIBLE REPRESENTATION

The paradox of *The Virgin of Saint Sixtus* and its singularity (by which Vasari was struck) are closely connected with the essence of representation. A dramatic spectacle, although it can be *repeated*, is something ephemeral and accessible only to a limited public. A painting, on the other hand, however fragile, can be said to possess a relative permanence and can be contemplated by ever increasing crowds. But what we have here—and the case is by no means exceptional—is a canvas designed to be an integral part of a spectacle that for its part and for obvious reasons was destined to take place but once. As Wölfflin has pointed out, on its own the painting retains in its organization something of a theatrical and therefore fleeting representation: the curtains are raised, as in a theater, to reveal the actors; the divine characters descend onto the stage, here given material form by a simple plank; and finally, there is the cloud, whose function, at the level of the image, duplicates the function imparted to it within a theatrical space. In the context of a representation, its importance is measured by the very special role that is assigned to it: the cloud, which now reveals the spectacle that it conceals, can here clearly be seen to be one of the signs most favored by representation and one that shares its essence.

The representational functions of cloud, and the kind of complicity —to put it briefly—that becomes apparent between dramatic spectacle and figurative representation, make it possible to understand the paradoxical treatment that clouds are given in so many of the paintings that have been mentioned so far. For the banks of cloud that accompany miraculous visions and transports of rapture in the paintings of Raphael, Correggio, and Zurbarán are not altogether immaterial. They may blur into the vaporous, golden distance ("The heavens part; through myriads of cherubim we glimpse infinity"),[78] but at the same time they provide a solid seat for the figures arranged upon them, and thus even provide the framework for two-tiered compositions, of which the sixteenth and seventeenth centuries produced so many examples, ranging from *The Dispute of the Holy Sacrament* (more accurately known as *The Triumph of the Church*), also by Raphael, to Tintoretto's *Paradise* and Zurbarán's above-mentioned *Apotheosis of Saint Thomas Aquinas*. All are "Triumphs," organized in a hierarchical fashion, with the earthly level, where the high and mighty of the hour are grouped, surmounted by one or several aerial levels in which, surrounding Christ, or even God himself, a whole celestial crowd is gathered. Raphael's *Dispute* makes it crystal clear that the sign designated as /cloud/ cannot be referred to a natural reality, but instead must be recognized to be an element in a setup that is both symbolic and illusionist: the Church Fathers and authorities are seated upon a semicircular bank of cloud that provides a kind of bench for them, floating just above the ground and surrounding another cloud—of more subtle appearance—which carries Christ, the Virgin, and Saint John. Such a composition does not refer to anything ever seen, any miraculous apparition or a real spectacle. Here, symbolism and illusion interplay on a different level: the level of a *possible* representation, or staging.[79]

Representation/Repetition/Substitution

> A signifier is, from the start, the possibility of
> its own repetition, its own image or resemblance.
> —Jacques Derrida, *Of Grammatology*

Decors

The Virgin of Saint Sixtus is a substitute for a *tableau vivant*; *The Dispute* is constructed on the model of a conventional representation: those are statements that need to be supported. At this point, let us simply recall a few facts that justify likening the pictorial "series" to the theatrical "series" and that also illuminate the remarkable destiny of the element "cloud" in the context of spectacular representation. The Middle Ages produced many paraliturgies: dramatic representations of the Christian legend that unfolded in a church or in front of it. The tradition was maintained right down to the Quattrocento, even as artists were introducing a new figurative order.[80] The organizers of these paraliturgies often arranged a "glory" or a celestial throne above the Church porch in front of which the spectacle was played out. Long before the appearance of the first street theaters, city monuments, in particular church facades or the town gates, served as a backcloth, if not a frame, for such representations; and it was common practice to use the upper parts of those edifices as places in which to position celestial characters and from which to lower an announcing angel at the appropriate moment.[81] In this respect, Veronese's *Triumph of Venice* can be seen as a sumptuous secular climax to a formula for which the Quattrocento provides many illustrations. One of the most characteristic of these is probably *The Death of the Virgin*, a mosaic frequently attributed to Mantegna, which adorns the ceiling of the Mascoli Chapel, in San Marco in Venice: surrounded by the Apostles, the Virgin lies beneath a triumphal arch through which can be seen a street in perspective where Christ, seated in a mandorla on a throne of clouds, occupies the tympanum.[82]

Installations of this type were also to be found in the secular spectacles of the sixteenth and seventeenth centuries and, in particular, in the *tableaux vivants* that constituted an essential feature of street parades laid on to honour a sovereign or important visitor: these "entries" were organized according to a characteristic structure of duplication, for the official procession in the honor of which the spectacle was performed was itself a spectacle or *representation* that had a part to play, as the crowd looked on, in a wider, comprehensive entertainment (although for a long time the processional element was regarded as of secondary importance in the areas where these kinds of street spectacles were particularly popular: northern France, the Low Countries, and England). The booths erected on these occasions to shelter groups of actors or painted allegorical scenery, often behind closed curtains, were usually arranged on a number of different levels, with elements that were at once symbolic and emblematic, such as a tree or a pillar, holding up an upper story wreathed in clouds. Zurbarán's *Apotheosis of Saint Thomas Aquinas* fits that model perfectly. It shows, on one side of the pillar, Charles V clad in an ermine cloak and a crown, on the other Archbishop Deza, the founder of the College of Saint Thomas, in Seville, both of them kneeling with their attendants beneath a cloud in which the top of the pillar is lost. Upon this cloud, Saint Thomas, in his Dominican robes, occupies the central position, flanked by the Church Fathers, while above this group are Christ, the Virgin, Saint Paul, and Saint Dominic.[83] Street stages, *tableaux vivants*, and entries were introduced into the Iberian Peninsula at a relatively late date, but artists there had long been familiar with the principle of compositions on several levels, as illustrated by stone and wooden altarpieces and the stage facades of Spanish theater. But the *Apotheosis*, for its part, bears comparison with the temporary decors that were improvised by the designers of the day, one of the most remarkable of which was that organized in Lisbon in 1619, for the entry of Philip II.[84] Here, a three-storied facade displayed a whole collection of constructed, painted, or sculpted motifs borrowed from the traditional repertory. The structure was decorated and made more imposing by architectonic elements—architraves, cornices, and pillars—some of which were topped with "what seemed to be clouds." In the center of the principal arch stood a pilaster that rose right up to the third story, where its top was lost in a cloud that supported a scene of heavenly glory. This contained a cross that stood out against a background of blue fabric strewn with gold and silver clouds and surmounted by other clouds upon which were positioned figures who appeared to be holding the cross aloft. From these heights a messenger wafted by a cloud descended to the stage at the bottom. Various niches in the facade contained an assortment of symbolic accessories (thrones, rocks, etc.), and a succession of other objects (monsters, chariots, boats) appeared on the stage.

The Nuvola

ENGINES

Francastel has shown that the figurative repertory of the Quattrocento contained a whole set of "signs" that correspond, term for term, to the materials used by the organizers of popular paraliturgies and parades. And it is true that, even where cloud is used for descriptive and illusionist purposes, it is also depicted, in many frescoes and on many panels, as a stage prop. Francastel suggests that the pious imagery of the Quattrocento was designed not so much to illustrate the sacred texts, but rather to reproduce certain familiar spectacles;[85] and it is perhaps in the work of Mantegna that such borrowings are most clearly evident, if it is fair to speak of *borrowing*— in the sense that the word takes on in discourse designed to trace sources— in circumstances where the reference to the theatrical series is manifest, or even deliberately emphasized, and the figurative element has only such signaling and symbolic meaning as is assigned to it in the context of a theatrical representation. Thus it is not good enough to say that the figurative work *signifies* to the extent that it borrows its signs from an institutional level of reality and from preestablished orders of signification (even if such a remark is important in that it illuminates the "makeshift" [*bricoleur*, as Lévi-Strauss might say] side of artistic production). What is borrowed has a *signifying* function in itself, precisely as something *borrowed*: it constitutes a sign in its role of representation, and an image, in its turn, insofar as it represents a representation.

In the right-hand panel of the triptych in the Uffizi, Mantegna depicted *The Ascension of Christ* (Plate 3) in unequivocal fashion, as a "flight" in the precise sense that this term had in medieval theater. Against the background of a cloudy sky, treated in an illusionist manner, Christ ascends, borne aloft by a machine covered with theatrical clouds and cherubs' heads. Just a few examples will suffice to point out how very respectful of the material aspect of the theater prop painters were—at this stage, at least. The rocks that shelter Uccello's dragons are cardboard caves, and the clouds of Mantegna and those of Signorelli, too, in his multistoried compositions in the museum of Cortona, look wooden or resemble painted canvas or even the wadding used to cover the bodies of theatrical machines. Mantegna makes no effort at all to conceal the fact that he has borrowed his cloud from the contemporary stage. Indeed, the simultaneous presence, in the Uffizi *Ascension*, of atmospheric clouds and theatrical clouds, painted in such a way as to rule out any confusion between the two modalities of the sign, explicitly draws attention to the opposition (in itself significant) between the two ways of using this element, two ways in which we should perhaps recognize two distinct signs, one that refers to a natural phenomenon, the other that refers to a cultural object that belongs to a different order of signifying from that of painting. As Francastel shows,[86] the central panel of *The Adoration of the Magi*, the same triptych, contains a remarkable image of an engine that was frequently used in the street spectacles of the Quattrocento, namely the *nuvola*, believed to have been invented by Cecca (Vasari's collection of drawings included several models drawn by his hand).[87] It was a kind of rack in front of which an actor would stand and upon which a variety of symbolic accessories could be placed. With all the precision of an ethnographer, Mantegna carefully represented the vertical metal rod, the tip of which upheld a star and a foursome of angels in a cloud of wadding. It is as if he wanted to emphasize the ambiguity of the status of the painted image, for the illusion here operated at the juncture of reality and the imaginary and by referring to a theatrical representation rather than to a natural reality or an intellectual notion.

SPECTACLES

It matters little whether or not Cecca was the inventor of these devices, which could take the form of a lily or a tree just as well as that of a cloud, and which were very much in evidence in front of the baptistery of Florence at the time of the Festival of Saint John. By the time Vasari was writing, however, they were seldom used, since, as he points out, the street spectacles had in the meantime changed a great deal both in their nature and in the means that they employed. The author of the *Lives* was anxious to provide as accurate a description as possible of these engines, in the hope that they would not be forgotten. He also took care to describe rather more ambitious "engines," the invention of which was also attributed to Cecca and from which he first got the idea of fabricating *nuvole*. For the Feast of the Ascension, held in the Carmine church, Cecca had designed a wooden mountain, where the Apostles were lodged, while Christ was lifted into the air on a *nuvola* swarming with cherubs and swept up into a "heaven" at the top of the vaulted ceiling, while two angels descended into the theater where the festivity was being "recited"[88] or enacted (*dove si recitava la festa*). Vasari recounts that the whole thing was covered with an abundance of wadding, arranged in such a way as to represent a cloud full of cherubim, seraphim, and other angels, all of different colors and beautifully arranged.[89] It would be hard to find a better description for Mantegna's *Ascension*, which clearly constitutes a reference to this type of spectacle.

Cecca was credited with the invention of this large machine, as it was he who brought it to an unprecedented peak of perfection. But Brunelleschi had already designed similar contraptions and was sometimes likewise credited with their invention (although, as Vasari notes, some people claimed even then that there was really nothing new about them at all). Brunelleschi had constructed a "Paradise" at San Felice, in Piazza San Felice, for the *representation* of the Feast of the Annunciation, celebrated in this same place and in the same manner as it had been ever since long ago (*anticamente*); and in his life of Brunelleschi Vasari gives an astonishing description of it.[90] This "Paradise" consisted of a machine in the form of a celestial disk that operated in conjunction with a complex device that made it possible to engineer a whole series of flights into the air, movements from one spot to another, descents and ascents. Between two of the beams that held up the roof of the building, the architect had positioned a hemisphere shaped like an inverted shaving dish (Vasari's description), which could be moved by an iron star, itself suspended from a large ring, around which it could be made to revolve. This hemisphere was large enough to hold not only three rows of lanterns, strung together, but also a dozen children, all equipped with wings and golden locks, who represented angels and who appeared to be dancing when the machine rotated it. Seen from the floor, this cupola-like structure, the sides of which were decorated with imitation clouds of wadding (*che parevano nuvole*), resembled a heaven, the doors of which could be opened or shut at will, with loud rumblings of thunder. To it was fixed yet another apparatus, composed of eight branches, on each of which a child was perched, and operated by a winch. This "bouquet of angels" surrounded an empty copper mandorla decorated with tiny lights that could be made to appear and disappear by means of a spring. When the bouquet reached the right height, the mandorla slipped sideways onto the platform (*palco*) where the festivity was being "recited" and which supported a four-storied structure designed to hold various actors (*a uso di residenza*) and into which the mandorla slotted. At this point a child climbed out onto the stage, saluted the Virgin, and gave her the news that he brought, then climbed back into the mandorla, while the angels lifted their voices in song and the "heaven" really did seem to be a paradise (*quello pareva propriamente un paradiso*). Vasari adds that the illusion was reinforced by the fact that God the Father stood alongside the hemisphere, surrounded by angels and accommodated in similar fashion.

The Representational Function of the Sign

AN ACCESSORY IN A STAGE SET

The above description testifies to the existence, in the Quattrocento, of a form of "mechanical" spectacle that Brunelleschi, and Cecca after him, may simply have perfected and *systematized*, which associated with a "paradise" a *celestial cupola* (where the image of God, it should be noted, did not find a place) a whole complex of machinery designed to simulate miraculous movements. All the elements of a spectacle such as this formed a single whole, out of which it would not be possible to pluck any particular prop or device in order to consider it on its own. But that remark applies to painting just as much as to the theater: the mandorla on which Mantegna's Christ is positioned is itself backed by a *sky*. And the time would come when Correggio, in the cupolas of Parma, would combine in the unity of a single structure all the until then separate elements in the system. It is clear enough that, in the context of representation, cloud has more than a symbolic or discursive meaning: far from serving simply to designate the sky, the canvas or wadding clouds also made it possible to conceal the workings of machinery that had been elaborated long before the appearance of stage sets in the strict sense of the term. We find such clouds mentioned in most of the inventories of materials belonging to brotherhoods in northern countries, as well as in France and Siena,[91] and that is because they were not so much a means of hierophany as an indispensable accessory in stage sets of many different kinds, which were by no means used exclusively in fifteenth-century Italy (even if it was there, significantly enough, that the prop and the machine were both referred to by the same word: *nuvola*).

The representation of an ascent or descent of Christ or of any other figure, in a Glory or a Paradise, posed many problems both in the theater and in painting. The organizers of the French-style Mysteries did not always respond to those problems in the same way as those who laid on

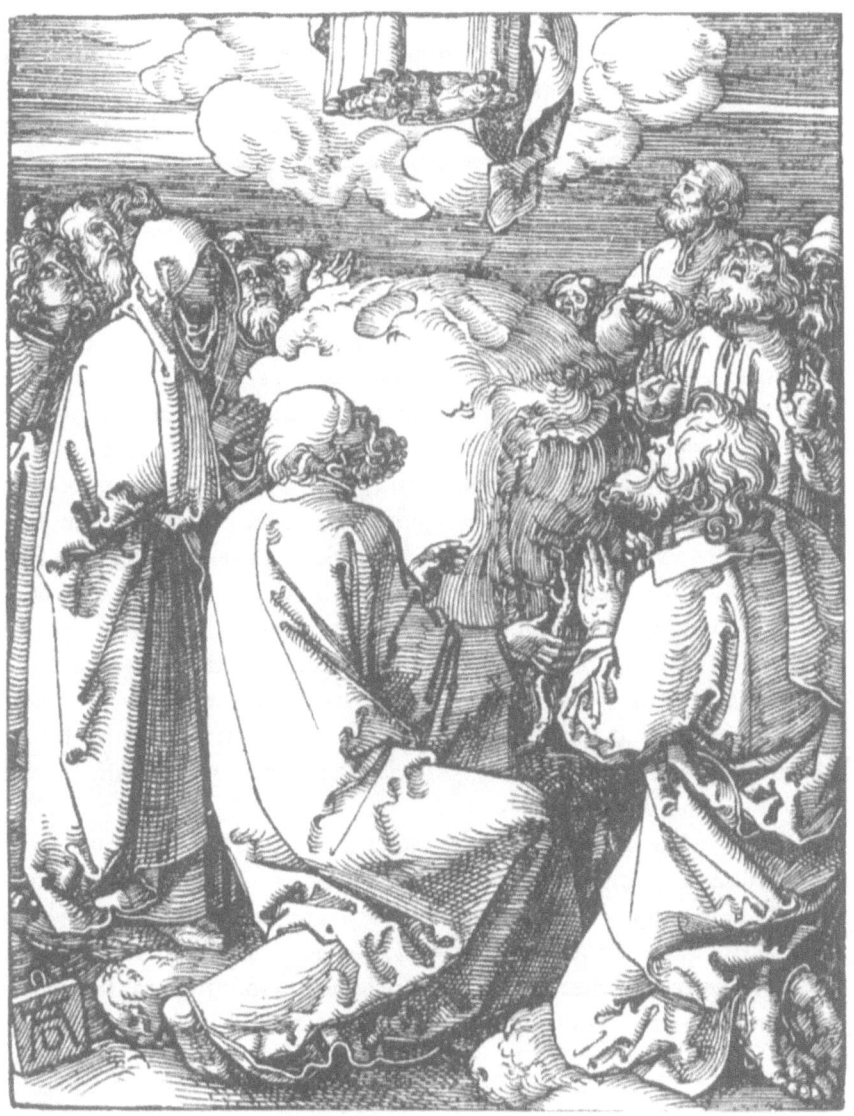

FIGURE 2. Dürer, *The Ascension of Christ*. Detail of a small woodcut entitled *The Passion*.

the Italian form of *sacra rappresentazione*. Yet for both groups clouds seem to have served the same purpose, according to modalities that are echoed in the painting of the time. According to Gustave Cohen, *The Mystery of the Acts of the Apostles* constitutes the peak of the genre, in which all the specific features of religious drama are seen to their best advantage.[92] In this, the Apostles, scattered in various places (the *loci deputati* characteristic of the fragmented space of the medieval theater), were each transported to the mansion where the Virgin lay and from which her body was then borne away on a cloud.[93] And the manuscript of the *Resurrection* provides an amusing description of the ascent of Jesus and the souls delivered from limbo. The latter climb up to heaven by a hidden path, while a piece of machinery seizes Jesus by the waist so that "his legs can be seen below the engine . . . , the cords that haul up the instrument where Jesus is must be concealed by clouds made of fabric."[94] That is the fashion, surprising to say the least and certainly very different from the Florentine mandorla, in which the Ascension of Christ was regularly represented in the art of northern countries across the board from Martin Schongauer's cycle of the Passion (in the Colmar museum) to Albrecht Dürer's little *Passion*, on wood, and from one particular Flemish tapestry preserved at La Chaise-Dieu to the small panel painted by Juan of Flanders, recently acquired by the Prado Museum of Madrid:[95] no part of Christ is visible except the feet that have just left the little hill, where their imprint is clear and is surrounded by the Apostles, the rest of his body being hidden in a cloud cut off by the framed edge of the composition.

Emile Mâle thought he could trace a connection between this type of incongruous-seeming iconography, which appears in so many fourteenth-century ivories and illuminations, and the stage machinery mentioned in the manuscript of the *Resurrection*, and he goes so far as to declare that "it is almost always the theater that should be held responsible for the somewhat down-to-earth realism that sometimes shocks us in the iconography of the fifteenth century."[96] But in a scholarly essay in which the *question of the cloud* is posed in all its aspects—theological and theoretical, semiotic and/or illusionist—Meyer Schapiro has shown that this representation, wrongly described as "Gothic," could not be regarded simply as a plastic translation from the theater. For the fact is that as early as the tenth century it features in numerous English manuscripts, and can be linked with a whole vernacular tradition that—from a description of holy places produced on the Isle of Ilona in the seventh century by Abbot Adamnanus, pronouncing on a report from the monk Arculfe, down to the Blickling homilies, a tenth-century Anglo-Saxon text—constantly emphasized the subjective, phenomenal aspect of the vision of Christ disappearing into the heavens.[97] The description of the church built on the site of the Mount of Olives, the rotunda "open to the sky" erected around the place where the imprint of Christ's feet remained in the earth, emphasizes the fact that pilgrims were admitted to the place where the Apostles had stood and were encouraged likewise to raise their eyes to heaven:[98] the heaven into which Christ had risen, disappearing into a cloud. The Blickling homilies on the Ascension (dated 971) refused to acknowledge that the cloud had served to support him or to carry him aloft: it was by his own propulsion, solely by his own power, that the Lord and master of all things had risen and disappeared into the cloud in which he first wrapped himself.[99] The text in this way raised a theological question relating to the process of the Ascension and thus also—indirectly—to its visualization, a question which, as we have noted, still reverberated in the paintings of the fifteenth and sixteenth centuries. Meyer Schapiro rightly stresses the role played by the terrestrial observer, which the image emphasizes, and also the representation of the phenomenon of the disappearance of Christ as it was seen by the Apostles who witnessed it. According to him, herein lay the originality of the English invention, and it was this that distinguished it from the traditional solutions that presented Christ in a mandorla or being wafted up to heaven.[100] The idea was not borrowed from the theater; nevertheless, right from the start it had been founded upon a representational structure, an idea of the visualization of scenes from Scripture, a repetition of moments from the sacred drama, a "dramatic re-enactment," as Meyer Schapiro so aptly puts it, from which imagery borrowed its norm and which was to triumph in the most spectacular way in the religious theater of the late Middle Ages.

THEATRICAL REALITY AND FIGURATIVE REALITY:
PAINTING AS REPRESENTATION

The use of cloud for representational ends must thus be understood in a spectacular, as well as a symbolic or discursive, sense. On the figurative level, this element (cloud) has a part to play as one term, among others, in a vocabulary that in part refers to an order of meaning distinct from that of painting: its appearance, its use value in the pictorial context, depends upon the functions assigned to it in the text of the representation to which the image refers, albeit in an imaginary fashion. But this does not mean that, even if the theatrical source is attested, the image serves solely to *fix* a fleeting spectacle. That would be to do as realist prejudice claims, namely to grant primacy to the referent of the painting, which would downgrade the true mainspring of what appears to be a transposition of elements borrowed from theatrical spectacle into the pictorial order. In any case, the communication between the two series of representation is not a one-way affair: contrary to the opinion of Emile Mâle, who regarded the sacred theater of the late Middle Ages as the source of very many iconographic schemata, we should perhaps agree with George Kernodle, who believes that, in the absence of a continuous tradition, the art of the Middle Ages bequeathed to Renaissance theater many of the conventions of the ancient stage, and that the elaboration of modern theatrical forms went hand in hand with a new figurative order, in many cases with the same individuals being involved simultaneously in both kinds of enterprise. But although their respective modes of expression constantly influence each other, they nevertheless do constitute two distinct *orders*, which do not communicate as directly as communicating vessels do. A painter may well have recourse to meaningful elements, or even to principles of theatrical design elaborated beyond the realm of painting. But as soon as an object is removed from the order of theatrical spectacle to be introduced, as a pictorial element, into a signifying system sui generis, it moves from one plane of reality into another: from the plane of theatrical reality, where it is defined in operational terms, to that of figurative reality, where it fulfills functions of a different nature.[101] The fortunes of *cloud* in sixteenth- and seventeenth-century painting testify to an extension of its functions, an extension of the purely plastic value assigned to this element, which cannot be explained simply by investigating its "sources." Thus, the problem ought at this point to be formulated as follows: what happens to painting when it presents itself as a representation of and the equivalent or substitute for a (theatrical) spectacle from which it probably borrows part of its vocabulary, in no way concealing that debt but on the contrary making it appear to be exactly that, even to the extent of drawing from that *production* specific signifying effects?

REPETITION, SUBSTITUTION

Perhaps we are beginning to glimpse a possible link, at the figurative level, between two modes of representation, the one discursive, the other theatrical. The logical space of representation is constituted in such a way that the positions of the signified and the signifier are strictly reversible. And, in truth, the representational function of the sign is unequivocally confirmed as soon as what is signified by the painting appears in the signifying guise that it assumes in the context of a theatrical spectacle, and as the representation of a representation; the pictorial sign reproduces (represents) a sign of a theatrical nature (and the painted *cloud* represents a theatrical *cloud*), just as the image reproduces (represents) a theatrical spectacle that is itself organized as a re-presentation, a *repetition* of an ancient scene. But representation may, equally, operate as *substitution*. In the figurative regime instituted by the Renaissance, the representativity of the iconic sign, like that of the representation itself, is governed by a characteristic structure that is determined by a combination of the various meanings that enter into the definition of the word representation: (re)production, evocation, substitution, and so on. This structure is clearly detectable in many works of indefinite status. For example, the museum of Rheims contains an important series of fifteenth-century paintings that somewhat crudely portray the Mysteries of the Old Testament, the Passion, the Ascension, the Resurrection, and the Vengeance of Christ.[102] Were these compositions used as backcloths for performances of the Mysteries? Did they serve to make up for the dearth of tapestries to adorn a church on the occasions of certain festivals or ceremonies? Or were they just banners carried in processions, designed to be exhibited (produced) in the streets on a number of fixed dates? Whatever hypothesis is favored, the question indicates the possibility of one order of representation being substituted for another, or even (where theatrical scenery is involved) of the representation itself being duplicated, with the paintings playing a theatrical role and the Ascension of Christ, to pick out that particular example, obeying the principle adopted by Dürer in his little *Passion* of 1511, a work whose "popular" character has often been emphasized:[103] the very principle upon which the representation described in the manuscript of the *Resurrection* was founded.

THE SYSTEM AND ITS RELATION TO BORROWINGS

It is as if, at a time when, having reached its climax, liturgical drama was about to give way to new forms of theater, painting had absorbed its heritage even as it presented an image of street processions and parades of a "modern" type. This is not the place to show how the two systems co-existed within the more general framework of the figurative order of the Quattrocento, as is cogently illustrated by the division in Mantegna's work between, on the one hand, the Christian drama (the triptych in the Uffizi) and, on the other, the triumph of Antiquity (the panels in Hampton Court). The fact remains that the quest for sources can never suffice to explain claimed borrowings and archaisms, let alone the fortunes that some of these encountered in the painting of the time. Cloud played no more than the role of an accessory in the art of the fifteenth century; but in the sixteenth and above all the seventeenth, it was to invade cupolas, domes, and ceilings, and assume increasingly extensive illusionist and pictorial functions. Before seeking to discover the possible ideological motivations for this development, we should assess its implications precisely, and the impact that it made upon the order of the system, in conformity with Saussure's formula according to which "everything that changes the system in any way is internal."[104] (Similarly, in the theater, the taste for aerial effects, elevations, descents, and paradises did not disappear altogether with the advent of the cubic, unitary stage in the Italian manner.) The problem of the arrangement of the upper parts of the stage was to continue to exercise the imagination of theater designers and directors, and the flights of the Baroque theater and opera echoed the descents and ascensions of medieval drama. But it is misleading to describe these effects as survivals and to seek for their sources within the theatrical series of representations of the past.[105] The "borrowings" and "archaisms" should be seen as stylistic ploys, which can only be judged in relation to the system in which both their principle and their justification are to be found.[106]

Research into sources is such a limited exercise that it actually prevents one from seeing the process of artistic production—both in the theater and in painting—as anything but a more or less fortuitous and haphazard sequence of borrowings and survivals, or even innovations, which are always "one-offs" and the end purpose of which is not apparent, a sequence that throws no light at all upon the real mainsprings of communication between these different series. Perhaps the relations and interchanges between painting and the theater were not limited to their vocabulary, and the elaboration of a syntax of representation may have been the work of painters as much as or possibly more than of people in the theater world. And, as we shall see, it is in relation to that very syntax that the fortunes of cloud in modern art and theater can best be explained, while at the same time that syntax manifests the underlying structures of an order in which, paradoxically, this "sign" may not have had a place.

3

Syntactical Space

Nihil est in lingua quod non prius fuit in oratione.
—Emile Benveniste, *Problèmes de linguistique générale*

A Problem of Reading

The problems, both technical and theoretical, posed by the depiction of an apparition or a miraculous ascension, an aerial transportation, or even a celestial gathering on a two-dimensional plastic screen—those problems are by no means secondary or subsidiary in relation to the enterprise that, according to a strongly established tradition, epitomizes the Renaissance's contribution to the development of Western art and civilization: namely, the institution of the system for representing space that is known as linear perspective. As we shall see, the above problems constitute that system's accompaniment, if not its necessary countersubject; and this counterpoint, in its turn, has a place in a far vaster field that determines the very principle if not structure of linear perspective: namely, the field of *representation*, in the most general sense of the term, for theory is definitely less concerned with the problem of the representation of space than with that of the space of representation, pictorial space seized upon in its relation to representation, which was initially—as early as the beginning of the fourteenth century—conceived as spectacular, that is to say theatrical.

In the introduction to his *Studies in Iconology*, Panofsky formulated the problem of apparitions in terms that reveal its theoretical relevance: when presented with an image—*The Vision of the Three Magi*, the right-hand panel of a triptych by Rogier van der Weyden, now housed in the

Berlin museum[1] — in which three figures are seen kneeling in a hilly landscape, with their eyes raised to the heavens where a little naked child is suspended, how is it that we know (or decide) that we should regard this as an "apparition"? In this picture there is no bank of cloud or machine to serve as a support or vehicle for the divine figure, such as are to be found in so many works — above all Italian ones — of the period. Instead, there is a golden halo that isolates this figure from the cloudy sky in the background, a halo that — as Panofsky correctly points out — differs in no way from the halo in Nativities that surrounds a child who is represented as being truly present. And as for the attitude in which the child is portrayed, it is that of a baby seated on a cushion rather than suspended in midair. From this, Panofsky draws the following conclusion: "The only valid reason for our assumption that the child in the Berlin picture is meant to be an apparition is the fact that he is depicted in space, with no visible means of support."[2]

As Panofsky immediately recognizes, it is by no means a decisive conclusion, but it does shift the question: before we pronounce on the actual nature of the apparition, what is the evidence on the basis of which we declare that this child occupies a position that is contrary to all the laws of nature? The reason for our conclusion is, of course, the fact that this composition is organized according to the norms of a representation that is founded upon a perspective view of the figurative space and upon the subjection of the figures and objects that occupy this space to laws analogous to those that govern the physical universe. In one particular medieval miniature, in contrast, the fact that the image — or "hieroglyph" — of a town stands out against an abstract background above a group of figures set on a conventionally depicted earth does not imply any infraction of the laws of gravity, but simply denotes where the scene is set: if we interpret that sign as a kind of ideogram, it is because we more or less spontaneously refer it to the medieval code of figurative representation, the key to which has been learned and is still possessed by the age in which we live.[3] The image painted by Rogier van der Weyden belongs to a different figurative order, one that began to be established at the beginning of the fifteenth century and that Alberti's "window" was to provide with its most systematic, but also most narrow and confining, definition. Alberti assimilates the surface of the picture to an open window set in the wall, through which can be seen (*per-spicere*, to borrow Dürer's [anachronistic] etymology)[4] a portion of more or less extensive space. The implication here is that laws analogous to those that govern empirical space also apply in figurative space, so any flouting of those laws takes on the *aspect* of a miracle or a supernatural event, and the "realism" of the system is thereby confirmed, if not strengthened, by any contradiction created at the symbolic level.[5]

One could challenge the fundamentally "realist" or even "naturalistic" connotation that Panofsky assigns a priori to the perspective code of the Renaissance and show, equally well, that the apparition by which the Magi are honored in the panel that we are considering is not so much an apparition of the infant Jesus himself as an apparition of an image of him, since the depiction is governed by the duplication that is characteristic of the representational system. For the child set in the sky is simply a variant of the one positioned in the *Nativity* that is represented on the central panel of the same triptych, but one that, through being detached from that context and in particular from his customary support (the crib or bed of straw on which he lies), here assumes the character of an icon—in the very same way as does the picture of the Virgin and Child enthroned on an altar (*haec est ara coeli*: this is a heavenly altar) that the Sibyl of Tibur is presenting to the Emperor Augustus in the left-hand panel of the triptych.[6] In these circumstances, the absence of any support, cloud, or machine, far from denoting an apparition (in the way that even an absence can carry a meaning), signals that what we have here is a (premonitory) vision, which is represented figuratively by a simple commutation of signs, rather than through means imitated from the theater. That being said, the question posed by Panofsky nevertheless retains its relevance as soon as it is situated in its true context: that of figurative production and comprehension, and the mechanisms by means of which a semantic interpretation can be attached to a complex of iconic signs.

Writing and Representation

An Incidental Remark

To prevent any confusion arising here, it is important to explain that, for reasons of principle as well as of a methodological nature, it seems impossible to separate a study of the rules that govern the elaboration and deciphering of images, in the context of a given figurative order, from a study of the way that they are, quite literally, put to work, actually, in concrete works. Linguists draw a distinction between language and speech or—to make use of the fundamental concepts of generative grammar, the discourse of which the present text is, for now, imitating—between the tacit system of aptitudes that define the linguistic *competence* of a subject and their application (or *performance*) in real speech acts. But such a distinction would have no operational impact in the field of the semiology of art, a discipline in which it probably makes more sense not so much to construct hypothetical models of different systems of plastic expression, but rather to address, in its own terms, the very question of *the* system.[7] The creativity of articulated language, linked as it is to the recursive power of the rules on which its system is founded, is expressed, in current usages, by an unlimited proliferation of forms, with the code of the language as assimilated by speaking subjects *determining* the semantic interpretation of an indefinite collection of real phrases, either expressed or understood.[8] However, where painting is concerned, the problem of "creativity" cannot be posed in the same terms (even if the—at least relative—independence of expression in relation to reflex activity, which Chomsky claims for articulated language, is the rule in both cases).

On the one hand, art, in the form assigned to it by Western society, by no means belongs to everybody and, as a result of the selectiveness that it implies as a condition for its existence, the number of its productions is, in principle, finite. In this respect, it perhaps presents more analogies with writing than with language: for while language, linked as it is with the *speaking masses*, is at every moment something that concerns everybody, and—of all social institutions—offers the least opportunities for individual initiatives,[9] in contrast writing, in the current sense of the term, for a long time remained, like art itself, the prerogative of a small group of specialists or privileged social groups,[10] and its development, indeed its very institution, have frequently been responses to some deliberate scheme. In language there is no private property;[11] and that is why, as Saussure remarks, language can never be revolutionized. But that is clearly not true of art, which is still a long way from being collectivized; nor—to a lesser degree—is it true of writing, which constantly seems to offer material for "reforms." On the other hand, insofar as the distinction mentioned above is founded upon the linguistic field and does not itself depend upon the structure of representation (for a linguist certainly always has to produce a *representation* of language), where the products of art are concerned no one could claim that "language" (competence) somehow takes precedence over "speech" (performance), or that artistic production, in all the relative variety of its forms and configurations, can be reduced to the use of a finite code that determines it a priori. If such a thing as a pictorial system exists, that system has no reality, even of a theoretical nature, outside the products in which it can be instituted in various strict forms. To borrow (and adjust) Saussure's metaphor, the "treasury" into which artists dip is not that of a language that has been deposited with them as a result of their using it; rather it is a treasury consisting of "masterpieces," works that are of *capital* importance by reason of what they have achieved and their consequent authority. As I am here opposing language (*langue*) to "taste" rather than to "speech," as Saussure does, it will be helpful to set out the system at the level of those realizations that have acquired the value and force of models. For (as we shall see by studying one particularly prestigious example) they reveal not only the quality of inventiveness and a concern for style, but also *a strong desire to impart to painting the force of a language* that has no equivalent in works of the common run (which belong simply to the order of *taste*), and this confers upon their quest for innovative detail an unprecedented emphasis not to be found where such a determination is lacking, in second-grade works or those all produced from the same mold.

It should also be pointed out that "tasteful" figures ("taste" as opposed to *language*, in the sense explained above), inserted in the lexicon with the same status as linguistic metaphor, are no less deceptive than the latter. Both have the merit of drawing attention to the system that supports the historicity of art. But what is at stake in the question of the system is definitely easier to think about in terms of theory rather than in terms of expression (speech), sense (taste), or organ (tongue). In this respect, great artistic revolutions are closer to the revolutions encountered in scientific thought than to the slow evolution of linguistic structures. Artistic revolutions are not so much a consequence of the introduction of new forms and new methods, even less of new codes; rather, they stem from a theoretical break that engenders a new definition of the system and its internal articulations. (The fact that such breaks occur at identifiable moments should not mislead us: art is never, under any circumstances, the product of a single individual, and the names that the Renaissance recognized as those of its heroes—Brunelleschi as well as Giotto, Masaccio as well as

Uccello—are the effect, not the cause, of the historical project known as the "Renaissance.")

Depiction and Representation

On the theoretical level, the question posed by Panofsky about the depiction of an apparition or—to put that better—about the mechanism of reading as a result of which a particular panel painted by Rogier van der Weyden can be described, as a first approximation, in the terms of a proposition ("The image of the infant Jesus appears to the Three Kings")—that question has the merit of providing us with a model. As we have seen, the association of a semantic interpretation with a complex of iconic signs operated through the mediation of a syntax, which, although figurative, nevertheless provided the material for a structural description, composed according to the ways of articulated language and in a mode at once inductive and comparative. We may assume, for methodological purposes, that recognition or identification of iconic elements (personages, objects, etc.) stems, in a representational context, from a reflex action. However, the interpretation of images *in terms of* figurative propositions, articulated with a view to expressing a meaning (as is possibly registered by the *title* given to them) in contrast brings into play semiotic mechanisms that differ according to whether one is dealing with an Ottonian miniature or a fifteenth-century Flemish altarpiece. There is nothing "natural" or "universal" about the rule that governs the formation of an image and its interpretation. It only has the force of a rule within a particular order, a system that can under no circumstances be reduced to a simple formula, whether the formula be that of what is claimed to be perspective realism, or that of an ill-defined medieval symbolism. It is not a matter of indifference that Panofsky chose to approach the question of the apparition starting with an image borrowed from the corpus of early Flemish painting. At a very early date that school of painting, albeit by means of trial and error rather than on a theoretical basis, elaborated a tradition that involved constructing a perspective space within which representations that contradicted the laws of nature were the exception: the exception that proves the rule? It was a rule that definitely seems to have been observed more systematically in fifteenth-century Flanders than in the place usually associated with it historically, that is to say the Florence of the Quattrocento.[12] But before moving on to consider how it was that the problem came to be posed just at the time when an explicit connection was established between the art of painting, scientific optics, and the work of "perspectivists," it is worth considering the solutions that were produced within a decorative corpus traditionally recognized as one of the monuments (in the Nietzschean sense) of the new style, if not as one of its founding cradles (the toponymic pun seems unavoidable).[13]

THE ASSISI STORY

We do not know for certain how much Giotto really contributed to the execution of the cycle of paintings representing the life of Saint Francis, in the upper church of Assisi. The many differences of both style and technique have led specialists to detect at least three different "hands" at work.[14] However, a formal analysis of the cycle reveals that it was certainly organized according to an overall plan, as can be seen both from the unity of the decor and from the symmetry observed between the two opposite walls of the nave, and also from the lighting of the various scenes, the arrangement of the painted edifices, and so on.[15] But such is the movement, "the impetus of the narrative,"[16] that even the least informed observer is impressed by a sense of evolution, or even progression. Leonetto Tintori and Millard Meiss have made a chart of the successive stages of the work of the fresco painters on the basis of a meticulous study of similarities discernible in the *intonaco* of the frescoes and the delineation of the zones or portions painted in the course of the "same day" (*giornata*). It provides striking confirmation of the conclusions reached by stylistic analysis. The scenes from the Old and the New testaments, which are positioned above the Giottesque cycle, follow the order that was usual at that time; the work was carried out from west to east on both walls of the basilica simultaneously. In contrast, the story of Saint Francis (with the exception of the very first scene: *The Saint's Entry into Assisi*) was painted in the order of the episodes of the story itself; and the stylistic variations that are discernible occur at identifiable points where, it is claimed, a particular artist can be named, but always within the framework of a homogeneous narrative sequence.

Cennino Cennini was to declare that Giotto changed the art of painting from Greek into Latin,[17] and it would be impossible to overestimate the importance of the fact that this monument of the new art is presented in the guise of a *story* systematically told by the means peculiar to painting. As a whole,[18] the Assisi cycle departs from the level of painting understood as a system of signs and enters that of painting conceived as the *instrument* of a narrative. In contradiction to the principle stated by Peirce, according to which it is not possible to construct propositions on the basis of iconic signs,[19] a language, or rather a figurative writing, is established through a "discourse" that links together, in an ordered progression, a sequence of images each of which possesses its own unity and is self-sufficient, but all of which mesh together following the thread of a story, a "continuous narrative." Furthermore, in its linearity, the sequence of frescoes obeys a necessity that is external to the order of painting, namely the demands of the hagiographical account which inspired the *tituli* that originally accompanied each scene.[20] This account provided not only a number of descriptive elements, but also the indispensable reference on the basis of which the communication operates and the images become readable as units in a discourse that it is their function, not so much to illustrate, but rather to actualize in a pictorial fashion.

Some of the images seem decipherable at a glance: *Sermon to the Birds*, for instance, and to a lesser degree, *Saint Francis Receiving the Stigmata*. Others are more enigmatic and bring into play more complex reading mechanisms. The theme of *The Death of the Knight of Celano* can only be explained (*dichiarato*, as the iconologists would say) in the light of the legend in the *titulus* that accompanies the image ("The Saint obtains the salvation of the soul of a knight of Celano who had piously invited him to dine; this man, making his way home, following his confession, while the others were sitting down to table, suddenly gave up the ghost and fell asleep in the Lord"). The picture is organized partly as a double composition: on the left, the saint and one of his companions are seated at table in a small, open chamber; on the right, the knight lies surrounded by a group of weeping women. The caesura, which is in effect also the suture that links the two moments in the story, is marked by a figure clad in red positioned, significantly, at the junction of the two "places." His raised right hand signals that he is listening to the saint, while his left hand points to the moribund knight lying on the ground. The same double schema is also found in *Saint Francis Dreaming of a Palace Stacked with Weapons* and in *The Dream of Pope Innocent III*, both of which juxtapose within a single image a representation of a man sleeping and of the dream that comes to him in his slumbers.

But the important thing is not so much to indicate the "subject" of these images (which, given the existence of the legend, does not pose a problem) and to situate them in the chain of a story told by means of a language. Rather, it is to draw attention to certain characteristic figurative elements and procedures, and this obviously requires deciphering the paintings bit by bit rather than producing an overall interpretation.[21] From a semiotic point of view, the Assisi cycle can be summed up as a set of pictorial signals distributed over a given surface. But the grammar and the syntax that make it possible to associate these signals with a semantic interpretation should not be understood in a merely decorative sense. The cycle translates into images a story the telling of which involves pinpointing a number of logical relations that painting, considered as a *visual* art, has in principle no means of denoting: the relation between dreams and reality, between two distinct yet simultaneous moments in the story, and also, as we shall see, between the present and the past, or even the physical register and the metaphysical one. Like dreams, which Freud regarded as a kind of regression from the linguistic order to the perceptive, painting is required to depict many conceptual relations by formal procedures, the discovery of which had the effect of considerably extending the domain of "that which is depictable" (*Darstellbarkeit*, to borrow Freud's concept from his *Traumdeutung*). The analogy that it is tempting to establish between art and dreaming is not all that well founded at the level of meanings and functions (artistic production is not resolved by a substitutive satisfaction of buried desires). Art presents, rather, an analogy at the level of what Freud called the "dream work," for the work is effected at the meeting point of what is said and what is perceived, and the products of this must fulfill one imperative condition: they must render the thoughts involved by essentially visual marks.[22]

THE GIOTTESQUE DRAMA AND ITS ACTORS

Let us return to the main theme of the present work and consider the means by which, in the cycle of the life of Saint Francis (described, for convenience's sake, as "Giottesque"), apparitions, oneiric or miraculous visions, and even one levitation (*The Ecstasy of Saint Francis*) are represented. It seems possible to reduce these to a few simple principles, some of which (a) can be described by and large as formal, the rest (b) as expressive.

(a) The surface is divided into two distinct zones, either vertically, as in the cases of the above-mentioned dreams, or horizontally, as in the cases of miraculous apparitions (*The Apparition of the Chariot of Fire*, *The Vision of the Celestial Thrones*). In the last two examples, the opposition heaven/earth is reinforced by the divergence of the respective vanishing points at the two different levels.[23]

(b) The connotation between the two zones or registers is ensured by the network constituted by the attitudes, gestures, and gazes of the actors, or (*The Death of the Knight of Celano*) by a figure positioned at the hinge connecting the two moments or levels of the story, whose function it is to coordinate or *conjoin* them. There may appear to be no intermediary between Saint Francis in his chariot of fire and his disciples, whose hands and eyes are lifted up to heaven, but in fact there is one brother who is leaning over a few of his sleeping companions, to draw the prodigy to their attention. In similar fashion, an angel shows Brother Leone the throne in the sky reserved for the saint, and Christ points toward the palace filled with weapons, an allusion to the future Franciscan militia (just as, in Rogier van der Weyden's altarpiece, the sibyl allows Emperor Augustus a glimpse of the enthroned Virgin). But in *The Dream of Innocent III*, the mere juxtaposition, within the limits of a single image, of two distinct elements of representation (the small chamber where the pontiff is dozing, and the church upheld by the saint) is enough to depict a relationship that the painters of the sixteenth and seventeenth centuries were to indicate by means of a cloud that would allow them to integrate into a unified spatial framework figures belonging to different and complementary registers (dream and reality, heaven and earth, and so on).

Would it be fair to say that the juxtaposition of distinct representational areas, with no real attempt at any illusionist coherence, corresponded to the medieval forms of figurative representation? Not really, for the schema of *The Dream* was to reappear in the Quattrocento, when this characteristic described as archaic took on the air of a stylistic ploy where the perspective code was concerned.[24] No doubt the Assisi cycle does not break totally with the scriptural means of establishing liaisons between narrative elements that characterize medieval figurative representations. But that does not mean that it does not simultaneously establish new rules of integration the essentially dramatic, *representational* nature of which needs to be underlined (representational in the theatrical sense of the word). The cycle is presented as a frieze; but the episodes that the painter chose to illustrate are set out like so many separate *scenes* that correspond to the chosen moments in the story, separated from one another by fake columns, and each with its own coherence. Not that the painter (Giotto himself or some other) right from the start defined the principles of a modern unified scene. Even where the composition appears to tend toward a geometric organization of the space (which happens particularly in the indoor scenes: *The Approval of the Rules, The Apparition in the Arles Chapter House*),[25] the unity of the representation is nevertheless above all dramatic, for the relations between the actors—as conveyed by their respective positions, attitudes, gestures, and even the direction of their gazes and the meetings of their eyes[26]—suffice to turn a relatively indefinite spatial framework into a stage.

An analysis of the topographical structure of the *intonaco* confirms what can be learnt from a rapid syntactical study. For the division of the surface to be covered, dictated by technical considerations—such as the need for the painters to work on scaffolding on several levels, set up before the wall—had quite a lot to do with a division that is not so much figurative as semantic. The Assisi frescoes were executed scene by scene, working from the top downward, the succession of levels being particularly noticeable on the southern wall, which was the last to be painted: the painter first covered (usually in two stages) the surface that corresponded to the top part of the space and the sky, then the central part of the composition, which was the most complex and the most closely packed with figures, and lastly the bottom part. In itself, such a division does not make for any particular spatial construction beyond that of a separation of the painted surface into horizontal bands one above another, as was also practiced in the Middle Ages.[27] And although it may seem particularly suited to the structure of compositions with several stories or registers, one placed on top of another, such as *The Apparition of the Chariot of Fire* or *The Vision of the Celestial Thrones*, it only derives its formal impact through the way in which it operates in relation to the figures in the drama and the elements that are strictly part of the representation. In *The Ecstasy of Saint Francis* (Plate 4a)—which deserves particular consideration since it represents the saint being carried aloft on a cloud—it is worth noting that one whole day was devoted to the execution of the saint's head, which occupies more or less the center of the image; another to the depiction of his hands raised heavenward in a gesture conveying the reception of grace. These were drawn with particular care (in contrast to the rendering of Christ's hands emerging from his divine circle). One further day was devoted to the painting of the faces of the spectators, treated as an independent unit, separate from their bodies, while their feet were painted at the same time as the ground on which they

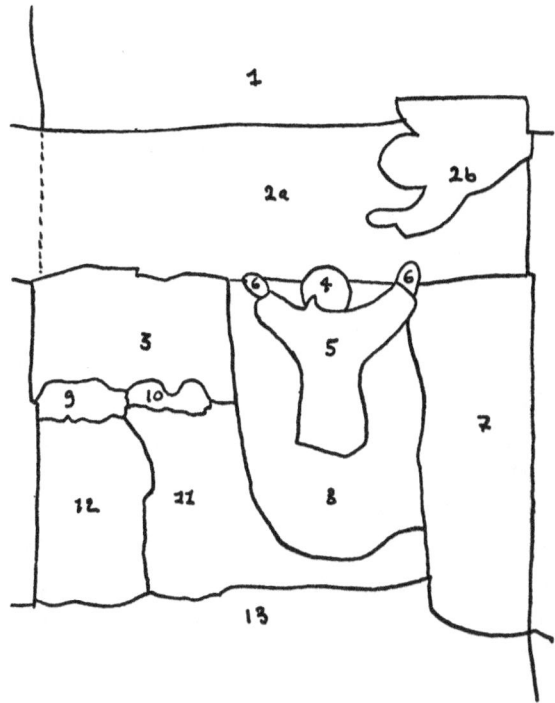

FIGURE 3. Giotto, *The Ecstasy of Saint Francis*. An analysis of the structure of the *intonaco*. From Leonetto Tintori and Millard Meiss, *The Painting of the Life of St. Francis of Assisi* (New York, 1962).

were placed. Other pictures in the cycle would produce similar observations. The emphasis placed on the expressive parts and the elements of a symbolism of gestures that is indicated by a technical analysis of the frescoes is altogether in keeping with the above observations and also with the qualities that Alberti, one and a half centuries later, was to ascribe to the artist of the *Navicella* (in terms that anticipated the teaching of the Academies, which very much favored the depiction of character and passion). Giottesque drama is played out between actors, each of whom conveys the psychological impact by which he is affected through a whole ordered mime that includes the expression on his face, his body language, and his attitude, and that takes into account the degree of importance of the role that falls to him in the organization of the representation.[28]

Giottesque representation is theatrical; and it is in terms of dramatic spectacles, if not of the direction of the actors, far more than of the perspective organization of space, that we should assess the so-called realism of this art and also an "expressionism" that owes more to theatrical skills than to purely pictorial effects. The tendency toward a "present reality" that Hegel considered to be characteristic of Giotto's art should be understood primarily in a theatrical sense: the Assisi painter wants human gestures to define the space of representation, a space in which the dimension of transcendence is no longer shown except by means of small machines relegated to the periphery of the composition. (As for the chariot of fire and the celestial thrones, they—like the child in Rogier van der Weyden's triptych, are no more than images, icons, and take their place as such in the context of a representation that is essentially carried by dramatic relations.) It is thus fair to say that Giotto confirmed the role of the human figure as an instrument of figurative thought,[29] but only provided that that thought—and with it the figures that it uses—was itself recognized to be here placed at the service of representation. The figures that Giotto's art sets in play cannot be dissociated from the action in which they are engaged. They are figures in a dramatic spectacle staged by painting rather than by a theater.

THE AMBITION TO FIND A FIGURATIVE WRITING

The Assisi cycle introduces gesture into the pictorial art of the West, by allotting it a coefficient of corporeity, of *relief*, that has often been declared to be the most innovative feature of Giotto's art (cf. the "tactile values" noticed by Berenson, the interest, demonstrated by John White, in objects, geometric volumes, to the detriment of spaces between bodies, etc.). Gestures seem to escape from the surface into which symbols are inserted, and to open up a new space, connected with movements and the interplay between the actors, whose function is not so much to signify a story as to establish it by the means of mime. The evolution that is noticeable in the structure of the *intonaco*, as the cycle proceeds, confirms the decisive role assigned to the human figure and to the effects that stem from its representation in a manner that is often described as "sculptural."[30] In the first part of the cycle, the feet of the human figures were painted at the same time as the ground, not when the figures themselves were painted. But on the south wall,[31] they were painted at the same time as the figures that they support, figures that become increasingly numerous, as if to satisfy the principle that Alberti was to set out: the greater the number of bodies included in the composition and the more varied they are, the more attractive the *istoria* will be.[32]

In the context of Western civilization, which is dominated by the model of the sign and verbal thought, dreams are not alone in making play with a form of writing (*Bilderschrift*, as Freud puts it) and figurative ploys that, despite owing nothing to the linguistic norms of expression, are (as we have learnt from the *Traumdeutung*) nevertheless capable of expressing relations that plastic thought, as such, has no means of representing.[33] Panofsky, limiting himself to one ploy out of many, has shown how in early Flemish painting the opposition between left and right operates in such a way as to make this a signifying pair capable of expressing the opposition between the before and the after, the past and the future, the old law and the new, and so on. Now, the Assisi cycle contains at least one scene that poses a problem of interpretation, and the only way to explain how the image equates to the episode in the story to which it refers is to resort to a syntax that is altogether foreign to the categories of language. This is a *representational* syntax that defines a space in which this same left/right opposition seems to play a determining role. The scene in question is entitled *The Apparition of Saint Francis in the Arles Chapter House* (or, as the *titulus* puts it: "While the blessed Antonio was preaching before the chapter of Arles on the theme of the Cross, the blessed Francis, although physically absent, appeared before the brothers, whom he blessed with uplifted hands"). There is nothing in the image as it is presented to our eyes to justify one declaring (*dichiarare*) that it represents an *apparition*: the saint stands in the doorway to the chapter chamber, which is treated as a unitary cube, with his hands indeed raised in a gesture of benediction, but his body in no way contradicts the laws of gravity. The only feature that distinguishes this image from the one that precedes it (which represents Saint Francis preaching before the pope, in a comparable architectural setting)[34] is that here the saint, instead of entering from the left, enters from the right. Now, the Assisi cycle as a whole unfolds in a progression moving from left to right, a progression that does not result solely from the movement of the narrative. This orientation is linked, as that of a hieroglyphic text might be, to the movement of the figures representing living beings, in particular that of the central figure in the story, Saint Francis himself. It is as he moves steadily from left to right that the saint makes his entry into Assisi, offers his cloak to a poor knight, sees in a dream a palace filled with weapons, chases the demons out of Arezzo, addresses the birds, and so on. In this context, any infraction of this dominant direction takes on a significance, a specific *meaning*. Sometimes the saint does turn back toward the left. In one instance he does so to remove his clothing (and here it is worth noting the gesture of the man who restrains his father from interfering, for it interrupts the movement that was carrying the latter *toward the right*, and by doing so leaves an empty space at the center of the scene in which that suspension or restraint of a movement constitutes the hidden mainspring of the narrative, a blind spot that underpins the figurative movement of the cycle). In another, he does so to hold up the tottering church that he will set about restoring (in *The Dream of Pope Innocent III*); in yet another to unmask the magicians of the sultan (who depart through an exit to the left, while the saint, already, turns his gaze to the right and toward whatever is to follow in the story). When he does turn back in this fashion, thereby contradicting the general orientation of the cycle, the movement is clearly designed to signify a break with the past: the march of history proceeds from left to right. In this context, the saint's appearance *from the right* into the hall of the chapter of Arles can be seen as an infraction, not of the laws of perspective and the physical universe, but of the rules of the narrative conducted according to the similar ways of painting and dramatic art. It is in accordance with those rules, which correspond to a higher figurative level than that of the images taken one by one, that the figure of the saint should be read as the apparition that Panofsky says it "is meant to be."

A study of the distribution—which, as has been pointed out above, is systematic—of the lighting and the colors (which, as technical analyses have shown, have retained a remarkable freshness) would no doubt prompt similar observations and reveal the extent and complexity of the figurative procedures followed in the Assisi cycle. It would make it necessary considerably to extend the field of depictability, or *representativity*, of this art that is so openly representational, and also to establish the limits of both Giotto's so-called "realism" and also his "expressionism," which does not seem to be an end in itself but is rather one, among many others, of his means of dramatic organization.[35] In this respect, it should be noted that the stylistic differences detectable within the cycle have less to do with the narrative syntax than with a syntax of "taste." The figurative story, for its part, obeys very strict principles, which govern its development throughout the entire cycle and which only tolerate deviation when this is deliberate and meaningful. (A structural analysis, corroborated on this point too by technical studies, reveals that the very first scene, *The Saint's Entry into Assisi*, in which the orientation from left to right of the pictorial text is *from the outset* clearly established, seems to have been executed after all the others, once the rest of the cycle was completed and as if to emphasize the unity of its conception.)[36] Certain theoretical implications stem from the fact that differences of style are revealed at a level below that of the story, where on the contrary there seems to be a clear intention to found the whole figurative representation, in the widest sense of the expression, upon a body of rules that is at once coherent and open-ended. The assimilation of painting—like the subconscious, as mentioned above[37]—to a "language" is valid primarily at the level of a "code," and it is at that same level, at which all individual differences are canceled out, giving way to the unity of *the* style, that *an ambition to find a figurative writing*, irreducible to the institution of a finite code, is confirmed. This is particularly clear in the case of the cycle of the life of Saint Francis.

GESTURE AND SPEECH

Considered both as a whole and in its various parts, the Assisi cycle shows that the process of visualization of which it is the theater cannot be reduced, even if we do regard it as a transformation, a theatrical production, to the association of a text and an image: reduced to its own components, the image—as Francastel points out[38]—presents to the eye a text that speaks for itself. But the fact that these images *speak* to the eye and that their text justifies a linguistic approach constitutes a sure enough indication that they remain dependent upon *phonè*, upon verbalized thought, or rather upon representation entirely organized so as to produce a meaning designed to be *understood*, just as if it were spoken aloud, whatever the scriptural procedures that are employed. For those who commissioned the work were no doubt not particularly concerned that the crowds who would be seeing these images should be able to understand what it was that made them work. You only have to read Boccaccio to appreciate the extent to which Giotto's art must have seemed *alive* compared to that of his master Cimabue: the "figurative self-evidence"[39] that has been adduced to characterize the works credited to Giotto is supported by the fact that these are images that immediately attract the eye, so much so that the mechanisms at work seem, as Vasari put it, to be borrowed from "Nature" itself. And those mechanisms, however much they are linked in their very principle to a regime of gestures that in itself is alien to the circuit of linguistic communication, are from the outset part of a representational system that makes them serve ends that essentially remain the aims of Words.

Giottesque "realism" is that of a representation to which every figure is subordinated (as is denoted by the meaning that the word *realism* has acquired in the language of the theater). The figures that this representation set in play lack speech: words are integrated into the pictorial text in the form of an absence that the *titulus* that accompanied each image merely serves to underline. But, by the same token, the gestures that constitute the most visible mainspring in the depiction of these figures serves to produce a meaning that is both played out *and* recited (or that might be either *played out* or *recited*, as the Italian language would have it, thanks to the equivocality of the term *recitare*). In this context, gesture assumes semiotic functions that are very different from those that might be conferred upon it by cultures alien to the schemata of Western civilization, cultures that would not grant speech the primacy that is assigned to it in the Greek tradition. Taken as it is usually understood, the use of gestures, in particular that of pointing, designating (which, as we have seen, plays an important part in the Giottesque story), opens up a relational space that disintegrates the entities and dichotomies that form the basis of linguistic communication (subject/object, word/idea, signifier/signified, and also image/concept, depicted/signified).[40] It is a theatrical space that Antonin Artaud hoped would equate to life, "not the life of individuals, not that individual aspect of life in which characters triumph, but a kind of liberated life that sweeps aside human individuality and in which man is no more than a reflection."[41] However, in Assisi we are still a long way from theater of that kind: Giottesque representation developed out of people's gestures and their effects (although, as Artaud, again, asked, "is theater really designed to describe to us man and what he does?").[42] Faithful at least to the teaching of the saint who decided to live "according to the Holy Gospel" and whose ardor to model his earthly existence upon that of Christ was rewarded by the stigmata by which he was the first to be honored, and above all respectful of the desire of the early Franciscans to contemplate God-made-man in the form of the man of God,[43] Giottesque representation organized the metaphysical drama to suit the measure of the man who was the instrument of that representation: the sacred representation *repeats* the "life" of the saint just as that life, in its day, repeated the life of the Son of Man in whom the Word was made flesh.

The images of *The Life of Saint Francis* do not actualize the *original scenes* of which a first representation is proposed by the Franciscan legend; but no more do they transmit a memory of real spectacles.[44] Nevertheless, they quite clearly do constitute a response to a preconceived program. And the image, along with the speech that duplicates and reduplicates it, by virtue of the threefold role that it has in relation to the program that it obeys, the *titulus* by which it is accompanied, and the commentary that it prompts, is part of a relationship characteristic of the general structure in which every element is linked, through representation, to all the rest,[45] the *tituli* themselves being so many representational texts that are subordinated to a representation.[46] By the same token, the gestures, in the form of icons, seem to be simply a means placed at the service of the representation that demands and governs the way that they are used. Giottesque gesture is not that of the theater of, for instance, the Far East, in which individualities are effaced in a sacred epiphany. It is gesture that is attached to one particular *character* among those that the representation puts into play, each of whom enacts, recites, and *repeats* the role allotted to him in this theater. And there is no gesture, including those of indication or designation, that is not a part of the context of a spectacle in which gestures assume strictly signaling functions—a spectacle governed by speech and totally enmeshed in a circuit of communication that works relentlessly to obliterate the presence of signifiers in the pictorial text.

SAINT FRANCIS'S CLOUD

Additional proof of this is provided by a fresco that is part of this same Assisi cycle but is often neglected by critics. It is, however, very important to the present discussion.[47] This particular fresco, already mentioned above, shows "how [to quote the *titulus*] the blessed Francis, one day when he was engaged in fervent prayer, was seen by his companions with his hands raised at a moment when his body was lifted from the ground; and a dazzling cloud shone all around him."[48] Clearly, this image could prove disconcerting for, in contradiction to the figurative design evident throughout the whole cycle, does not the transcendental make a reappearance here, in a most obvious guise?

In truth though, a comparison between the figurative means employed in two images as apparently different as this *Ecstasy* and *The Apparition in the Arles Chapter House* clearly reveals that both are instances of one and the same kind of writing. In the earlier of the two, the figure of the saint erupts into the circle of brothers without any sign of the miraculous movement that is implied by the apparition. All the painter needs to do in order to draw attention to the exceptional nature of this intrusion is have the saint enter from the right, contrary to the rule that governs the cycle as a whole. In this case, he did not need to resort to any signs other than those of a mime in which the positioning and expressions of the characters reveal that only a few of them were witnesses to this prodigy, which, at the figurative level, implies no infraction of the laws of the physical universe. In *The Ecstasy*, in contrast, the painter has paid less attention to the external manifestations of the vision and its effect on the spectators, being more concerned to translate into an image the actual phenomenon of the saint's rapture. But that translation is nonetheless still founded upon the unity of the representation as it is established by the Assisi cycle, that is to say by dint of theatrical rather than geometric means. The Middle Ages produced certain representations in which cloud played a *determining* role and the figures affected by it were designated either as heavenly or as occupying a position midway between heaven and earth. But the sculpted tympanums that show the resurrected Christ half emerging from a fringe of undulating clouds are not composed as unified scenes. Only in *The Ecstasy of Saint Francis* in the upper church of Assisi does the unity (in the theatrical and dramatic sense of the term) of the representation, extended to the cycle as a whole, for the first time provide the *systematic* mainspring for the depiction of an incident of rapture.[49]

In *The Apparition in the Arles Chapter House*, the representation of the prodigy is played out within a perfectly closed network of coordinates, both spatial and dramatic, in which the figure of the saint is introduced as something extra that is not directly denoted to be such. *The Ecstasy*, in contrast, is visually signified by a figurative operation designed to dissociate Saint Francis from the circle of his companions and also from the ground upon which the entire representation rests. Yet mime, even if it sufficed to define the overall scene that was affected by the saint's levitation, in itself offered no means of representing it: the painter was thus perforce obliged to resort to a *sign*, one that was certainly borrowed from the traditional repertory but that, because it was used within the context of the Assisi cycle, took on a new force and was assigned syntactical functions that were to undergo considerable developments in the centuries that followed. This cloud introduces a break into the fabric of dramatic and theatrical relations: it removes the saint from the common space and makes transcendence appear as an antithesis in a representation conceived in strictly "human" terms. But that break is only "readable" to the extent that the intervals that generally separate the protagonists in the mime (intervals upon which Giottesque writing confers a decisive significance) are here replaced by a discontinuity as a result of which the ordinary means of the use of gesture are here suspended— in this picture at least.

In this perspective (and the word here takes on an unexpected resonance), it is not hard to see how ambiguous is the Assisi painter's use of the /cloud/ graph in *The Ecstasy of Saint Francis*. On the one hand, it manifests the coherence of the system of figurative writing that is based essentially on dramatic means and semiotic mechanisms, some of which operate within the limits of a single scene while others do so at the level of the narrative sequence, that is to say the cycle considered as a "whole." On the other, this use (of a cloud) underlines the limits of representation that remains dependent upon the problematic of the sign and of the division that it imposes between the levels of what is signified and what signifies. As a consequence of the introduction of a perfectly conventional and arbitrary sign into this dramatic tissue, two things are revealed: first, the flatness that threatens the representational scene—flatness that cries out for its counterpart, "relief," and for an exaltation of "tactile values," and an insistence on the volumetric effects sanctioned by geometry; and second, the way that gestures can be reduced to the dimensions of the surface within the framework of which they are inscribed. A rupture can be introduced into one of the images in the cycle without tearing it apart or destroying its coherence; that is because all that mime ever ensures is a kind of expressive liaison between the figurative elements that are simply juxtaposed (the gate to the town, the group of brothers, the saint in his cloud, the little hill, etc.). Never does the pictorial depiction of gestures on its own open up any spatial depth for a stage no longer subject to the tyranny of speech: whatever its particular kinds of prestige, which are irreducible to those of articulated language, the theater of painting whose *bases* were immediately established by the cycle of Saint Francis was right from the start strictly subordinated to, if not made to serve, a representational structure that it reinforces with its signs. That structure is one of duplication, based on repetition, and it encompasses firstly the Giottesque enterprise (the aim of which, by dint of dramatic means, was to introduce a new, figurative writing, which tradition was soon to reduce to the dimensions of a "taste": the "Giottesque" taste). Secondly, it encompasses the Quattrocento's endeavors to use the means of perspective with a single vanishing point to construct the theatrical framework for a spectacle to which "Giotto" was the first to try to impart a unity, but which could only acquire such a unity at the price of an infinite duplication, the very duplication that is the basis for thought about signs and the system of representation.

Surface and Signs

The Scriptural Space

CONSTANTS

To speak of representation in terms of "writing" is to pose the question of the system that underpins the figurative process and of the constants thanks to which that process can be described and analyzed. However, the reference to writing does not imply that this question must logically be resolved by revealing a finite repertory of elements and by an exhaustive inventory of the possibilities for their combination. Even where writing of a phonetic, syllabic, or alphabetical kind appears to be a simple duplication of the speech from which it is derived and of which it offers a kind of more or less thorough analysis, it cannot be reduced to a finite series of procedures of notation. Even before tackling the *literal* meaning of a written text, reading is guided by a whole collection of marks that manifest the fact that the text belongs to the space in which it is inscribed and that this inscription constitutes and produces in the guise of a scriptural space. Some of those marks — their succession, the spacing between the letters, the separation between the words that results,[50] and even the punctuation — depend, either directly or indirectly, upon the system of phonetic notation and so belong to the order of signs; others correspond, rather, to the order of inscription, in the strict sense of the term, which is not solely linear: the disposition of words on the page, the arrangement of the lines and possibly of the paragraphs, the margins, and so on—not to mention features that may be described as external, such as the quality of the handwriting, the nature of the material upon which it appears, the format, the grain, and so on, and all that requires to be read "between the lines."

The fact remains that phonetic writing, insofar as it forms a system, has an external reason: however fleeting the speech that it immobilizes, that speech provides the text that it is writing's function to fix; and if it allows itself to be deciphered, it is by reference to the linguistic order, a language (or a type of language) to which it may provide access. The history of this writing, despite being repeatedly marked by progress (in the sense of simplification), is also a history of degradation: from being an autonomous means of expression, writing has sunk to the level of a mere substitute for speech,[51] although the phonetic model's superiority lies only in its efficiency and a transparency that borders on effacement. Pictorial writing, in contrast, even in the representational guise, does not necessarily aim for simplicity, let alone for effacement. And as for efficiency, the effects that it aims for are not those of language: how could they be, when this kind of writing, for the same reasons as apply to the writing of dreams, strictly speaking has at its disposal only *visual* means to render concepts and relations that language is able to express in words, conjugations, inflections, and so on? Even where figures seem to be organized by a text, a story, and where the "subject" of the image can be developed and explained by means of discourse, that image and those figures are neither a duplication nor an adornment for preexisting speech. A pictorial account has laws and principles of its own; but apart from a possible narrative perspective, the grammar that governs the organization of the figurative elements on a two-dimensional surface borrows nothing from the mechanisms of language by means of which an interpreter tries to understand the painting's image and then to describe and analyze it. Nor, once again, does there seem to be any reason why a systematic approach to the pictorial phenomenon should imply, as its necessary condition or end, the production of a code. Here, the position of the interpreter is close to that of a psychoanalyst who must try to discover the grammar of dreams as he proceeds in his or her deciphering, without hope of ever being able to lay bare a system on the basis of which it would then be possible to read and translate the whole of the dream.[52] But where the analyst knows that a "writing" that stems from the nonhistorical order of the subconscious probably exists, no one could deny the diversity or heterogeneity of the syntactical mechanisms by virtue of which a semantic interpretation might be historically associated with a complex of iconic signs. Pictorial writing, unlike the writing of dreams, does have a history: the problem, for theory, is to resist submitting to the tyranny of humanism that is disinclined to see in the products and periods of art — or of any other human achievement — anything apart from their singularity or individuality, and that would condemn as illegitimate, or even inadmissible, any quest for historical and/or transhistorical constants on the basis of which plastic art could be defined in terms of its general characteristics and its fundamental structure. A theoretical appreciation of such constants would seem to be the precondition for any rigorous, if not scientific, history of art.

PERSPECTIVES

The question of the *place* of a pictorial experiment, or rather the question of pictorial writing as it relates to its substratum, and of the signs that it brings into play as they relate to the space that they create through being arranged there in a certain order and with a certain positioning — that is a question that is clearly part of a generalized problem. Pictorial writing itself *produces*, either positively or negatively, its own substratum. It either requires a canvas, a panel, or a wall to function as a mediating or immediate component of the image, as is the case in so-called "surface" arts; or else it denies them their substantiality, their very materiality, by replacing the perception of a flat surface by that of a three-dimensional illusionist space: when this happens, we say that the painting "makes a hole" in the wall, or that a cupola "opens" onto the sky, and so on. In both cases — each of which lends itself to variations that may differ radically both in principle and in their effects — the treatment of the pictorial space fulfills truly syntactical functions, over and above any decorative or illusionist aims. Again there are two alternatives: either the substratum accommodates apparently heterogeneous figures and signs, the assembly of which on a single surface is governed, even in its stylistic modalities, by imperatives of a semiotic nature; or, alternatively, the fake space suggested or constructed by painterly or geometric means appears to hold beings and objects whose positioning, spacing, and interrelations seem governed by laws that are themselves illusionist but are conceived in imitation of the laws of nature.

Formulated in these terms, the question seems bound to exclude any comparative research into the *disjecta membra* of systems whose existence theory would merely postulate without ever producing them explicitly. On the other hand, there are some systems that are nonfinite, not closed, and perhaps nonsystematic, but which nevertheless appear to obey a number of constants. Some of those constants are particular and guarantee the identity of a given historical system throughout the diversity of all its manifestations, while others are general and fundamental to the pictorial phenomenon as such, and if these were revealed, they would make sure that a scientific history of painting had the basis that it lacks. Where those nonfinite systems are concerned, theory, at least at a preliminary stage, should endeavor to eliminate the particularism of the concept that constitutes the main stumbling block for a purely descriptive history of art conceived on a philological model. That particularism can be expressed in many ways and take on a diversity of forms when it is not obliterated by some misleading generalization of the kind that is usually invoked at the level of "style" (as we have already noted in connection with the very lax extension of the concepts of "Baroque" and "Mannerism" in the discourse of art history). And, on a more strictly formal level, the confusion is just as great. Let us restrict ourselves to a single example, albeit admittedly one that is particularly typical at this point. Panofsky was obliged to undertake a great deal of work (the results of which, even then, were far from conclusive) in order to show that the classical notion of linear perspective with a single vanishing point does not suffice to explain the predecessors of a symbolic form (*Symbolische Form*, as Cassirer put it) that corresponds to a far more general definition: it is possible to depict (*Darzustellen*) several objects, as well a portion of the space in which they are positioned, in such a way that the representation (*Vorstellung*) of the material substratum to the image is replaced by that of a transparent surface through which we think we can see those same objects disposed in an apparent sequence in an imaginary space that is limitless, yet is delimited by the edges of the picture.[53] If that is so, perspective appears as

one of the symbolic forms through which a spiritual content is linked to a perceptible concrete sign [*an eines konkretes sinnliches Zeichen*] in such a way as to be identified with it. For that reason, it becomes essential, where different periods and provinces of art are concerned, to decide not only whether they were familiar with perspective, but also with what kind of perspective they were familiar.[54]

There could be no clearer formulation of the relation that links theory to its concrete applications, nor any better way of illuminating the classificatory function of this concept (of symbolic form): the definition of perspective as a symbolic form leads us to distinguish between the periods and provinces of art that knew nothing of it and those that used it in one form or another. Furthermore, the basis for such a distinction is not so much theoretical but rather historical, marked as it is, in the way that it is stated, by the stamp of cultural prejudice according to which all pictorial systems, however foreign they may be to the ideological space of the Western Renaissance of the fifteenth and sixteenth centuries, are given negative characterization from the outset, based on the absence of any kind of perspective order. Meanwhile, systems where that link is attested are assessed according to a progression that stretches from the first attempts at illusionist depth, in the Hellenistic period, all the way to the geometric method of constructing representative space, as developed by the contemporaries of Brunelleschi and Alberti. The constants that any theory would have to recognize would be of an essentially historical nature and linked to the slow conquest, birth and rebirth, after the Middle Ages, of a projected three-dimensional space that was consummately defined by the likes of Paolo Uccello and Jan van Eyck. Such a teleological view should not be justified by subordinating the developments of Western art to the progress made in techniques of planimetric reduction and the achievements of the perspectivists of the sixteenth and the seventeenth centuries (achievements that were often paradoxical if not aberrant, as is shown by the many anamorphoses and other "curious perspectives," to which we shall be returning). Rather, it should be justified by dialectical articulation from the starting point of a far more general problem: namely, the problem of the figure and the background, the scriptural space within which signs are arranged. What Hellenistic painting would then be found to have in common with Renaissance painting would be the way that it replaced the view of the pictorial surface, as a material reality, by a fake depth in which all references to the surface from which the figures stand out are abolished. Such references are, in contrast, decisive in Byzantine and medieval art (at least up until the point when the stained-glass window began to supplant the wall fresco. The stained-glass window was quite literally the first painted "window," but one that, it must be said, was designed not to allow the view to open up onto the beyond, but rather to play with and make the most of the light that passed through it).

Syntaxes

AN ANCIENT PROCEDURE

Whether or not they involved the establishment of a perspective order and a negation of the material substratum of iconic signs, the various systems of pictorial writing that were respected, at one time or another, in the Christian West made room, in various ways and to varying degrees, for figurative units that analysis labels, proceeding according to the means of articulated language, using the word *cloud*. Of course, the assignation of a name is not enough to guarantee the identity of the units thereby denoted. As has already been pointed out, even if they are identical from the point of view of an analysis of the pictorial text into its iconic components, from the point of view of their formal texture and the functions that they assume in different contexts and at different levels, the units denoted as *clouds* constitute so many distinct objects. Yet, over and above the delimitations introduced by a consideration of stylistic periods and over and above the differentiation of levels imposed by analysis, the presence of those units corresponds, at least at the level of that which is signified, to a common factor: *cloud* intervenes in a figurative text wherever it is a matter of relations not only between the earth and the sky but also between down here on earth and the beyond, between a world that obeys its own laws and a divine space that is unknowable to science. We have noted the syntactical functions fulfilled by *cloud* in Giottesque pictorial writing, and how the painter of Assisi resorted to the use of this sign in order to extract the saint from common space and place him in a space of contemplation and mystic rapture. But similar uses of *cloud* are attested in a much earlier period. Studying the mosaics in the nave of Santa Maria Maggiore in Rome—which dates from the reign of Sixtus III (432–40)—André Grabar noticed the interesting use of cloud in certain panels illustrating the most striking episodes in the stories of Abraham and Jacob, and in those of Moses and Joshua. It was in *The Vision of Abraham* and *The Stoning of Moses, Caleb, and Nun* that, for the first time, a divine cloud was introduced as an accompaniment to the appearance of God, and that cloud was also used to protect a figure from the sight and blows of his adversaries. Grabar remarks that "this way . . . of drawing attention to the providential nature of the story may be ancient in its form, but was only really successful in Christian iconography." But all the same, he does consider that the Roman mosaic artists were "reworking" the Hellenistic modes of representing epic.[55] A reworking it may be, but it certainly testifies to a recognition of both the radical discontinuity and — at the same time — the possibility of communication between the human world and the divine order, and of the consequent need for the painter to elaborate figurative mechanisms that serve to represent the intervention of celestial beings into human affairs.

Does *cloud* appear in Giottesque writing merely as an archaic survival?[56] In the case of the Assisi *Ecstasy* and that of the Roman *Stoning*, what is signified is perfectly congruent. In both cases cloud is the instrument that carries a figure away from the earthly scene, and in both cases it introduces a differentiation into the pictorial field, a sort of fault line, a rupture that is a figurative way of manifesting the precarious nature of the human order, constantly exposed to being torn apart by a miracle. But what is signified there does not suffice to account for the functions of this element, which assumes a noticeably different importance in the two contexts that we are considering. In Santa Maria Maggiore, as in the upper church of Assisi, the images have a wall as their material substratum. But in Rome, the decoration, far from negating the church wall, on the contrary enriches it with its brilliance; the series of illustrations is organized overall according to ornamental principles and their sole unity is conferred upon them by the enclosed area within which they are positioned. In contrast, the Assisi story, even in its trompe l'oeil architectural framework, involves the definition of an autonomous and unified space, in which it can deploy its figures. Here, each *scene* has its own unity. But it is a unity that gains from the coherence of the whole, which is conveyed at very first sight by the intensity of the colors, the scale of the figures, the impetus of the narrative and — more covertly — by the rigor and homogeneity of its *dramatic* principles. For the fifth-century Roman mosaic artists, the cloud is simply a punctuation mark (a kind of parenthesis) that derives its meaning from the position that it occupies in a linear sequence. With the Assisi painter, the cloud accedes to the rank of a truly symbolical tool, the significance of which is linked with the spatial structures in which it is used, and by means of which the human aspects of the drama are at this point suspended.

Without leaving the terrain of Rome (which constitutes a special field of reference since it is only here that the continuity of Christian art is attested from late antiquity right down to the modern period), we can see *cloud* functioning in a wide range of very different conditions. Take the case of the clouds scattered here and there in the mosaics that adorn the apses and arches of Roman basilicas dating from before A.D. 1000. In the church of Santi Cosma e Damiano (sixth century; see Plate 4b), and in that of San Prassede (ninth century), Christ is isolated from a blue background by a kind of screen, a discontinuous grid composed of small blue and red strato-cumulus clouds, clearly delineated and regularly aligned. Similar clouds are to be found on the arch of the church of Santi Nereo ed Achilleo (eighth century), where a triumphant Christ is flanked by two groups of the faith-ful at prayer. In Sant'Agnese Fuori le Mura (seventh century) and in San Prassede, clouds surround the hand of the Creator, which is to be seen at the top of the vaulted ceiling, ringed by a series of concentric circles, in conformity with a traditional symbolic schema that is found again in the twelfth century, in the apse of San Clemente and also in Santa Maria in Trastevere. Perhaps it would be fair to say that the Roman mosaic artists were not concerned to establish their figures in an imaginary place but were content to organize the surfaces that they were given to cover by defining qualitatively differentiated zones in which they could position their sacred effigies. (It is as if, by rejecting all illusionism, they set out to affirm the spiritual value of the image as such—that is, the painted or decorated sur-face—at a time when an iconoclastic frenzy was being unleashed in Byzan-tium.) Certainly, the organization of these icons is governed by rules that are above all conceptual and decorative: *cloud* is not used for its pictorial value, but as a marker, a conventional determinative.

Panofsky has pointed out that in early Christian art, and even more in Byzantine art (which never managed to cut itself off altogether from the ancient tradition), the place where figures are depicted, far from being re-duced to an opaque, negative surface, appears rather as a luminous film that denotes—but does not "reproduce"—a spatial environment.[57] The clouds that separate the figure of Christ from a uniformly blue background intro-duce a modulation into the colored depth: the figure is placed *in front* of them, on this side of a transparent screen beyond which color (light) reigns over all. Cloud—or a cloud of light, which fulfills the same functions[58]—*identifies* as the apparition of a god the divine icon that it isolates from the rest of the image and from the "sky" against which it stands out, and also from the "ground" upon which are set the witnesses whose mediation seems indispensable to any manifestation of the sacred. Already, in Santa Maria

Maggiore, one of Abraham's three celestial visitors is surrounded by an oval of light, while horizontally the three are separated from the patriarch greeting them by a few clouds. In *The Meeting of Abraham and Melchizedek*, above the group of horsemen that the priest is moving forward to welcome, a figure — possibly the Old Testament God?[59] — is half emerging from a row of clouds; the separation between the two levels is emphasized by the opposition between the gold background, which corresponds to the human level in the story, and the blue background against which the divine figure, here in the position of a witness, is depicted. But in truth these representations stem from a very elementary ploy, which goes back to the earliest centuries of the Christian era: at Doura-Europus, the hand of God, along with a small cloud, is introduced into the representation of the sacrifice of Abraham, just as it was to be a few centuries later in San Vitale, in Ravenna.[60] The same ploy is regularly used, in just as elementary a fashion, throughout the Middle Ages, in Giotto's paintings, and even later; but by then it was regarded as a deliberate archaism without consequence at the level of the system, the coordinates of which it even enhanced. On the other hand, the functions assigned to /cloud/, considered as a syntactical tool, were to make an ever increasing impact.

MASOLINO

A small panel painted by Masolino, now in the Naples museum, provides a good point of reference both historically and from the point of view of systematic analysis. Through the implied reference to a building, the decoration of which marked an important stage in the development of Christian iconology and even, as we have seen, in the problematic of /cloud/, the title of this panel manifests the continuity of the cultural tradition to which it belongs. *The Foundation of Santa Maria Maggiore* (Plate 5) illustrates the legend according to which the Virgin appeared to Pope Liberius to command him to build a church on the Esquiline hill, in a place that would be found to be covered by snow. In the lower part of the composition, between two rows of buildings depicted in exaggerated perspective, like a piece of stage scenery, the composition shows the much foreshortened figure of the pope tracing the plan of the future church in the snow. Echoing the pronounced perspective of the lines of the buildings, a series of regularly distributed clouds, growing increasingly small as they approach the horizon, separates this scene from the upper level, where busts of Christ and the Virgin are represented inside a medallion and in larger scale than the other figures. The clouds, set in perspective but not treated at all "realistically," belong to both levels, the one descriptive and illusionist, the other bearing the seal of a sacred effigy.[61] They do not so much effect a transition between the two levels of the composition, but rather divide it into distinct zones, each governed by a different set of figurative principles. Christ and the Virgin, in their medallion, are presented not so much as an apparition (not one of the figures in the principal scene is looking toward them), but rather as a seal, a stamp that confirms the miraculous nature of the event. Christ, who is pointing at the scene, is here in the position of a witness. The composition is thus in principle analogous to that of *The Meeting of Abraham and Melchizedek*, described above, except that in its lower level it is governed by the principles of perspective, in an almost outrageous fashion, only to ignore them, equally openly, in the upper level. The contrast, through its very exaggeration, draws attention to the miracle supposed to have been at the origin of the foundation of the basilica. But it is only seen as an opposition thanks to the introduction, at the join of the two levels, of a hinging element that ensures that the two interact. If one imagines the picture without the mass of clouds that divides the panel into two more or less equal parts, it loses most of its *depth*, in the literal as well as the figurative sense of the term. For the clouds, in their regulated flight, on the one hand compose a kind of "roof" or covering, which reinforces the perspective effect sought in the lower level, while, on the other, it is thanks to them that the literal message, denoted by the title that accompanies the image, serves to carry a second meaning, of a theoretical nature. The opposition modulated by the cloud signifies something quite different from what it denotes: it conveys a series of questions about the status of the perspective "code" and the exclusions that it determines, that is to say the contradiction which—possibly—constitutes its very mainspring.

History and Geometry

THE TASK OF THE PAINTER

Do the rules of perspective lend themselves to the representation of phenomena that break through the ordinary bounds of the human order? Or, to put that another way, do divine interventions, which open up this world to the beyond and, more generally, mystical—or even physical—exchanges between the earth and the sky provide matter suitable for representation despite the fact that depiction seems to be subject to an organizational principle for the pictorial field that seems to imply as its corollary that the illusionist space constructed by geometrical means is governed by rules analogous to the laws that operate in the empirical universe: bodies and objects are subject to gravity, and this imposes limitations upon the manner in which they move around, and so on? Can such representations even have a place in this system, unless supernatural manifestations or miraculous events allow themselves to be reduced to the common norms of perception, or vision? Alberti's *Della pittura*, a book that—as its author tells us—is not a history of art but an *art*—a theory—of painting, and the first of its kind,[62] states that a painter should attempt to *feign* only that which can be seen: things that cannot be seen are no part of his remit ("Delle cose quali non possiamo veder, niuno nega nulla apartenersene al pictore. Solo studia il pictore fingiere quello si vede [No one would deny that the painter has nothing to do with things that are not visible. The painter is concerned solely with representing what can be seen]").[63] His job is "to describe [*descrivere*] with lines and to tint [*tigniere*] with colour on whatever panel or wall is given him similar observed planes of any body so that at a certain distance and in a certain position from the centre, they appear in relief and seem to have mass."[64]

It is important to emphasize at the outset that painting, thus understood, is only meaningful in relation to the *istoria*, which is the main preoccupation for a painter (*summa opera del pittore*). The *istoria* means the composition of the bodies that belong to it according to their respective sizes and functions: therein lies the genius of the painter.[65] The greatest care should be devoted to placing each thing in its rightful place (*a suoi luoghi*). In an *istoria*, what is immediately striking and pleasing is the abundance and variety of the figures, animals, objects, buildings, and even landscapes that fill it.[66] But all the same, blame is due to painters who do not manage to accommodate any *empty spaces* in their compositions, but instead cram them with so many figures, disposed in such confusion that the *istoria*, taken over by tumult, no longer possesses the dignity of an action.[67] That is quite an important observation, for it implies that the organization of the pictorial space cannot be defined in purely geometrical terms: there is also a dramatic quality to the space, and because of that the scene that frames the figures should not seem to be simply a receptacle for the bodies;

rather, it should be created by their composition, their disposition, their relations, their interaction (and by the space that separates them, which may seem to play an intrinsically insignificant part in their articulation, constituting a merely negative condition for the permutation of the figures in the painting, but which nevertheless, in the context of a unified illusionist representation, in truth acquires a positive value). Painting is a matter of empty spaces as well as full ones (and theory ought to be able to take both into account and define them, with their respective statuses, on an equal footing);[68] and it is also a matter of the weight of things. Each figure should play its part in its proper place (*a suoi luoghi*), where it should be in proportion,[69] and the place should suit the nature of the particular figure and that figure's situation. A diversity of gestures and attitudes and a variety of body movements constitute some of the major attractions of an *istoria*. But if visible movements — the only ones that a painter should concern himself with[70] — are defined in relation to a place and can be reduced to a change of position (of which several kinds are possible: [a] upward, [b] downward, [c] to the right, [d] to the left, [e] moving away, [f] coming closer, [g] turning in a circle), then some transgress the bounds of reason,[71] that is to say the rule according to which every figure should be satisfactory in respect of the balance between its parts and, with all acrobatic postures excluded,[72] should also satisfy imperatives analogous to those of gravity in the empirical universe. But there is one nuance that, by contrast, emphasizes still further the laws by which the system is marked: in every movement, grace and beauty should be aimed for. And the most agreeable and animated movements of all are movements upward, up into the air, that is to say movements that contradict weight and manifest the freedom of bodies.[73] The same goes for hair, similar to flames (*simile alle fiamme*) and garments, drawn naturally downward by their weight: it is good when the wind lifts and moves these things, that same wind that blows *through the clouds* (*che soffi fra le nuvole*).[74]

PERSPECTIVE AND REPETITION

In painting, the impression of weightlessness sometimes possesses a charm of its own (and here it is Botticelli's *Venus* and the figures in his painting of *Spring*, or Lippi's *Salome*, and so on, that leap to mind). *As an effect*, it only exists thanks to the illusionist basis of the painting and the perceptible, "visible" relation between the body and the space within which it moves. This clearly means that the composition of the bodies, which constitutes the *istoria*, presupposes a geometrical construction of space. Linear perspective is a means that is placed at the service of the *istoria*, a means that it cannot do without or attempt to supplant. However great the emphasis that Alberti lays upon the *istoria, summa opera del pittore* ("the greatest achievement of the painter"), there is no trace in *Della pittura* of the "humanist perspective" that Pomponius Gauricus was to describe in his *De sculptura*, in the early sixteenth century, in Padua. The *Della pittura* proceeds according to the terms of (medieval) optics: Alberti recommends that artists should construct a checkered floor, a volume in perspective, before going on to consider foreshortening and the composition of surfaces and bodies. In Gauricus, in contrast, perspective teaches no longer how to paint a scene but, right from the start, how to compose a story. On the same grounds as Aristotelian verisimilitude, to which Robert Klein, with reason, compares it, it now appears not so much as a principle of illusion, but rather as a rule of unity, of perspicuity (of legibility): the clarity of the *istoria* depends not only upon the gaps between the figures but also upon the quantity of figures that predetermines those gaps, and their disposition.[75] Gauricus claims that the number of figures is associated with the perspective;[76] so does he not replace an illusionist perspective with a dramatic one? Thereby he would reaffirm, in contrast to what the Quattrocento had to say, the drama's priority over the setting in which it takes place. Does he not avoid posing the question of representation in relation to the place or setting in which it takes place and which defines its fundamental structure, thereby reducing it to a problem of expression, if not of the direction of the actors (who are the "interpreters" of the drama and its "protagonists")?

The distinction between geometry and history that *Della pittura* introduces makes a quite different point. The rule to follow in the construction of a checkered floor and its corresponding vertical walls is a rule that affects the composition:[77] it is an indispensable preliminary that is not only "practical" (the perspective construction provides the scale for the foreshortening of the figures), but also, and above all, "theoretical." In *Della pittura*, Alberti's condemnation of the technical method that made it possible to determine how to space out the transversal lines by means of mechanically reducing the intervals between them comes immediately after his description of the figurative surface as an open window through which the painter is supposed to see what he is about to paint.[78] But that definition in itself presupposes defining a painting as a representation, on a given surface, of an *intersegatione* (section) of the visual pyramid at a predetermined distance from the eye and in a position calculated in relation to the central point.[79] The painter must impart to his figures an appearance of the reality that he is endeavoring to imitate. But in order to do so, he needs to be equipped for the task. It was to this end that the method of perspective construction, the invention of which is traditionally credited to the architect Filippo Brunelleschi, was elaborated. In principal, there is nothing empirical, or even experimental, about this method—unless experimentation is understood as a systematic enquiry. The break that it introduced into the development of art (a break of which contemporaries were very much aware) was of an essentially theoretical nature.[80] Just as Galileo was to do for the science of his day, Brunelleschi presented the painters of the Quattrocento with a "language" that would enable them to interrogate nature, the external world, and to interpret its responses.[81] This language or, if you like, "model" (in the epistemological sense of the term), in which the task of the artist as defined by Alberti is allotted its status, is none other than that of *perspectiva artificialis* (the perspective of painters, as opposed to the *perspectiva naturalis* of medieval optics), as it was instituted at the beginning of the century by Brunelleschi's famous optical machines in which the principle of the new system, by having the vanishing point coincide with the viewpoint, seems to have found its first theoretical illustration. In the first of the two experiments described in the *Vita* (Life) of Brunelleschi attributed to the mathematician Manetti, the observer was supposed to place his eye against the back of a panel upon which the baptistery of Florence and the square surrounding it were depicted, at a point that corresponded to the vanishing point of the construction, and to look through a peephole made at this point at the reflection of the image in a mirror placed parallel to the panel, at a suitable distance. In this way the observer obtained a view similar to the view he would have had of the square from the door of the cathedral, that is to say the very spot in which Brunelleschi had positioned himself, ideally at least, in order to set up his model.[82]

Alberti's *Della pittura* has nothing to say about the decision that was the veritable intellectual breakthrough and with which Brunelleschi's name is associated (the confirmed coincidence, in the mirror, between the vanishing point and the viewpoint). But the definition that it produces for *the task of the painter* only makes sense in the light of that revolution. The painter's function is to delineate and color the surfaces that he constructs in imitation of the surfaces of the bodies seen at a predetermined distance and from a predetermined point of view: clearly then, both the concept and the execution of the *istoria* depend directly upon the representational system for which Brunelleschi produced the theory. And the reason why the mirror is such a good guide for the painter and helps him to judge the qualities and defects of a painting,[83] is that it introduces between the eye and the painted image (just as Brunelleschi's machine did) the distance achieved by reduplication, repetition. But the duplication that is the principle of representation is not just a matter of mirrors. In Alberti, it takes on the aspect and value of a dramatic reduplication: through a kind of *repetitio rerum*, the painter starts by constructing the scene in which the story, once he has done so, will be inscribed and in which everything will find its place.[84]

Intarsia II

> La prospettiva è di tale natura ch'ella fa parere il piano
> rilievo e'l rilievo piano.
>
> Perspective is of such a nature that it makes what is flat appear
> to have depth and what has depth appear to be flat.
> —Leonardo da Vinci

THE PARADIGM OF WRITING

It is thus not hard to understand the importance ascribed to geometry in the training of a painter and also the primary place assigned to drawing—assimilated to delineation—among the various parts of painting.[85] Anybody not well versed in geometry would understand nothing of the elements of painting or any of its rules. A painter needed to know only of visible things, that is to say things that occupy a place.[86] If he referred at all to scientific optics, it would be to borrow concepts necessary for his own demonstrations: the mechanism of vision, the role and nature of the eye, and so on mattered little to him. His "I speak as a painter" (*Parlo come pittore*) was to be taken literally: for a painter, whose reasons are not those of a mathematician, the field of vision was the equivalent of whatever could be inscribed (projected) onto a surface and could be reduced to an interplay, a regulated construction of surfaces seized upon from various angles, each of which was defined, designated by its outline, and all of them interacting according to the point of view adopted for composing figures and scenes, in the same way as letters are arranged so as to form words and sentences.[87] The work of painting is a work of inscription, and its apprenticeship is not unrelated to that of writing: "I would have those who begin to learn the art of painting do what I see practised by teachers of writing. They first teach all the signs of the alphabet separately, and then how to put syllables together, and then whole words. Our students should follow this method with painting. First they should learn the outline of surfaces."[88]

POINT/SIGN/SURFACE

The introduction of the paradigm of phonetic writing is decisive here: it establishes representation (what is visible being here assimilated to what is representable) as directly dependent upon language, the same language that gives things their names and by which *stories* are told; also the language that Western tradition links *in general* to the voice, to hearing, to speech, in relation to which writing has a merely derived function—derived, in that it is representational.[89] And, in *fact*, the same applies to the perspective model as to the phonetic model: *model* being the right word here, as Jacques Derrida insists, rather than *structure*; for "it is a matter not of a constructed system, functioning perfectly, but rather of an ideal that explicitly governs a functioning which *in fact* is never phonetic through and through,"[90] any more than it is ever "perspective" through and through. It would be interesting, at this point, to show that the paradigm of phonetic writing, far from having a purely pedagogic value, in fact governs the entire pictorial culture of the Renaissance, right down to the symbolic developments that the classical age was to classify under the rubrics *emblematic* or *iconology*. The reference to a hieroglyphic model—a reference already to be found in Alberti's text, where it does not, however, imply any break with the phonetic model, rather the reverse[91]—does not correspond, either in fact or in theory, to a remanipulation, let alone an undermining, of the fundamental structures of representation. It simply illustrates an extra level of articulation at which the painting operates when, to borrow Ripa's formula, it aims to signify something other than what it *presents to the eye*. No more does iconology, even in its reference to the hieroglyphic model, break with the strictly representational notion of the written sign that is imposed by the phonetic model: it is up to the painter to create a figure all of whose parts correspond term for term to those of the thing that is signified, and these should be disposed in an order in conformity with the order of the elements of representation (see Chapter 2, pp. 51 ff.). In their visible interaction, allegory and figurative metaphor remain part and parcel of the pictorial syntax the elements of which Alberti indicated.

It is revealing, in this respect, that Alberti's *Della pittura* refers to the outlines of surfaces — those same outlines at which novices should practice "as if they were the first elements of painting" — under the rubric *sign*. "I call sign anything which exists on a surface so that it is visible to the eye."[92] The point will be defined as a sign that cannot be divided into parts; a line will be defined as a sign that can be divided in its length but is too fine to be divided in its width.[93] But what about a surface, or plane (*superficie*)? "If many lines are joined closely together like threads in a cloth, they will create a surface. A surface is the outer limit of a body which is recognized not by the depth but by width and length, and also by its properties."[94] That is a proposition of great consequence, since it implies that for a painter all surfaces must be *projective*: drawn lines, the composition of surfaces on the flat canvas, aim to produce an illusion of real bodies seen from a certain distance and a certain point of view (a definition that, it is worth noting, coincides with that which Peirce produced for a sign, or *representamen*: "A sign, or *representamen*, is that which replaces something for someone with a certain aspect or position"). Several assembled surfaces compose a body. But the composition of bodies, in its turn, presupposes that between them there should be intervals, empty spaces. What of these with regard to the problematic of signs and the visible surface of the picture upon which they appear? This question is one that a painter may choose to ignore, pretending that he looks right through (*per-spicere*) the screen upon which he is working, denying it any reality or value of its own. An interval would only present a problem if the outline were nothing but a limit between what is filled in and what is empty, and belonged neither to the figure nor to the background.[95] But Alberti declares specifically that the outline is the edge, the extreme limit of the surface that it encloses and to which it belongs.[96] The line that surrounds them provides the reason for the surfaces that it designates[97] and that in part derive their name from it,[98] just as it provides the reason for the colours and the very composition. But even if the *istoria* is *woven with lines* in this way, it will still be necessary, as has been said above, for those lines not to be too definite, for otherwise the image would appear to break up into a constellation of juxtaposed fragments (see Chapter 1, pp. 25 ff.).

If a painter had to construct his picture piece by piece—as the paradigm of writing would appear to suggest—his task would be infinitely painstaking. Fortunately though, there were ways of cutting through the problem. One was the *intersector*, the transparent veil that Alberti advised the painter to place in between himself and the spectacle that he wished to paint, in order to trace the outlines of objects directly upon it.[99] The day would come when Vasari would write that it was not a good idea for a painter to become too involved with matters of perspective: anyone who paid them too much attention would lose his "naturalness" and, through excessive attention to detail, would fall into a dry, angular "constrained manner, . . . full of profiles" (*una cattiva maniera . . . secca e piena di profili*). One example was Paolo Uccello who, according to Vasari, could have been a painter as original as Giotto if only he had paid as much attention to the drawing of figures as to this art, the most difficult of all, which caused him to put everything in his compositions into perspective, everything from the ground to the roof tiles, "by means of intersecting lines [*per via del intersecare le linee*]."[100] Uccello's friend Donatello, who watched him devoting his time to drawing a ducal cap (*mazzocchio*) from the most varied of angles, and polyhedrons with diamond points, and all sorts of other "bizarre things," apparently often took him to task for "losing the subject for the shadow" and devoting his time to things that were of no use to a painter and could only be useful to *the makers of marquetry* (my emphasis).[101]

A SPACE WITHOUT DEPTH

The work of a painter was, however, not unrelated to the skill of *intarsia* (marquetry), at which Alberti himself was to try his hand on a monumental scale.[102] It is well known that marquetry specialists owed a great deal to the theorists of perspective, including both Piero della Francesca and Brunelleschi.[103] But their work simply carried the paradox of perspective to its limit, thereby revealing the position of painting, in its relation to space, once it became dependent on the idea of the sign and became affected by the limitations and the exclusions that this dictates. A sign operates at surface level and in a linear fashion; it functions primarily in extension: its meaning is conferred upon it by its relations to other signs with which it may or may not be associated and with which it enters into composition or opposition.[104] A painter knows only of things that occupy a place: in terms of painting (that is, in terms of an art of painting governed by the idea of the sign), he knows only of figures (surfaces) that stand out against a background (a plane). To delineate the area within which they are inscribed, he is guided by the contour that is circumscribed or signified by a line.[105] An illusion of depth depends upon a division, the correct adjustment of the plane of projection. In other words, strictly speaking, a perspective construction implies, in the first instance, the elimination of depth—primarily in the form of color—and the representation's reduction to the graphic dimensions of Euclidean geometry. Simultaneously, to the extent that the construction must extend to the limits of the plane and articulate it in its totality, the division from which it proceeds is no longer a response to technical imperatives (as was the case with the division of the *intonaco* in the Assisi frescoes): it immediately involves a descriptive articulation that confers upon intervals a positive value equal to that of the figures. This has been shown by the example of Giotto: the theater of painting (in which the classical stage was to establish its influence) was a place where something, one dimension—that of bodies and gestures, that of color—was forgotten and negated. Here painting was confirmed primarily as a practice of interplay under no obligation to produce any recital of speech.[106] For space, it simply proposed a substitute, a model without volume that the painter would then try to animate by means of a "story," or an account. All that Alberti says about movement (and the movements of bodies being the visible expression or translation of the movements of the soul) and all that he says about color and the light and shade that confer *relief* upon figures is part of a rhetoric designed to compensate for the reduction,[107] the flattening and linearity that are fundamental to signifying representation, representation based on signs and their operations.

THE MYTH OF NARCISSUS

The paradox of perspective is that it aims to captivate the eye and introduce it into an illusory depth by means that depend solely upon a surface and descriptive geometry. It is fundamentally a matter of mirror images, as is also shown by Brunelleschi's experiment, which reveals the equivalence of the real spectacle and its iconic *representamen*, from the point of view of their mirror reflections. Alberti was given to declaring that the inventor of painting, the flower of every art (*fiore d'ogni arte*), was none other than Narcissus, whose story is apposite at this point: "What is painting but the act of embracing by means of art the surface (*that surface*) of the pool?"[108] The formula, by naming the spring, draws attention to the formal narcissism upon which representation is based and recognizes the flat surface (the original nature of which is connoted by the word *spring—source*, in French) that is its corollary. To paint is to grasp or embrace a surface entirely, the surface upon which the painter works (*that* surface) or, equally, the surface of the mirror that serves to guide him: the distinction between the two is of no importance, given that the two surfaces are part of a single system and are seized upon through the same structure of reflection. The reference to Quintilian, which follows that "definition" in Alberti's text, has more than a purely erudite resonance here. Alberti, it must be repeated, was writing not a history, but an *art* of painting, a study of a theoretical nature. When he reminds the reader that the earliest painters were accustomed to delineate the shadows produced on a wall by the light of the sun, an art that then developed and became enriched,[109] he was making the point that the formal precedence that drawing was recognized to possess was dependent upon the (specular) structure of representation. The surface is seized upon by means of lines, and color operates merely as a *supplement* in relation to drawing (which gives it its reason and marks out its limits). But that structure also dictates other exclusions, and these will throw more light upon the contradiction that is its mainspring. It is a contradiction that, given that it concerns the "sky" and "clouds" and, more generally, the whole aerial element, relates directly to the question that the present text aims to elucidate.

The Mirror for Clouds

The image used by Brunelleschi in the first of the experiments described in Manetti's biography (written after Alberti's *Della pittura*, by which it is clearly deeply influenced) presents one particularly noteworthy feature. On a small panel (*tavoletta*) about half a meter square, Brunelleschi had produced a painting in imitation (*assimilitudine*) of the baptistery of San Giovanni, in which he had depicted whatever his eye could seize upon from one particular vantage point (the porch of the principal door of the cathedral). The text explains that he took such care with this painting and rendered the mosaic of black and white marble so precisely that no miniaturist could have done better.

Vasari, for his part, adds a number of important details. Although

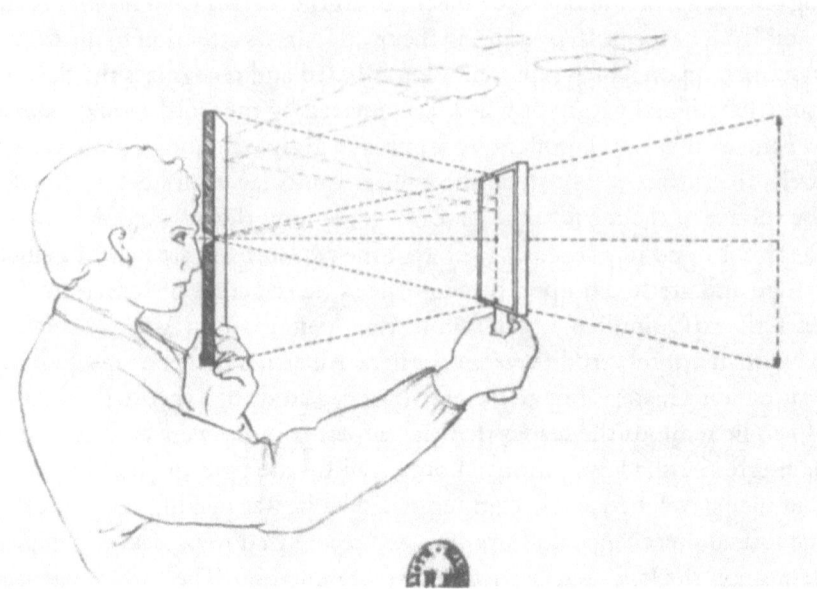

FIGURE 4. A reconstruction of Brunelleschi's first experiment. From Alessandro Parronchi, *La dolce prospettiva* (Milan), fig. 91.

he does not mention the mirror in which the painting was reflected or the peephole through which it was supposed to be seen, he does stress the fact that the perspective had been established using the *intersegatione* (intersection) method, "tracing with the ground plan and profile and by means of intersecting lines [*che fu il levarla con la pianta e profilo et per via del intersegatione*]" all the compartments of black and white wherewith the church was encrusted, "which he foreshortened with singular grace [*che diminuivano con una grazia singolare*]." [110] This was a most ingenious procedure, in Vasari's opinion, and very useful for the art of *disegno* (*cosa veramente ingegnisissima, ed utile al arte del disegno*), and one from which a painter such as Masaccio would later profit. Vasari goes on to say that Brunelleschi made a point of showing his constructions to specialists of *intarsia* (defined as the art of combining woods of different colors); and he stimulated them to such effect that this was the starting point for the many excellent things in this domain that were then produced in Florence. [111]

Judging by this text, the definition of perspective, in the modern sense of the term, and the new technique of marquetry appear to be logically contemporary and strictly complementary; and that complementarity and contemporaneity in their turn point to an important historical interaction.[112] The authority of Brunelleschian perspective stemmed from the totally coherent structure that it conferred upon the pictorial space, insofar as it reduced both the forms distributed within that space and the intervals between them to an interplay of identifiable and substitutable elements. Reciprocally, the juxtaposition of a number of two-dimensional geometrical figures was enough to produce the illusion of a third dimension. As André Chastel remarks,

Following the Ancients who had happily made the most of the technique in their "Greek" works, for example, people now discovered or rediscovered that through an amusing illusory effect, which is at the very origin of trompe-l'oeil, an interplay of articulated surfaces summons up depth. In short, the unifying function of perspective certainly reflects coherent mathematical thinking, but the procedure of analysis and construction that resulted was the technique of marquetry.[113]

And yet the work of analysis and construction governed by perspective was not *all inclusive*. Brunelleschi had depicted the part of the square around the baptistery that corresponded to the viewpoint he adopted, stretching from the Loggia dei Pescatori to the Canto alla Paglia. So far, so good: all perfectly justifiable, even in the terms of *Della pittura*. But that ceases to be the case once we move from the buildings (and the ground upon which they are built) to the sky, against which they stand out. The painters of the Renaissance are often credited with the elimination of the golden backgrounds that were characteristic of medieval painting, with lowering the line of the horizon, and with opening up the perspective composition to the sky. But Brunelleschi made no attempt to depict that sky; he merely showed it (*dimostrare*). And in order to do so, he resorted to a subterfuge that introduces into the representational circuit a direct reference to external reality, and at the same time a supplementary reduplication of the specular structure upon which the experiment was founded. When the observer applied his eye to the back of the panel and, through the peephole made at the vanishing point of the composition, looked at the reflection of the painted image in a mirror placed at the appropriate distance, all that he in fact saw of the "sky" was the reflection of a reflection. Whereas Brunelleschi had to take account of and show the space where the walls depicted in perspective space were imprinted (*stampassono*), he had, in contrast, covered the corresponding part of the panel with a surface of dark silver in which the real air and heavens were reflected, and likewise any clouds that appeared there, propelled by the wind, when it happened to be blowing.[114]

Both this text and the optical and scenographic experiment that it describes testify to the limitations imparted right from the start to the theoretical perspective system (for it truly is a matter of theory, not of "painting"). Perspective only needs to "know" things that it can reduce to its own order, things that occupy a place and the contour of which can be defined by lines. But the sky does not occupy a place, and cannot be measured; and as for clouds, nor can their outlines be fixed or their shapes analyzed in terms of surfaces. A cloud belongs to the class of "bodies without surfaces," as Leonardo da Vinci was to put it, bodies that have no precise form or extremities and whose limits interpenetrate with those of other clouds.[115] So the procedure to which Brunelleschi resorted to "show" the sky, that version of a mirror which he introduced into the pictorial field like a piece of marquetry and in which the sky and the clouds could be glimpsed—that mirror was considerably more than just a subterfuge.[116] It has the force of an epistemological emblem (in the twofold sense of a symbolic figure and also a foreign body, an inserted, interlocked element: *emblēma, embolē, symbolē*), to the extent that it reveals the limitations of the perspective code for which the experiment provided the complete theory. It reveals perspective as a structure of exclusion, the coherence of which is founded upon a series of rejections, and yet which has to make room for the very things that it excludes from its order, just as it does for the background upon which it is imprinted. And Brunelleschi's mirror device also has a formal, if not a stylistic, effect in that it seems to impose a limitative, strictly graphic notion of the pictorial space and the figures positioned within it.

4

The Powers of the Continuum

Infinity is the genus of which a continuum is a species.
—Jules Vuillemin, *La Philosophie de l'algèbre*

"Style" and Theory

A Structure of Exclusion

WHAT IS IGNORED IN THE SYSTEM

Even though Brunelleschi, right from the start, produced a *theory* for it,[1] the institution of perspective was nevertheless not defined all in one go, at one particular point, and its fortunes, initially at least, were not linked to any act of authority. Linear perspective with a single vanishing point has a history, and in this respect the same can be said of painting as is said of language: "Nihil est lingua quod non prius fuit in oratione [There is nothing in language that was not first in speech]."[2] The distinction between language and speech is not relevant to the pictorial field (and that implies quite a lot about the definition of a so-called "code," which, in truth, encodes nothing but itself, and the function of which is, as we shall see, not so much productive, but demonstrative). Nevertheless, what is worth remembering about the above formula is that it makes the "code" dependent upon *what is said* and upon *phonè*, thereby giving a negative connotation to that which is not said (not thought), which is its reverse, the *scriptural* surface against which it stands out, and the inarticulate backing that surrounds it and gives it an outline and its coherence as a figure. In other words, the "code" appears as a structure of exclusion whose balance and necessity stem from the very things that it designates as being external to its own order—that which is unsayable (undepictable) but which nevertheless, as such, continues to "work within" the system.

What we need to ask is: did Greek and/or Roman painting produce representations of clouds or cloud formations?—a question that is clearly not unrelated to the ever thorny problem of ancient "perspective."[3] In the preface in which he describes the genesis of his *Cupido cruciatus*, the poet Ausonius, who was active in the second half of the fourth century, compares that poem to a cloud painted on a wall (*nebulam pictam in pariete*). Panofsky, in his commentary on this text,[4] suggests that Ausonius may have been thinking of an illusionist painting opening onto a trompe l'oeil sky, in the manner of so many late Roman paintings. However, the background against which the architectural perspectives of Pompei are painted are always of a transparent blue, as are the cloudless skies of the mythological landscapes in the Vatican, which some art critics nevertheless persist in regarding as one of the first examples of a "pictorial" treatment of space. Yet Ausonius refers explicitly to a cloud as to an object that painting is perfectly capable of depicting, of representing by its own particular means: "Most people have seen painted clouds and remember them."[5] So far as we at present know, his assertion has not been corroborated by any archaeological remains. In fact, on the contrary, the vague masses that edge the upper border of two golden goblets discovered at Vaphio (Laconia), which Aloïs Riegl thought he recognized as clouds, for him constituted an extra reason to deny those famous pieces a Mediterranean origin. Today they are thought to be Minoan, dating from the end of the sixteenth or beginning of the fifteenth century B.C. According to Riegl, cloudy skies were never to be found in ancient art, and it was not until the late Middle Ages that painting paid them any attention.[6] Yet, as Ruskin pointed out,[7] even if clouds were excluded from the pictorial field, they were certainly not absent from literary texts: quite apart from in Aristophanes' *Clouds*, they have a part to play, in many forms (as vehicles, shelters, threatening signs, masks, etc.), in the *Aeneid*. However, it was only with the advent of Poussin, an assiduous reader of Virgil, that Venus was shown, in all her splendor, descending earthward through the clouds—"At Venus aetherios inter dea candida nimbos [Venus, the shining goddess among ethereal clouds]" (*Aeneid* 8.l.608)—to present Aeneas with the weapons forged by Vulcan.[8]

How was the question posed in the Quattrocento, at the time when increasingly in-depth theoretical work was proceeding—work that was to produce a veritable revolution in pictorial art? Brunelleschi's experiment had the merit of presenting the terms of the problem perfectly clearly. Cloud cannot be depicted by means of geometry; such a body "without a surface" cannot be "described" or reduced to the coordinates of an experimental setup that only reproduces objects as the clearly delineated, outlined shapes that are apprehended by an observer positioned at a particular spot (that is to say, whose view must be defined as from one specific point). Yet, even so, cloud does find a place in Brunelleschi's representation: the mirror image accommodates it by means of a supplementary duplicating ploy, as a reflection of a reflection. Similarly, when Mantegna was to seek to represent an Ascension, he would begin by reducing it to a piece of theatrical machinery, complete with the festoons of wadding by which aerial flights were accompanied in the theater: the duplication or repetition that is at the origin of signs and representation makes it possible for the painter to treat such subjects without contravening the principle laid down by Alberti, who declared that painting should deal only with visible things. (All the same, clouds other than theatrical ones did appear *outlined in profile* in Mantegna's skies, just as they were to later in Bellini's, and their appearance was no more "realistic" than that of the theatrical clouds: "Ever since signs appeared, that is since forever, there has been no chance of coming across the purity of reality anywhere.")[9] The significance of the plaque of darkened silver that Brunelleschi inserted into his painted panel in place of the sky did not merely have the role of a parody or even a criticism.[10] However perfectly it was adjusted to the other elements in this perspective division of the picture, this plaque introduced into the network of the *intarsia* as it were an alien element, which manifested the closed nature of the "code" and, at the same time, the fact that it was impossible for representation to remain within the limits of the field as it was defined.

The theoretical principle of the perspective regime excludes the representation of cloud formations. But pictorial creation cannot be reduced to the implementation of a code; and it is not hard to show that, in fact, only very few of the works of the Renaissance satisfy the demands for a "correct" perspective construction.[11] Vasari, already, delighted in noting—with false naivete—that Uccello himself, in his *Drunkenness of Noah*, had committed what he (Vasari) pretended to regard as an "error" of the gravest kind with respect to doctrine, since the composition was governed by two vanishing points instead of one.[12] But it may be that the perspective "code" (and likewise the color code subordinated to it) occupied a paradoxical position in the representational system of the Renaissance, assuming very different functions from those assigned to a symbolic code, in the modern sense of the expression. The so-called perspective code provided the elements for a geometrical construction of space and rules for combining those elements. But it is as if painting, ever since the revolution inaugurated by Brunelleschi and right down to the break subsequently introduced by Cézanne, thrived, at least in part, on repeatedly transgressing the law that it had given itself. The "code" functions in the system not so much as a matrix to which all productions derived from it should relate, but rather as both a guiding and a regulating device (Leonardo was to define perspective as "the rein and the rudder of painting [*briglia e timone della pittura*]").[13] It was a *theoretical* device, regarded as a model (in the epistemological sense of the term), which related to the *real* practice of painting in the same way as the theoretical models elaborated by science relate to the reality that science aims to understand. We have already insisted on the term *model* rather than *structure* (see Chapter 3, pp. 116 ff.), and it is a model that governs a way of painting that is never "in perspective" through and through. The historical successors of the system themselves in the last analysis show that this model never ceased to have a normative value. Even though the painters of the Quattrocento may seem to have taken the greatest liberties with the principles established by Brunelleschi and then codified by Alberti and Piero, the perspective regime gained increasingly strict authority as a paradigmatic code, once the system inherited from the Renaissance began to be fundamentally threatened in its very *order* (which was defended by the academicians). But if that is the case, and if this single spatial ideal was maintained (and continued to function) for close on four centuries, in the course of which painting experienced considerable stylistic fluctuations (fluctuations for which Wölfflin tried to provide a theory, and at the limit of which Correggio's work seems to lie), this certainly proves that it was not limited a priori to one style in particular, and first and foremost to linearity. The perspective regime excluded clouds insofar as it seemed it had to confine itself to graphic interpretations. But what would happen when the hierarchy of functions that conferred upon a style its particular look and configuration was questioned—in the first instance when the preeminence of line was challenged?

If we accept Panofsky's interpretation of the panegyric of Dürer that Erasmus saw fit to include in his *Dialogus de recta Latini Graecique sermonis pronuntiatione* (published in 1528, the year of the painter's death), this work confirms the relevance of our question both from a historical and a systematic point of view. Erasmus's text declares that Dürer is the Apelles of modern times; he is even greater than the ancients in that—and this is the important point—he knows how to express everything without the aid of colors, solely by means of lines: shadows, light, dazzling brilliance, relief—everything. Taking things situated in a predetermined place, he can show more than just the profile that they present to an eye obeying the rules for constructing the visual pyramid precisely. But above all, he paints the *unpaintable*: fire, rays of light, storms, lightning, even—some say—clouds on a wall (*nebulas in pariete*), and much more besides: every kind of feeling and affection, the whole human soul as manifested by the appearance of the body, even the voice itself (*ac pene vocem ipsam*). And, to repeat, he does it all by means of the most felicitous of lines, black lines to which no color could be added, for fear that the work would suffer from it:

Although Dürer is also admirable elsewhere, what can he not do in monochrome, that is in black lines? Shade, light, splendor, highlights, depressions! In this respect, from a single viewpoint, it is not just a single view that offers itself to the eyes. He exactly observes symmetry and harmonies. Indeed, he depicts what cannot be depicted, fire, rays, thunder, blasts, lightning, or clouds (some say) on a wall, the senses, emotions, and finally all the human soul that appears in the body, and almost the voice itself. This he does in black lines so well judged that, in visual effect, were you to add colors you would harm the work. Is this not more miraculous, to achieve without the allure of colors what Apelles did with their assistance?[14]

This is another crucial text. Even though the essential elements of it are borrowed from Pliny (and from Ausonius, as Panofsky shows, where the *nebulae in pariete* are concerned), the way that it is put together and proceeds, and all the additions for which Erasmus is responsible, serve to produce a theoretical thesis the rigor and acuity of which are somewhat surprising. Given that it relates to the question of both linearity and depictability, the problematic of cloud is addressed and is most accurately declared to arise at the point where the visible meets the invisible, the representable meets the unrepresentable.

Quae pingere non possunt: that which cannot be painted (and at this point how can we not evoke Goethe, Goethe who, only fifty years before Monet and van Gogh, was declaring that painting was not suited to portraying trees in blossom, because these *could not be captured in an image*?).[15] That which cannot be painted: fire, rays of light, storms, lightening, clouds; but also feelings, affections, the voice itself even; and also (according to Panofsky's interpretation) the reverse side of an object, the side that is excluded by a planimetric reduction. In the context of a dialogue devoted to *pronunciation*, the reference to the voice is significant. (The panegyric opens with the assertion that children ought to learn to paint and draw at the same time as they learn to write, so as to acquire more skill in the writing of letters, just as those who know how to sing will have better elocution.[16] This is a variation on Alberti's paradigm of writing: painting or drawing is to writing what music or singing is to speech.) The writing, black on white, of a line does not duplicate *phonè* but it evokes it as what is missing in what is shown, as something that supplements the representation, proof of it that can always be demanded ("Speak and I shall baptize you"), just as, in the calculated effacement of its outline, the visible profile of an object holds the promise of the profiles that it excludes and reveals what it conceals.[17] Painting needs to know only of what is visible (representable). But the field of representation does not coincide with the field of what is denoted: suppressed depth tears at the flat representational surface through which the dazzling effects of fire and light, the shattering effects of storms and lightning, the aerial, impalpable effects of clouds, the fixing of which on a flat surface (*in pariete*) immediately carries an oneiric connotation (since a cloud—as Erasmus himself points out—is something too immaterial to be expressed by colors, a painted cloud resembles a dream, in other words, it resembles *nothing*).[18]

In this text, the demonstrative structure of which did not escape Panofsky, Erasmus designates most precisely the limits of a mode of representation in which the principles of perspective with a single vanishing point are interpreted in strictly graphic terms, to the exclusion of any way in which color might determine the objects. By so doing, he forestalls and repudiates Wölfflin's declaration that a style, even if it is linked with a particular idea of "beauty" and possesses specific characteristics, nevertheless constitutes "a language, capable of expressing everything."[19] It is not so much that, with painting, one is dealing with a *restricted language*, in the sense understood by Hjelmslev—that is to say, a language that, unlike "philological" language, cannot be used to express every possible kind of meaning, whatever it may be, but only lends itself to certain specific uses.[20] In Erasmus's own terms, and taking into account the principle according to which painting only needs to know of that which is visible, the idea that a painter can paint *everything*, even *that which is unpaintable*, rules out defining a style as a code of which works of art are the product. It implies that a system or representation cannot be reduced to a particular interpretation, be it linear or pictorial, nor to any particular one of the codes, however productive, that enter into its definition. Style—even the style which, thanks to the preeminent place that it assigns to drawing, seems the best suited to the demands of the perspective code—is no more than a formal moment that corresponds to a specific level of analysis, and that does not contain in itself its own determination. Intent upon praising the painter of his portrait as he was, Erasmus nevertheless did not fail to perceive the theoretical conflict by which Dürer's work was marked. But of course Dürer was not to be the only painter to wish to paint the unpaintable: fire, storms, brilliance and light, and clouds. Two years before his death, Correggio set out to decorate the cupola of the Parma cathedral, where the Assumption of the Virgin takes place amid a great gathering of clouds. And twenty years later, Pontormo set out to imitate, no longer with lines, but *with colors (co' colori)* all the things produced by nature (*tutte le cose che ha fatto la natura*), fires (*fuochi*), brilliance (*splendori*), clouds (*nugoli*), and so on, and to do so on the same flat surface (*e farla a piano*) where painters strive to animate their figures and give them the appearance of life.[21] No one succeeded better than Pontormo in expressing the unease that may grip a painter conscious of the disproportion between the effects of painting and the means available to it, between the intensity of the illusion and the precarious, meager nature of his materials. But that unease was not simply a historical harbinger of an imminent change of style, for it had deep and distant roots. It was fed by the paradox that constitutes the mainspring of the system and that became manifest long before the so-called Mannerist period, or even the period of Dürer.

From the Linear to the Pictorial

MANTEGNA'S LUNETTE

As we have seen above, the work of Mantegna—well known to have been an important influence on Dürer—displays an almost ethnographic precision in its rendering of certain machines used in the contemporary theater through the duplication of which the painter managed to get around some of the prohibitions linked to the institution of the perspective code. He thus represented the Ascension of Christ on a theatrical cloud or on a swathe of wadding that surmounted the *nuvola* upon which the actors would be positioned, and meanwhile filled the sky with heavily delineated compositions of stratocumulus. But, as we have also seen, his work also presents other features that have a direct bearing on the subject of the present work. In the first place, his *Saint Sebastian*, now in Vienna, and his *Triumph of Virtue*, in the Louvre, contain clouds in which the silhouettes of figures in the round can be glimpsed. They constituted a significant "aberration" in their context, and one that shows that the bizarreness for which Vasari criticized Piero di Cosimo, on account of the latter trying to make out strange figures and constructions in cloud formations—that so-called bizarreness was by no means exceptional at the time and so did not deserve to be labeled pathological.[22] The second feature of Mantegna's work is the "spatial unease"[23] that "animates" his paintings and that is expressed by essentially graphic means, as can be seen from the fragments of the predella of the altarpiece of San Zeno (1457–59). In *The Crucifixion*, now in the Louvre, the vanishing lines of the ground, *treated in curvilinear perspective*, are abruptly interrupted: a path, along which horsemen and soldiers are advancing, slopes down across them toward a valley, the view of which is blocked by the stone paving, treated like a theater stage, upon which the crosses stand, filling the foreground. In the distance, the same path leads up to a town perched on the top of the hill. It seems to be at another point along this path, closer to the town, that *The Agony in the Garden* (London, National Gallery) is set. Here, too, the space, defined by a complex network of arabesques, curves and "streaks away" at the bottom of the picture, while fluffy clouds "float across" the sky. This description (necessarily metaphorical, as any description of a fixed image must be if it introduces any reference to movement) would serve for most of Mantegna's landscapes (in which the rock or *rocky mountain*, with the linearity of its cracks and crevices, fulfills functions, at once plastic and thematic, if not fantastical, as widely diverse as those assigned to *clouds* in Correggio's works). It seems that Mantegna was trying, *through his use of lines*, to liberate himself from the rectilinearity of the spatial coordinates, even as he endeavored to make his compositions open onto the sky. The Correggio of the Camera di San Paolo was to remember the leafy canopy affording glimpses of the sky, beneath which *The Virgin of Victory* shelters (Paris, Louvre). And in similar fashion, in the Ovetari Chapel at the Eremitani in Padua, and the Spouses' Chamber in the palace at Mantua, the vanishing lines converge at points situated *beneath* the image's frame. Their arrangement obeys an illusionist principle that was later to enjoy considerable success, according to which the floor of the scene represented on the wall was positioned higher than the eye of the spectator. *The Crucifixion* presents a variant of the same procedure, one that Mantegna was to exploit to the limit in the lunette of the ceiling of the Spouses' Chamber in Mantua (Plate 1b). This shows, if not the first perspective *dal sotto in su* in the history of Italian painting, at least the first perspective constructed strictly vertically, following an axis that is rigorously perpendicular to the plane of the floor upon which the spectator stands.[24] The lunette, painted entirely in trompe l'oeil, does not open directly on to the "sky"; a fake balustrade intervenes, upon which girls and servants are leaning while putti flutter around them and a peacock stretches an unusually long neck toward the sky—a faintly misty blue sky, in which the eye loses itself.

CHIMERAS: ANAMORPHOSIS AND CLOUDS

These effects of accelerated or reversed perspective, linked with the position that the spectator occupies in relation to the painted surface, are not so very different, in principle, from other ploys in which the code is used explicitly as what it is: an artifice, the instrument of an illusion. In "normal," "scientific," "rational" practice, perspective functions as a technique for appropriating reality and setting it in order, a technique that has been said to correspond to the establishment of an "adult" relationship with the world.[25] But that same technique, those same principles can be used in precisely the *opposite way*, for hallucinatory or disconcerting purposes (for *delusion* or deception, as opposed to illusion). The history of perspective cannot be solely identified with that of artistic "realism." It is also the history of a dream, the arcane nature of which has been probed by Jurgis

Baltrušaitis.[26] Anamorphosis was known as early as the sixteenth century (although the word did not appear until the seventeenth) and was widely used in, among others, circles close to Dürer. (One of his pupils, the engraver Ehrard Schön, left some famous examples of it.) And what anamorphosis did was turn the perspective order against itself. This *bella e secreta parte di perspettiva* (Daniele Barbaro, 1559)—the art of secret perspective, *die Kunst in geheimer Perspektive*, into which Dürer may have been initiated in Bologna,[27] which shows "the power of perspective in its deceptions" (P. Accolti, *Prospectiva practica*, 1625), is based upon the systematic deformation of regularly constructed figures, by mechanical or geometric means. In this way one obtains either an image in which, if one adopts the appropriate point of view, one can recognize a number of secret objects or portraits, or "a chaos of lines" that only become recomposed when the image is viewed obliquely or through a hole pierced in the screen that hides it (the day would come when anamorphosis would appropriate catoptrics and would use specular reflection as an extra means). Elements of Brunelleschi's experiment are detectable in all this, right down to the *hole*, a viewing point strictly defined for this particular viewing from one specific point, against which the eye must be placed, and any repositioning of which would lead to not only breaks in the continuity of outlines, but also a kind of disorientation, reducing the system of linearity to nonsense: "Instead of reducing forms to their visible limits, they are projected out of themselves and are dislocated in such a way that they only fall back into place when they are looked at from a predetermined viewing point."[28]

It may well be that anamorphosis is no more than a curiosity, a "chimera." Nevertheless, through this variation in the writing, it reveals the limits that the code imparted to perspective construction, and the prohibitions that govern its "normal" functioning. Rather than annihilating the system, anamorphosis works at its systematic deregulation, and does so—it must be emphasized—using the means of the code itself, which is defined as a *regulator*. It is most remarkable that it should be that same Mantegna, whose work reveals a new spatial unease, who conceived the very first trompe l'oeil cupola. The fact that it depicted an evanescent cloud at the center of that cupola indicates that the games in which perspectivists were to delight had from the outset been linked with the assumption of clouds into the skies of cupolas and painted decors, and that certain uses of this element would probably not be unrelated to anamorphosis. That, at any rate, is what Galileo suggests in his *Considerazioni*, mentioned earlier, when he criticizes allegorical poetry, which he assimilates a little further on to *intarsia* work (see Chapter 2, p. 51). His criticism is that it subjects natural narrative to an oblique, hinted meaning and renders this obscure in an extravagant fashion, by dint of resorting to altogether pointless inventions. He goes so far as to compare it to one of

those paintings which, observed sideways and from a predetermined point of view, show us a human figure, but one constructed in accordance with a rule of perspective such that, when seen from in front, as is natural and usual for other paintings, presents to us nothing but a confused and disordered mixture of lines and colours in which, with much application, one may form an image of sinuous rivers and roads, deserted beaches, or *clouds or strange chimeras*. (my emphasis)[29]

LEONARDO I

Rivers, winding paths (impossible not to think of Mantegna), clouds, and chimeras, before which the eye begins to dream, as a painter might before stains on a wall: such are the things to which an anamorphic image is reduced unless the eye recomposes it by looking at it from the appropriate angle (as with an allegory, if the hidden, implied meaning is not grasped). It is as if the decentering, the possibility of which is recognized from the start by the "code," could only operate within very narrow limits beyond which the image will seem to disintegrate although it has, in truth, merely been *transformed*. In a similar fashion, today, it is possible to reduce speech to a sonorous magma that can be transmitted more economically and then recomposed by electronic means, restored to an articulation that conveys meaning. Leonardo da Vinci, who may have invented anamorphosis,[30] certainly saw how this "composed perspective" (as he called it) summed up all the problems of perspective, even as it subverted its function of making things understandable. Perspective is an instrument for setting in order the perceived reality that flows toward the eye in the form of images of the bodies that fill the air, images for which the eye is at once "a target and a magnet."[31] By referring painting to perception (rather than just to geometry), Leonardo managed to bring to light all that was implied by Alberti's system, and at the same time to criticize it. Given that the images of objects are all infused in the air as a whole, all of them in each of its parts, nothing could be seen, according to him, except through a tiny *fissure* through which the atmosphere passed and the rays of light could separate out, become concentrated, ordered, and composed.[32] Nothing could be more deceptive, in this respect, than the *specular* metaphor: for no object is defined in itself in the mirror; it is defined only by the eye that sees it there and that gives it a configuration.[33] Perspective *demonstrates* this operation, which is a procedure that is at once selective and organizational, in geometric terms. But whereas simple, "natural" perspective operates on a plane of intersection that is perpendicular to the axis of the visual cone, and leaves the spectator a relative freedom of movement, composed perspective, which is a combination of natural perspective and artificial (or even *accidental*) perspective, operates on oblique planes, where natural diminution is combined with geometric foreshortening and forces the observer to look at the image through a little hole: if several people are looking at a picture constructed in this way, all at the same time, only one will be in a position to understand the effect of perspective, while the rest will perceive it only confusedly.[34] Anamorphosis thus provides a kind of *reductio ad absurdem* of the *aporia* into which perspective leads when it is reduced to solely its linear components. But this demonstration, in its turn, leads to certain pictorial consequences.

All perspective relates to the plane on which it is inscribed. But linear perspective implies an extra reduction, the reduction of the visual faculty to a single point — from which the whole series of its elements are engendered: point/line/surface/, elements that, although they serve a rational construction, nevertheless have no value in reality. In effect, geometrical perspective does away with the very element of vision, the atmosphere, within which images and colors are conveyed. It reduces bodies to surfaces defined by the outline by which they are represented from a predetermined point of view, no account being taken of the fact that those "points," "lines," and "surfaces" have no more than a nominal existence. A surface is a *limit*: it is not part of those bodies, merely their common frontier, the point of contact (*contingenzia*) of their extremities.[35] When referred to the element upon which vision is operating, the surface has a name, but no substance:

Surface is the name given to the boundaries of bodies with the air or I would rather say of the air with bodies — that is what is enclosed between the body and the air that surrounds it; and if the air makes contact with the body there is no space to put another body there; consequently, it may be concluded that surface has no body and therefore no need of position. . . . A surface has existence and not space. Consequently, this surface is equal to nothing, and all the nothingness in the world is equal to the smallest part if there can be a part. Wherefore we may say that surface, line, and point are equal as between themselves, and each is of itself equal to the other two joined together.[36]

If painting presumes to an existence that is not solely nominal (presumes to be something other than a "nothingness"), it will thus have to resort to means other than those of linear perspective: to color perspective, which is linked with the fading produced by distancing, and to the perspective of diminution (which takes account of the blurring of outlines, and the imprecision of forms seen from far away), diminution itself being linked with the aerial perspective that is caused by the density of the air and the interposition of mist or fog of varying degrees of thickness between the eye and the object.[37] Linear perspective has an essentially demonstrative function: it makes explicit the function of visual lines and serves to define and characterize forms without regard to the effects produced by distance.[38] It thus contradicts the perspective order, of which painting claims to be the equivalent;[39] for, as is confirmed and demonstrated by experimentation, the image of the substance of a body is projected further from its source than its color is, and color in its turn is projected further than its form.[40]

   One can see here how attention paid to the "real" conditions of perception, and not just to the data of geometrical optics, results in a subversion of the hierarchy of the components of painting as laid down by the linear interpretation of the perspective construction. Form, defined by outline, is less powerful than color, which itself is less powerful than substance, or rather the image of substance (*la similitudine corporea*) — by which Leonardo means the visible mass of bodies, quite apart from any determining features of form and color.[41] Experimentation, in the sense of the word as used in Leonardo's text,[42] proves that the eye cannot make out the limits of any object and that it is impossible for it to verify precisely where in the field a particular object comes to an end when it stands out in the far distance. According to Leonardo, that is due to the fact that the visual faculty resides not in one point, as was held by other painters who had pronounced on perspective, but solely in the pupil, where the images of objects are integrated together.[43] It is also due to the fact that the limits of bodies are constituted by the line of the surface, a line of invisible thickness. "Adunque tu pittore non circumdare i tuoi corpi di linee [Wherefore, O painter, do not surround your bodies with lines]."[44] Leonardo's intention in stating this precept is not to set one aesthetic in oppositon to another, nor is it to praise "haziness" in painting. His thesis is essentially theoretical and in no way sets out to destroy the perspective order or even to call it into question. Instead, it is designed to widen the concept or, to take him literally, to *deepen* it. It leads Leonardo to assign to the system of linearity, based on a strict delineation of surfaces, a primarily demonstrative function, and to bring greater attention to bear upon operative elements which that system excluded (as is seen from Brunelleschi's experiment) or to which it granted no more than a subordinate position. Generated by distance (or by too great a proximity), by darkness, or by mistiness, *sfumato* (gradation) and the *mezzo confuso* (chiaroscuro) operate in such a way that the outlines of objects no longer stand out from the atmosphere,[45] which now appears as the most fundamental element in the representation. Leonardo may have invented anamorphosis (only to reject it); but he was far more interested in the representation of the atmospheric phenomena that upset the order of linearity most violently: the wind, storms, hurricanes, even downpours of rain (and it is interesting to note the rule assigned to clouds in these *representations*).

In representing wind, besides the bending of the boughs and the reversing of their leaves towards the quarter whence the wind comes, you should also represent them amid clouds of fine dust mingled with the troubled air [*li ranugolamenti della sotil polvere mista col la interbida aria*].[46]

Let the dark and gloomy air be seen buffeted by the rush of contrary winds and dense from the continued rain mingled with hail.[47]

The crests of the waves of the sea tumble to their bases, falling with friction on the bubbles of their sides; and this friction grinds the falling water into minute particles, and this, being converted into dense mist, mingles with the gale in the manner of curling smoke and wreathing clouds and at last it rises into the air and is converted into clouds.[48]

If you wish to represent a tempest, consider and arrange well its effects as seen when the wind, blowing over the face of the sea and earth, removes and carries with it such things as are not fixed to the general mass. And to represent the storm accurately, you must first show the clouds, scattered and torn, and flying with the wind, accompanied by clouds of sand blown up from the sea-shore, and boughs and leaves swept along by the strength and fury of the blast and scattered with other light things through the air. . . . Make the clouds driven by the impetuosity of the wind and flung against the lofty mountain tops, and wreathed and torn like waves beating upon rocks; the air itself is terrible from the deep darkness caused by the dust and fog and heavy clouds.[49]

Perhaps it will seem to you that you may reproach me with having represented the currents made through the air by the motion of the wind notwithstanding that the wind itself is not visible in the air. To this I must answer that it is not the motion of the wind but only the motion of the things carried along by it which is seen in the air.[50]

SYMBOL, ICON, INDEX

L'aria spaventosa per le oscure tenebre fatte in
nell'aria dalla polvere, nebbia e nuvoli folti.

The air itself is terrible from the deep darkness
caused by the dust and fog and heavy clouds.
—Leonardo da Vinci

These texts are doubly interesting. On the one hand they show that the limits that Alberti assigned to the representation of movement were, historically speaking, on the point of giving way. Alberti's *Della pittura* told the painter that, among "changes of place," the only ones that he needed to know of (*solo riferiamo di quel movimento si fa mutando el luogo*), there were some that were beyond reason (*in questi movimenti si truova chi passa ogni ragione*), and of these he should make no more than moderate use, as painting demanded gentle, graceful movements (*et conviensi alla pittura avere movimenti soavi et grati*). In contrast, Leonardo's writings introduced a far more dynamic concept of representation. And it was by no means by chance that, by way of illustration, the new concept called for images of perturbations, even of deluges of rain (in which all *stories*—incidentally— would be wiped out). The perspective space is not negated, simply tested, perturbed, as it was also to be by the flying experiments that Leonardo worked on ("The great bird will take to flight to the stupefaction of the earth and will fill all the annals with its great fame"). On the other hand, in so far as these texts explicitly aim to set out a program for representation (how to represent wind, or a downpour of rain, etc.), they introduce a decisive split into the field of representation: whereas classical discourse was to endeavor to reduce representation to a binary relation of substitution and reciprocity between the signifier and the signified,[51] Leonardo, through the plan that he adopts to show (*dimostrare*: the same word that Manetti, in his biography of Brunelleschi, applies to the representations of the heavens) phenomena and incidents in themselves invisible by means of visible *signs*, imposes the idea of a relationship, still admittedly triadic, but no longer in any way allegorical and which no longer owes anything to similarity. In Peirce's terms, you could say that cloud figures in Leonardo's text as a symbol (word symbol) that refers to an icon that is itself designed to be an index: the object, the word for which is the *representamen* while its pictorial graph is the *interpretant*—that object in its turn functions as a *representamen* and calls for an *interpretant*. Whereas an icon is defined by its immediate intrinsic representational quality, the word *wind* can have no iconic *interpretant*. The object with which it is associated cannot be seen, so can only be represented indirectly by means of an index that implies a physical connection between the sign and the thing that is signified: the movement of clouds serves as the index of wind (which, as we shall see, is its cause).[52] In other words, the functions imparted to *cloud* are, in this context, essentially semiotic: *cloud* is a sign, in the triple sense of a symbol (word), an icon, and an index.[53] But it was already as a sign that it appeared in Alberti's text (wind has no right to be in a painting except insofar as it blows through the clouds: *che soffi fra le nuvole*) and also in the text that Manetti devoted to Brunelleschi's experiment (the plaque of darkened silver reflected not the wind itself but the clouds that passed through the air when the latter blew, "e nugoli che si vedono in quello arieto essere menati dal vento, quand' e' trae").

This triadic relationship conditions all discourse on art, and on that account is fundamental to the present work. The graph marked as /cloud/ functions as a sign on several levels all at once. On the descriptive level, an iconic sign acquires the value of an index: just as a painter has no way of representing the movements of the soul other than through their bodily manifestations, he will depict wind, or a storm and so on, by means of their visible effects. On the expressive level, a sign takes on the value of a symbol ("the air itself is terrible from the deep darkness caused by the dust and fog and heavy clouds"). But it also functions as an index in another sense of the word and in another dimension: that of style, in its evolution and revolutions. Leonardo did not make room for /cloud/ in his paintings, and there is no evidence to suggest that he was thinking of devoting special passages to it in the treatise that he was planning to write. But the attention paid to texts that refer to this element in the editions of the *Treatise on Painting*, compiled in the sixteenth century by Francesco Melzi, and their collection together under a specific title in the part devoted to landscape indicate the emergence of new interests (and at the same time of a new "genre"). And how could one fail to think of the more or less fantastical landscapes, with skies filled with "threatening" clouds, so many examples of which are to be found in the paintings of this century, beginning with Giorgione's enigmatic *Tempest*? But we need to press on further and dig deeper into the objective texture of style. Associated as it is with the depiction of movement (movement which, as Braque was to observe, "upsets" lines), /cloud/, which appears so often and in such proliferation that it tends to disappear as an independent unit, is a symptom of a problem that stylistic analysis formulates in its own particular terms. As we remarked at the beginning of the present work, in the cupolas of Parma, the painting of clouds spreads to fill the dimensions of the whole field allowed to the decor and seems to open up the way for a style no longer founded upon the delineation of figures and perspective construction, but rather upon the dissolution of forms and the effects of light. But that description does not do justice to the truly theoretical aspect of the question of linearity any more than it makes it possible to assess the historical impact of Correggio's work: for the sign—as has been hinted in the analyses above—here serves ends other than stylistic ones, and lends itself to (and borrows from) quite different figures.

Heaven and Earth

Aesthesis

EL GRECO

If there is one oeuvre in particular that, in its very appearance, seems to satisfy the definition of style no longer based on delineation and trompe l'oeil but rather upon a specifically pictorial technique in the movement of which the geometrical coordinates of depiction seem to disintegrate, it is that of El Greco. At the level of the making of the painting, the characteristic figures that fill his paintings seem to be "bodies without a surface," in other words—as Leonardo would have put it—as figures with no definite frontier, always ready to sink back into the background from which they have emerged. But that background, in its turn, is not so much a place in which each thing finds its own place, but rather an indefinite, unstable site where it is impossible for the laws of nature to operate. Max Dvořák has shown, in a classic text, how El Greco's possibly most famous work, *The Burial of the Count of Orgaz*, despite resorting to a traditional compositional schema, nevertheless obeys principles that directly contradict those to which the painters of the Quattrocento had rallied.[54] The painting is divided into two parts. At the lower level, the painter has represented the miracle that occurred during the funeral ceremony (the apparition of Saint Stephen and Saint Augustine, who descended from the sky and themselves proceeded to bury the body of the count). The upper level depicts the reception of the deceased's soul by Christ and the Virgin, who are set within a semicircle. As Dvořàk rightly perceived, the novelty of the work lies in the interpretation of the schema on two separate levels, the meaning of which is completely transformed by the framing adopted by the painter, who allows no glimpse of the ground upon which the figures at the lower level stand. Those figures are so closely packed, squeezed one against another, that any allusion to the scene where they are congregated is abolished, to allow for an opening onto the "heavens" at the upper level of the composition where, in an upward spiral that is further emphasized by the round framework, the figures are intermingled with banks of cloud. It is as if the painter had been intent upon eliminating from the figurative field any reference to the terrestrial base of the representation so as to make it seem an inner vision that, as such, was not subjected to the objective conditions of perceptible certainty.

Historically speaking, El Greco's painting does not constitute a new point of departure. Michelangelo's *Last Judgment* had already presented the image of "a space without reality, without existence, the upper part of which is filled with human bodies which . . . , seen from a distance, float like trailing clouds."[55] In fact, many compositions of the period present a quasi-nebulous aspect that is most striking. Dvořàk with reason draws attention to the similarity between El Greco's *Burial of the Count of Orgaz* and Tintoretto's *Ascension of Christ*, in which the Apostles, the circle of whom traditionally constituted the pivot of the composition, are as it were pushed away, sucked into the background, of which only a few glimpses are afforded, while Christ rises amid a great mass of clouds. Here, the Ascension is not treated as an objective physical phenomenon, but as a mystical, subjective vision, seen only by Saint John, a figure who is relegated to the bottom corner of the composition (just as he is in Parma, on the rim of the cupola of San Giovanni Evangelista). This seems a far cry from the stable, clearly arranged, systematically organized scenic space that the Quattrocento had used as the theatrical setting for human action. But this evolution—or even transformation—should not be described solely in stylistic terms. When Max Dvořák chose El Greco as an exemplary figure for the period known—most unfortunately, in his view[56]—as "Mannerist," he wished to show that the rejection of the rules of both physical and dramatic verisimilitude and the abandonment of the objective frameworks of representation are, as principles, inseparable from the general trend of the culture of the day and in particular the mystical crisis that followed the Reformation. Saint Teresa wrote as follows: "What I see is of a white and a red color that is to be found nowhere in nature and is of a brilliance and power greater than anything man can see; these are images that no one has ever painted."[57] But they were images that El Greco did attempt to paint, according to Dvořák, never hesitating to sacrifice all preoccupations of objectivity and verisimilitude to the expression of inner rapture.

THE SYSTEM AS A FIELD OF PRODUCTION

But the most important thing here is not so much the "documentary" significance that may be attached to El Greco's work (documentary, in the sense that Panofsky, following Karl Mannheim, gave to the word: namely, the significance held by the work of art in as much as it offers access to the "spirit of the age," "the vision of the world" that the artist shared with other men of his time and class, or that of the social group for which he was working). Rather, it is the fact that that significance can be reconstructed from certain discrete features and partial aspects of the work.[58] So a rapid comparison between the procedures adopted by El Greco and Zurbarán in their efforts to represent miraculous visions may enable us to draw a contrast between the purely contemplative mysticism from which El Greco's images stem and a far more active concept of mystical effusion. El Greco tends to reduce to the minimum the space allowed for the ground, the terrestrial seat of his compositions, or to liberate such space from all natural constraints. Zurbarán, on the other hand, creates an opposition between the terrestrial level, which is treated as a closed architectural volume, and the celestial level, which is conceived as an open, indefinite cloud; and furthermore, his figures in a state of ecstasy are never represented as weightless but remain firmly fixed to the ground. Clearly, these factors can tell us something about the policies of the religious orders for whom the painter was working: the Church, under divine inspiration, has work to do in the world, and contemplation is not all that there is to religious practice. But the solution adopted by Zurbarán has a further significance. Insofar as it associates within a single composition two levels that correspond to different stylistic principles, one treated in the linear mode, the other one "vaporous," it makes the opposition, at once physical and metaphysical, between heaven and earth doubly effective and confers upon the opposition between the "linear" and the "pictorial" a resonance that is more than solely stylistic. It sets two stylistic configurations that seem to be contradictory within a single field, which, however, at the same time reveals that both belong to a more general structure with regard to which the opposition itself becomes significant.

This structure cannot be seized upon or even detected at first glance; nor does it become accessible through a description composed along the lines of a history of styles. Such a history recognizes only the successive configurations of art, and sets out to describe them in a philological way, in terms of evolution, affiliations, reversions, and remainders. But the aims of the system that underpins them call for a different approach to the historical material and a different focus upon the pictorial reality. In place of the periodization that corresponds to the level of manifest, visible figures, it substitutes a different type of breaks and discontinuities: figures that appear to be irreducible, incongruent, are found to communicate at a deeper level, that of structural constraints and regulatory principles, theoretical articulations, and practical options, and formal models and cultural and ideological affinities of the most general kind. The network and more or less stable and organized constellations of these interconnections define what the present text has above called the *aisthesis* of a period of art, that is to say the system that underpins its historicity. Thus the pictorial system of the Renaissance determined a *possible* field of production but at the same time imposed a number of prohibitions, rejections, essential structural limitations. By reducing the question of the space of the representation to that of the representation of space, the theorists of the Quattrocento gave the system a restrictive definition, which was, however, perfectly coherent and which made symbolic representation directly dependent upon theatrical representation. The term *spectacular* encompasses the term *specular*: the perspective code, through the assimilation that it engineers between spectacular (theatrical) repetition and specular duplication, establishes a structure of referral that is characteristic of the system of representation that stemmed from the Renaissance (the concept of which, in its turn, implies a duplication, a setting in perspective) and from the notion of the sign that is associated with it. All the same, the system cannot be reduced to the code in which it found—to borrow Leonardo's words—both its brake and its rudder; and even if the brake slipped or the rudder spun out of control, the functioning of the system, in its most general principles, would not be affected.

APOCALYPSES

That is exactly what happened in the first half of the sixteenth century, when the pictorial graph marked /cloud/, which until then had been limited to marginal iconic and signaling functions, progressively came to invade the whole plastic space, to the point where, in the cupolas of Parma, it came to "designate" the space itself. In the context of a system that *in theory* excluded the depiction of nebulous formations, such an invasion presents diverse if not contradictory aspects, with variable consequences. Sometimes /cloud/ provided the conventional basis for multilayered compositions (e.g., *The Dispute of the Holy Sacrament*, by Raphael) more or less directly inspired by a real theatrical spectacle or intended to be a substitute for it (see also Raphael's *Virgin of Saint Sixtus*), but in no way jeopardized the representational system the regulation of which was ensured by linear perspective. Sometimes, in contrast, it seemed (but was this perhaps a deliberate deception?) to serve to produce a pictorial space that owed nothing to geometry (e.g., Tintoretto's *Paradise*) or to open up a volume constructed upon a trompe l'oeil sky (e.g., the "celestial" cupolas of Mantegna, Correggio, etc.). In other instances, clouds more subtly introduced a contradiction into the very heart of the representation, by denoting a rent in the human space and a more or less brutal insertion of a dimension of transcendence into the system of depiction that depended upon geometrical coordinates. In one way or another, cloud connoted the closure of the system, revealing its limitations by operating on its margins, at the meeting point of what was depictable and what was not.

But the implications of that closure and the limitations that went with it, the prohibitions that stemmed from it, and even their transgression were more than just formal. As Machiavelli's text testifies in the field of political thought, the great social upheaval of the sixteenth century involved the end of the Florentine dream, the dream of a history in keeping with the dimensions of the city that for a time had been its theater. And it was not by mere chance that the kind of unease that, with Mantegna, took hold of the system of representation was first diffused in papal Rome and in regions in contact with the empire (Mantua, Parma, etc.) before eventually, with El Greco, descending upon the chosen land of the great mystics. Yet that unease in no way implied an abandonment of the system established in the Quattrocento: *The Burial of the Count of Orgaz*, with its figure of a young boy positioned as if at the front of the stage and looking in the direction of the spectator as he points toward the miracle, is still obedient to the principles laid down by Alberti, who wanted every representation to include a figure positioned on the edge of the story, as a *witness*.[59] Nor does Correggio's cupola of San Giovanni Evangelista transgress that principle: on the contrary, it subtly duplicates the effect, for here the Evangelist, whose vision the spectator is allowed to share, is represented *twice*. First he appears in the guise of an almost invisible figure, seated on the rim upon which the cupola rests (as if the painter intended to maintain—albeit covertly—the terrestrial basis of a representation apparently liberated from any architectonic support); and secondly, as another figure who occupies one of the lunettes in the transept and is looking up at the cupola. In this way the artist creates within the architectural volume a dramatic triangulation that defines the image not as an Ascension, but as the vision of Saint John on the island of Patmos, while at the same time making the trompe l'oeil decor of the cupola directly dependent upon the constructive organism the closed nature of which it is supposed to negate.

The art of Correggio provided an outlet for the unease detectable in the system of representation: a sentimental outlet, with strong pagan connotations. We should not forget that, in the imagination of the time, very little distance seemed to separate the figure of Ganymede lifted into the air by Jupiter from that of a Saint John carried heavenward. In the cupola of San Giovanni Evangelista, we are not confronted with an apocalyptic vision, in the catastrophic sense of the expression. Indeed, it is almost tempting to see this *reassuring* interpretation of Revelations as a response to another *Apocalypse*, which, for its part, certainly did deserve its title: the *Apocalypse* by Dürer, published in the form of a book—the first to be published by an artist at his own expense—in 1498, the very year in which, in Florence, the adventure of Savonarola came to an end when the prophet of the *Dies irae* and the end of time was burnt at the stake. The significance of

Dürer's undertaking was only too clear with respect to the crisis that was dividing Christianity and the struggle against Rome led by Luther;[60] and this version of an ironical pamphlet owed its essential force and power of persuasion to a formal paradox that Dürer managed to exploit systematically. Although Revelations was not recognized as a canonical text until the fourteenth century, it had long been providing material for commentaries and illustrations.[61] But it was not until Dürer's work that, for the first time in the history of Christian art, the rent in the cosmic order that the text of Revelations implies was not just the object of a *symbolic* and/or figurative *representation* and that the eruption of heaven on earth was conveyed, at the level of the system, by a rent in the very order of depiction, or representation. Max Dvořák comments that, in Dürer, the images of the Apocalypse appear as fantasies, the fantasies of a people without images (*den irrealen Phantasien und Meditationen eines bildlosen Volkes*);[62] Dürer had selected from among all the "undepictable" moments in the text (*im Höchsten unbildlich*) those that were "representable" (*Momente, die sinnlich darstellbar waren*) with such skill that this deserves to be described as a "resolution" of mental fantasies by means of perceptible expression—*einer Auflösung der gedanklichen Phantasie in bildliche Ausdrückmittel zu sprechen*. The great merit of Dvořák's formulas is that they pose the problem in terms (the very ones used by Freud) of depictability (*Darstellbarkeit*): how can one represent that which cannot be represented, since after all, in Revelations, all representation necessarily comes to an end?

Pursuing the analyses of Dvořák, Panofsky has shown that Dürer's *Apocalypse* derived some of its effects from the contrast between the naturalistic rendering of the figures and the resolutely antinaturalistic mode of presentation. This is confirmed by the ambiguity of the functions imparted to the figurative elements: the clouds, for example, which are often mentioned in the Evangelist's text and which, in Dürer's *Vision of the Seven Candlesticks*, provide a solid resting place for the candlesticks arranged in perspective, while at the same time being deployed so as to form weightless vertical columns of smoke, liberated from all the laws of geometric construction. The depth of the space is at once affirmed and negated;[63] but it is important to see that here affirmation and negation are indissolubly linked. Figuratively speaking, Dürer's *Apocalypse* is at once the first and the last of Apocalypses: the first because, with Dürer, the Apocalypse leads to the destruction of an order that, at the level of the system, is for the first time embodied by the perspective code; and the last (not excluding attempts such as that of Odilon Redon) because the order thus given material form at the very level of the signifier could not give rise to Revelations without disintegrating. The very same illusionist means that serve to construct a three-dimensional space are used to break its closure and impose the image of a vision that contradicts the very notion of representation. In short, to borrow the terms used by Panofsky, all the progress made in verisimilitude and animation strengthens rather than weakens the hallucinatory effect. The vision only appears as a vision thanks to the very order that it contradicts but that nevertheless opens up to receive it, at the cost of a seemingly irreparable fracture. It is a fracture in relation to which Correggio's *Vision of Saint John* takes on an enigmatical significance: what if Correggio, far from aiming to open up the Christian edifice to the Beyond, had instead been striving to plug the breach, soften its angles, and restore the system's equilibrium by fortuitous means?

Semiology and Sociology

PRODUCTIVITY/CREATIVITY

The examples of El Greco's *Burial of Count Orgaz* and Dürer's *Apocalypse* seem to show that there are some works that it is impossible to analyze or even to describe without returning to the historical context into which they fit. The sociology of art is concerned with concrete products: it aims to restore them to an overall context, attach them to definite social strata and there pick out the interplay of various points of view, beliefs, and interests, many of them contradictory, or even to explain such or such a stylistic transformation by relating it to a change in the attitude of the interested groups.[64] Conversely, over and above the superficial effects of signification and the ever attractive shimmer of correspondences and parallelisms, the semiology of art strives to reveal the general habits and principles that govern depiction in a given age. But that opposition (which echoes the opposition in linguistics between speech and language, diachrony and synchrony) is a specious one. What happens in Dürer's *Apocalypse* and El Greco's *Burial* only takes on its full resonance when related to the signifying work from which the artistic images proceed and, at a deeper level, to the question of depictability, the system by which what is stated is historically governed, and the constraints and possibilities that stem from this at a formal level.

The system that established the perspective code as "the brake and rudder" of pictorial representation, by associating it with a spec(tac)ular structure, thereby defines an organized scriptural space, one that lends itself to regulated variations and transformations that in their turn appear, with regard to the repercussions and remanipulations that they cause at the level of the system, as *internal* events. Now, basically such a system is of a social nature. Even if the collective practice to which they are linked, as to their condition of existence, is not as extensive as that which is at the root of language, in the Saussurian sense,[65] the system here called the *aisthesis* of an age, along with the "styles" whose configurations belong within its field, lead, as Saussure would say, "a semiological life."[66] To varying degrees, they elude the control of individuals and, as they develop and evolve historically, are governed by a sui generis logic the more or less constraining nature of which depends upon the extensiveness of the process of socialization of which they are at once the agent and the product. In other words, a semiology of art is bound to have sociological implications right from the start and, conversely, the sociology of art cannot do without that aforesaid semiology. As Tynianov and Jakobson were pointing out as early as 1928, "To consider the correlation of systems without taking into account the laws immanent to each system is, from a methodological point of view, a dire way of proceeding,"[67] and that observation is just as applicable to different cultural series, when considered from the point of view of their reciprocal relations, as it is to the various structural levels according to which social reality can be apprehended.

However, to assign to "styles" or, better still, to the system from which they stem, a status, if not a "semiological" existence, is not the same as assimilating them to a language, let alone a code, a finite collection of rules that would make it possible to engender an infinite number of works or images, just as generative grammar is used to engender phrases in a given language. Whatever language may do in this respect, the creative capacity of the semiotic practice known as "art" cannot be measured by the totality of works that that practice can in effect produce. "Works" are not all that "painting" produces. It provides those who make use of it, considered in all the diversity of their social and cultural allegiances, with a certain number of figures that play a role in their elaboration of the image that they maintain of themselves and the world in which they live, and the position that they occupy there. These are spatial figures first and foremost, as the most visible function of painting—or at least of representational painting—is to confer the force of an *institution* upon the particular notion of space and relations—geometric, dramatic, and so on—that define that image. Those figures—and others too, quantitative or qualitative, graphic or colored, descriptive or metaphorical, figurative or *nonfigurative*[68]—which provide a *theoretical* support for the pictorial process, are the product of a signifying practice the activity of which is marked and pinpointed by the works in question. They proceed from creativity constituted as a "network," with intersecting tendencies, for the works that it produces do not succeed one another in a linear fashion, as the pearls of a necklace do, but rather set out in the field to which they belong configurations that are variable, complementary, or even opposed, and that derive their meaning from their differences as much as from their common factors. In their respective finiteness, these works are indicators of a general productivity the deepest mainsprings of which can only be glimpsed where individual products come into contact with one another, as they interact and function in a reciprocal fashion. It is within that field, far more than on its periphery, that the question of the relations between the signifying practice known in our society as "art" or "painting," and other practices, or even the general historical context in which it operates, has a chance of being treated rigorously and systematically.

MEDIEVAL FIGURES

The figurative interaction between "heaven" and "earth" provides a good illustration of the relation between the internal destinies of the "system," in the sense in which the present text understands the term, and the transformations that come to light, historically speaking, in pictorial production. Considered in their details, those transformations are traceable to objective determining factors of a social or cultural nature. But that does not mean that the system evolves through an accumulation of contingent variations and determining factors. Adaptations (of which we have noted a few examples), borrowings (those that painting—among other things—derived from theatrical spectacles), and even transgressions (or what are presented as such) only have the impact allowed to them by the system that authorizes them, encourages them, or gives rise to them. And this is where any "reductive" sociology of signifying productions meets a stumbling block (for, as we shall see, we must push on even further, and seek within the system itself for the mainsprings, or at least the regulatory mechanisms, of evolution). The Saussurian principle according to which "anything that changes the system to any degree" must be internal is in no way antihistorical: on the contrary, it paves the way for historical analysis of the most profound kind when it explicitly contradicts the opposition long since abandoned by even the linguists themselves, between synchrony, in which the relative rationale of the system is supposed to reign, and diachrony, defined as the contingent dimension of acts of expression: a system has a history and that history, far from borrowing anything from chance, is itself systematic and interpretable as a system.

From a figurative point of view, and despite the difficulties of interpretation presented by some images, the Middle Ages seems to have conceived of the relations between earth and the heavens in the least problematical fashion: in those days, either a particular determining factor—the line of the ground, a rock, a cloud, a mandorla, and so on—was enough to identify sites and mark out characters or objects as either terrestrial or celestial, or to signal an eruption of the divine into the human field, or else some compositional artifice made it possible to reserve certain "celestial" compartments in the iconic field. In medieval painting there are countless instances of hands, angels, Christ, the Virgin, even God the Father in person, emerging from a circle or a crown of more or less closely packed clouds to present Moses with the Tablets of the Law, to dictate the text of an Evangelist's account, or to visit a sleeping pope or Saint Joseph in his dreams. The Utrecht Psalter (c. 820) provides the best example of the identification of sites and actors by means of a rapidly drawn line evoking the outline of the ground or the profile of a cloud.[69] As for separating off areas of the surface of the image by means of a series of frames, the Limoges Sacramentaire (c. 1100) provides an illustration of the Ascension that can be regarded as a model: Christ, seated in a mandorla flanked by two angels, occupies the upper part of the image, which is separated by a kind of lintel from the lower part, where the Apostles and the Virgin are positioned, their bare feet set on a narrow checkered band.[70] Another ploy used repeatedly throughout the Middle Ages, but only developed systematically in the fourteenth century, is an architectural setting in which the upper part, or upper story, is called upon to function as "heaven" or "paradise": a little chamber in which an angel or an Evangelist occupies the tympanum (a schema copied by numerous altarpieces), arches with the inner surface decorated, as in the

Psalter of Saint Louis, with an edging of clouds that designate as "heaven" the entire architectural superstructures that enhance the Parisian miniatures of the period.[71] All that was then needed was a few angels climbing a ladder to the upper reaches, to obtain a perfectly legible representation of movement between the earth and the heavens above (Plate 6a).

TWO SEPARATE ZONES: THE HEAVENS AND THE EARTH BELOW

But it is not the case that these configurations, which, in the case of the architectural settings, operate as veritable *tropes*, are purely formal. The strictly scriptural plane, apparently devoid of any spatial implications, where the characters in the Bamburg *Apocalypse* and the Mozarabic *Beatus*[72] are set, is not a neutral, indifferent surface: it is organized with a left-hand side and a right, a top and a bottom, and any movement depicted there acquires a definite connotation from the direction of its impulse, just as every figure does from the spot in which it is located. However, that is not all there is to it, as can be seen from numerous miniatures and frescoes, some of which date from the earliest days of Christian imagery. In many cases the surface is divided into a number of horizontal bands of different colors, and far from this being a purely formal feature, it often seems to correspond to a carefully elaborated cosmological concept. It is as if these apparently simple parallel bands, of variable width, correspond to different zones, places that are qualitatively different. In the lower basilica of San Clemente, in Rome, *The Ascension* (or possibly *The Assumption of the Virgin*, eleventh century) is represented as a composition consisting of a series of levels, respectively marked by bands of different colors. At the lowest level, the Apostles are positioned in two groups on either side of a stone relic built into a wall, against a brick-colored background; the figure of Christ (or the Virgin?) rises through the middle band, of a bluish hue, toward a "heaven" (or "paradise") that is greenish, where Christ(?) is enthroned in a mandorla held by four angels.[73] The positioning of the bands of color one above another certainly seems to be a conventional kind of signifying arrangement, as is confirmed by *The Meeting of Abraham and Melchizedek*, part of the cycle of mosaics in Santa Maria Maggiore, already mentioned above (see Chapter 3, p. 107). The protagonists in this scene are advancing along a green band of "ground," upon which their shadows are cast. Their silhouettes stand out against a background of uniform gold that looks like a wall upon which an ancient scene might have been painted. Meanwhile in the top zone God (or Christ?) is emerging, in a sky treated in a curiously "pictorial" fashion, from a cluster of blue and red clouds. The opposition here is clearly between the conventional golden background and, on the one hand, the "ground," on the other, "heaven," both of which are given a descriptive quality. The same opposition, reduced to a simple interplay of colors, is to be found in the *Book of Pericopes*, part of the Reichenau group (early eleventh century) in a scene depicting Saint Peter receiving his keys (in which Christ advances to meet the Apostles very much like Melchizedek confronting Abraham and his army). It is painted on a background organized according to the same norms: a green "ground," a gold "background," and a blue sky.[74] The significance of this opposition is made explicit in the Paris Psalter in the Arsenal Library (thirteenth century). This opens with an image showing three characters—a copyist, an astronomer, and a calendar expert—seated on a kind of chest. The three figures, flanked by two bushes, stand out against a brownish background beneath a starry firmament that is separated from the rest of the composition by a fringe of undulating clouds (Plate 6b).[75] The astronomer occupies the most important and dominating position, as is emphasized by the little stool upon which his feet rest, a sign of dignity, and by the subordination of his companions' book and calendar to the astrolabe that he holds up to the sky. The connotation of the opposition between the background to the figures and the vault of the heavens that the astronomer is observing is unequivocal: it corresponds to the Aristotelian division of the cosmos into a sublunary world that is subject to corruption, the center of which is occupied by the weighty and opaque earth, and the celestial spheres, in which the weightless, incorruptible, and luminous stars are fixed.

A PICTORIAL SERIES AND A THEATRICAL SERIES

There was nothing problematical about such representations, given a system that imposed no unifying organizational principle upon the figurative field. But when, in the Quattrocento, Masolino came to paint *The Foundation of Santa Maria Maggiore*, he too set clouds along the join between the two spaces: the human, sublunary space, subjected to weight and composed according to strict perspective, and divine space, the source of all light (as is connoted by the heavily accentuated relief of the clouds), in which are displayed the effigies of Christ and the Virgin, set within a perfect circle. To speak here of a contradiction would make no sense, since the opposition, quite apart from signaling through its very excess the miracle supposed to have been at the origin of the foundation, corresponds to the traditional division of the cosmos. Contradiction only appears at the point where linear perspective asserted itself so strongly as the brake and rudder of painting that any deviation from the order that it established was seen as an exception to the rule. Cloud itself would then, *in theory*, be excluded from representation, unless it took on the appearance of a theatrical prop (and, as we have seen, Vasari was to continue to regard anyone who succumbed to its attractions as a neurotic).

The *aisthesis* of a period can therefore not be reduced to an organized collection of means of expression the coherence of which stemmed from the existing state of the system. An *aisthesis* does not develop solely from use, nor in consequence of a quasi-natural selection that preserves only the results of plastic experiments that fit in with the general tendency of social and historical evolution. On the contrary, it provides the systematic basis in relation to which those experiments are made and become meaningful, acquiring a historical significance. Although one need not necessarily attribute any "intention" or a specified historical "plan" to it, it is imbued with a kind of finality that—in some cases—leads to spectacular reversals.[76] To cite Tynianov and Jakobson once again, you could say that the history of the system is itself a system: "Each synchronic system contains its past and its future, which are structural elements inseparable from the system."[77] In these circumstances, it is not hard to see that the use of cloud may appear, in both the art and the theater of the Renaissance, now as an archaism, now as an innovative feature, or sometimes as both at once: although inherited from medieval art and theater, it acquires new meaning in the representational structure that is linked with the institution of a unified scene. In the Middle Ages and in medieval representation, which was characterized by *discontinuity*, cloud served to designate one particular place among others (the place of the heavens, in both the divine and the meteorological sense of the term) and transition between one place and another. But in the perspective *continuum*, it introduces a kind of opening or detour that the system itself seems to call for and that seems to answer its needs and to serve figurative or iconographic functions of the most varied, or even contradictory, nature. Cloud makes it possible to introduce a dimension of transcendence a posteriori into a network of coordinates that appear bound to exclude it. (For instance, in *The Annunciation*, in the Uffizi Gallery, Veronese renewed a traditional theme by having a golden cloud invade a perspective of colonnades and porticoes of the type made popular by the art of Piero della Francesca.) But the very same cloud that serves as a vehicle for the Holy Ghost or Christ, and so on, may equally well transport the gods of Olympus; and the same cloud that opens onto paradise may, in a different context, provide the syntactical mainspring for a rhetorical figure. To borrow from Giulio-Carlo Argan the example that he chose in order to illustrate the notion of "allegorical transcendence," Rubens was to paint Henry IV, still clad in his armor and surrounded by a landscape of ruins, but already thinking of taking a wife, to mark the advent of a new era: Minerva appears in a cloud, to present him with a picture of Maria de' Medici, while in the distance, on another cloud, the marriage of Jupiter (the eagle) and Juno (two peacocks) is evoked.[78] A single *configuration* serves functions that are now representational (conferring a kind of verisimilitude, if not an illusionist quality, upon the mystical vision), now symbolic (introducing the distance of allegory into the iconic fabric).

In this respect, the radical break indicated by Georges Kernodle between modern theater and medieval spectacle illuminates the problem of representation, and at the same time situates at the right level the question of the exchanges and relations between the pictorial and the theatrical series. Medieval representation developed along a discontinuous temporal line, with successive episodes unfolding in different places (*loci deputati*), around emblematic objects. Not until the sixteenth century did men of the theater become interested in integrating all the different elements of a single spectacle within a unified framework and in returning to a dramatic form that had been abandoned centuries earlier.[79] It was a problem that theater directors had ignored so long as town squares, streets, and the thresholds of churches, or even huge sets of scaffolding served as a framework for the unfolding of the spectacle; but it was one that painters, for their part, could not push aside. Although they were free to regard the page, wall, vaulted ceiling, or panel as simply a substratum upon which to arrange various characters and symbolic accessories, right at the outset the very practice of painting posed the question of the systematic organization of the surface given them to cover. Linear perspective was to provide a solution both technical and theoretical to the problem. However, the geometrical construction of space did not totally resolve the problem of representation:

the specular duplication to which representation was linked according to the principles of depiction, and its signifying function that was the counterpart of that depiction, called for the reflective interplay from which a system of signs proceeds. Duplication, reflection: those terms should be taken literally. Far from fueling each other reciprocally, as do communicating vessels, art and theater both have their own respective semiological life. And even if art borrows from theatrical spectacle some of the latter's signs, what really matters is the semiotic mechanism upon which the sign depends and which prompts that borrowing: the element "borrowed" by painting from the theater no longer counts as the same thing once it is forced into specifically plastic functions and is governed by the internal demands of the system, in which it now represents a new articulating mechanism.[80]

Science/Painting

> La prospettiva, figliola della pittura.
>
> Perspective, the daughter of painting.
> —Leonardo da Vinci

CLOUD AND THE CLOSURE OF THE REPRESENTATION

Depending upon whether one chooses to compare the pictorial series with some other series—religious, theatrical, scientific, and so on—a given figurative structure will lend itself to multiple interpretations: the same images by Zurbarán that testify to an increasingly important role assigned —in the organization of faith—to mystical experience may be described in terms of dramatic settings and seem to propose a kind of compromise between the Italian stage, with its free-standing proscenium, and the "facades" of Spanish theater. But the space of classical representation, as staged by Quattrocento painting,[81] is a closed space, as is a cubic volume, chamber, or town square seen in perspective, in which the *istoria*—and, with it, every element that it brings into play—in principle finds its *place*. Quite frequently, the place of the sky is taken by the checkerboard that imposes its order and rationality upon any representation, marking out its closed nature; and it is not purely by chance that this happens.[82] Historically speaking, the first checkerboard surfaces that painters tried to depict in perspective may have been the coffered covers, ceilings, and arches of which Massaccio's *Trinity* (Florence, Santa Maria Novella) provides an admirable example, at the same time obeying a principle of construction that was then extremely "advanced." This was the principle that Mantegna, in his day, was to exploit to the full: with the vanishing point situated below the lower edge of the fresco, the composition is determined from "on top" and, in the absence of any visible "ground," its order stems from the compartmentalization of the coffered vault seen in perspective, against which background the figures of the Trinity stand out.[83] When, as a corollary to the institution of the perspective regime, the line of the horizon is lowered, this reveals the sky, which the painter may, if he chooses, organize in an architectonic or quasi-architectonic fashion. In Masolino's *Foundation of Santa Maria Maggiore*, the clouds set in perspective serve—once again to borrow Burckhardt's expression—to designate the space and to close off the terrestrial scene, while at the same time they mark the frontier between the sublunary world and the heavens where the divine figures are positioned. That is not at all the case in the small panel that Brunelleschi used for the first of his demonstrations, where the sky—with its passing clouds driven by the wind—was inserted like a wedge, in the guise of both a reflection and a body alien to the representational circuit, the linearity of which it upset. The corrosive function that Leonardo was to attribute to the representation of atmospheric conditions was thus justified right from the start: one and the same element—/cloud/, whose similarity to a stain has already been noted—may thus be called upon to perform a function that may be sometimes integrating, sometimes disintegrating, depending on whether it is used for constructive purposes or, on the contrary—both as a meteorological indicator and as a pictorial instrument—it causes *perturbations*.

LEONARDO II

In theory, cloud, a body without a surface and so also without outline, has no place in any reproduction based upon the reduction of opaque volumes to their linear projection on a plane. So if painting sets out to paint *what cannot be painted*, will it not have to modify, if not its projected purpose, at least its means? But Leonardo does not think and write solely "as a painter": for him cloud is not just a pictorial element or a theatrical accessory; it is first and foremost a meteorological phenomenon and it is as such that it has a role to play in representation. Scientifically speaking, cloud is a body without a surface but not without substance for, like mist, it is the product of a *thickening of the atmosphere*, a contraction of the humidity dispersed in the air. As well as playing a determining role in the creation of winds, by pushing air before it, compressing it and expressing it,[84] it also constitutes a kind of *site* from which sounds may be reflected and where rainbows may form. Being both an instrument and, at the same time, an indicator of the cycle of evaporation and precipitation,[85] cloud occupies a key position in da Vincian cosmology, at the junction of uranology, aerology, hydrology ("a large part of the sea will escape into the sky, and will not return for a long time: namely, clouds"),[86] and geology: in both its formation and its effects, cloud provides a good illustration of the universal liaison that links together all the parts of nature, and of the mixture and ceaseless permutation of the elements,[87] the separation of which dispersed chaos but which seem here to return to their primordial indistinctness and to change into one another, producing unstable masses the formation, evolution, transformations, and resolution of which are accompanied by perturbations and precipitations that upset the order of the world and its visible configurations: rain, lightning, snow, hail, and wind. Far from marking the limits of this world in which we live, beyond which the principles of mechanics and the science of weights are believed to cease to apply, cloud is thus a body that, like any other body, obeys the laws of movement and weight in accordance with which all natural action takes place.

Although it has no surface, cloud is *visible*.[88] Now, it is worth repeating that Leonardo thought that the functions of painting were equivalent to those of the eye and that his works *represented* those of nature ("l'opere del pittore rappresentano l'opere d'essa natura" — the works of the painter represent the works of nature).[89] Cloud is a part of the visible world; and clearly the latter cannot be reduced to a linear construction. And yet it seems that Leonardo, the theorist of painting, borrowed even more from geometry than Alberti did. Intent upon preventing any "confusion of genres," Alberti declared that he wished to speak solely *as a painter* as, for him, geometry's contribution to the art (theory) of painting — which calls for "a fatter Minerva" — definitely amounted simply to providing a principle for a reasoned construction of the scene of the *istoria*, or representation. For Leonardo, on the other hand, geometry assumed far more extensive functions and played a far more fundamental role. If painting is a science (*se la pittura è scientia o no*),[90] that is thanks to geometry. Or, to be more precise, no research merits the name of science unless it involves mathematical demonstration.[91] And what is perspective if not a rational demonstration that all things transmit their image in lines that form a cone?

That demonstration shows that painting, the *science* of painting, begins, as geometry does, with points, lines, and surfaces.[92] That is a theoretical point of departure if ever there was one: for perspective is the daughter of painting; and it was the painter who created it to meet the requirements of his art.[93]

Painting engenders perspective, which in return establishes painting as a science. The science of painting begins with drawing. But this generates another science, which involves shadow and light (*da questa nasce un altra scientia, che s'estende in ombre e lume*),[94] along with chiaroscuro, color perspective, and atmospheric perspective. Whereas geometry is limited to quantities, both continuous and discontinuous, painting for its part is concerned with the qualities of things, the qualities that produce the beauty of the works of nature.[95] And that, of course, is why painting has not always been counted as a science: painting, like the works of nature, has no need of words to attain its full potential; so painters have not thought it necessary to describe it and reduce it to principles.[96] Painting, which is a "semimechanical" discipline, since it is born of science but culminates in a manual operation, is not a science in the same sense as disciplines that begin and end in the mind (*che principiano e finischono nella mente*).[97] And therein lies the mistake made by those who seek to reduce it to geometry and who, confusing a natural point with a mathematical one, have spoken of perspective as if the eye were a point.[98] The "confusion of genres" that Alberti wished to avoid is inevitable, given that perspective is not solely a matter of geometry.

What is at stake here in Leonardo's text is fraught with heavy consequences, both epistemological and pictorial. Aristotelianism certainly proscribed all confusion of genres: a geometrician should not think as an arithmetician, nor should a natural philosopher (let alone a painter) think as a geometrician. As Alexandre Koyré pointed out, that is a perfectly legitimate demand: so long as "genres" subsist, they should not be mixed. But they may be destroyed;[99] and it seems that Leonardo set out to do exactly that— Leonardo, in whom the artist and the engineer coalesced and in whom the technologist rivaled the geometrician.[100] To what degree that endeavor ran contrary to the classical notion of "genres" imposed by a division of tasks is a matter that may be resolved by the question that Freud took as his starting point in his famous essay on Leonardo (to which I shall be returning elsewhere): in Leonardo the scholar, to what degree did the man of science

"smother" the artist? It is clearly a question that bears the stamp of ideological considerations that rule out any rapprochement, whether concrete
or theoretical, between art and science, between the image of the world that
art strives to impose and the image elaborated by science, and also that of
the relations between the procedures that a painter and a scholar of physics
respectively adopt to that end. But it is also a question upon which what
Koyré calls "the Leonardo miracle" insistently forces one to reflect.

Alexandre Koyré has shown that Leonardo's attitude in mathematics, that is to say geometry, was essentially that of an engineer, and that his
geometry, far from suffering from his concrete way of thinking (and the
"confusion of genres" to which this led), on the contrary benefited from it.
Leonardo, who was a born geometrician, possessed to the highest degree
the gift of an intuitive understanding of space (extremely rare, according
to Koyré); and this gift enabled him to overcome his lack of theoretical
training.[101] At the same time, in his case, the engineer's science was dominated by geometry, "so his geometry is usually that of an engineer and
vice-versa his art of engineering is always that of a geometrician." The same
could be said of geometry's relations with painting (after all, is not perspective "the daughter of painting"?). But it is manifestly clear that it is
thanks to the confusion of genres from which it proceeds that Leonardo's
geometry possesses the dynamic character that enables it to break resolutely with the Aristotelian tradition. Pierre Duhem, who was determined
to show that most concepts credited to modern science were stated, *anticipated*, as early as the Middle Ages,[102] claimed that Leonardo borrowed from
Ockham the notion of the point, the line, and the surface that he sets out
in his *Notebooks*.[103] But if one compares Leonardo's definition of the line
and the surface to the definitions with which Alberti's *Della pittura* opens,
it is impossible not to be struck by the way that Leonardo has transformed
mathematical entities such as the lines and figures of geometry. Whereas
Alberti regarded a line as being determined by a succession of points set
out one after another without interruption, and a surface as being determined by the meeting of a number of lines assembled together like the
threads of a fabric,[104] Leonardo conceived them to be *produced, described*
by the movement of points and lines through space. "A line is a length (extension) produced by the movement of a point, and all its extremities are
points. . . . A surface is an extension made by the transversal movement of a
line [elsewhere Leonardo was to write: 'by the movement of a line diverted

from its direction'] and its extremities are lines."[105] Now these definitions, far from stemming from the scholastic tradition, on the contrary, in the descriptive mode, "anticipate" the conceptual transformation that Newton was to accomplish when he brought mathematical entities closer to physics and submitted them to movement, considering them no longer in their "being" but in their potential "becoming," their "flux."[106]

Never mind, at this point, if this "anticipation" (which seems founded upon the heterogeneity of series and the possibility of relating one theoretical field to another) can only be interpreted as such, in Leonardo's text, by working back from the Newtonian synthesis, or if it only produces its effects in the logical time constructed by epistemology. The essential thing is to see clearly that, in Leonardo, the idea of submitting mathematical entities to movement is not based on theory but proceeds from that same "intuition of space" (which Koyré regarded as one of his most astonishing gifts) and depends directly upon his experience and practice as a painter (perspective is the daughter of painting, which itself originates from points, lines, and surfaces). But Leonardo's text goes yet further: whereas pre-Galilean, pre-Cartesian tradition assimilated movement to change, a process that affected the bodies subjected to it, the concept of movement that was to triumph in classical science would replace a physical, empirical notion by a purely mathematical one, devoid of any qualitative connotation: the notion of a relationship that has nothing to do with time, or rather of "a movement that unfolds in nontemporal time, a notion as paradoxical as that of change without change."[107] And what do we find in Leonardo's text?

Although time is numbered among continuous quantities . . . it does not altogether fall under the category of geometrical terms, which are divided into figures and bodies of infinite variety, as may constantly be seen to be the case with things visible and things of substance; but it harmonises with these only as regards its first principles, namely as to the point and the line. The point as viewed in terms of time is to be compared with the instant, and the line resembles the length of a quantity of time. And just as points are the beginning and end of the said line, so instants form the end and the beginning of a certain given space of time. And if a line be divisible to infinity, it is not impossible for a space of time to be so divided.[108]

The nature of time is different from its geometry.[109] In other words, the time known by geometry has nothing in common with the time of generation and corruption, the time of change and of movement in the Aristotelian sense, but is the nontemporal time in which "movement that does not change" takes place, and which makes it possible to speak in mathematical terms of realities such as the speed, acceleration, and direction of a moving object at any particular point of its trajectory. All these were questions that preoccupied Leonardo, although he lacked the theoretical training that might have enabled him to come up with anything more than the beginnings of answers.[110]

A STRANGE FACULTY

The strange faculty to which Koyré refers, which is supposed to have enabled Leonardo to reach astonishing conclusions on mechanistic and ballistic problems, even though he was ignorant of the premises upon which they were founded—that faculty necessarily derived as much from his experience and practice as an artist as it contributed to these. Leonardo's conception of physics, the model for which was provided by Euclidean geometry, coincides on one decisive point with the conception of painting with which his *Trattato* opens: physics (painting) must begin from a collection of primary principles and propositions that will provide the basis for subsequent developments.[111] The tendency to mathematize physics (following the example set by Archimedes), and which goes hand in hand with an attempt to make it dynamic, is also to be found in geometry and painting itself: "movement" and "rest" are among the functions of the eye (which are also the functions of painting), just as are light and shade, substance, color, shape, place, proximity, and distance. The confusion of genres, the constant interchange of roles, and the permanent intercommunication between series probably constitute one of the determining features of the "Leonardo miracle": Leonardo's persistent determination to relate a theoretically constructed scientific object to a perceived object presented in pretheoretical experience, a determination that is at the root of all natural philosophy, in his own case paradoxically operates contrary to the thinking of his day and makes it possible to free himself from the magical and poetic conception of Nature associated with the Renaissance. But the reason why, in Leonardo, the man of science cannot be separated from the artist, particularly the painter, is that in both cases something is involved that can only be seized upon where the two series join: here, where ideology and science vie with each other, a single question of an essentially theoretical nature comes to the fore in more or less rigorous and explicit terms.

The Taboo of the Infinite

> This thinking carries with it I know not what secret horror;
> one finds oneself wandering in a great immensity devoid of any
> limit or center, and therefore of any determinate place.
> — Kepler, *Opera omnia*

A STRUCTURE OF CONTRADICTION

The intuitions that shoot through Leonardo's text, the surprising perceptions and sudden premonitions that punctuate it, the desire to reform geometry, physics, and painting all at once and, possibly, each by the other, the very fragmentation of the text and its enigmatic, cryptic aspect may all stem from the irreverence that Leonardo manifested with regard to the separation of the genres and his low opinion of the social division of work. Alberti borrowed from geometry only that which was useful to a painter; Leonardo strove to define, on the basis of geometry, the conditions for a scientific practice of painting, expecting this, in its turn, to inform geometry itself. Where Alberti conceived of geometric figures in a static fashion, in terms of intercrossing threads, woven fabrics, stretched skins, Leonardo saw them as products of the movement of points and lines. But who precisely was speaking? The painter, the geometrician, or the man of praxis, "a man who constructs not theories but objects and machines and who, more often than not, thinks as such," and whose attitude in mathematics, that is to say geometry, is generally that of an engineer?[112] In truth, the confusion of genres and the intercommunication that it prompts between the mathematical (geometric), physical, pictorial, and other series make it necessary to treat Leonardo's text in all its theoretical depth, paying particular attention to some of the propositions that it contains: not so much in order to replace them in their cultural context or, as Pierre Duhem indefatigably strove to do, to interpret them in the light of Leonardo's predecessors, but rather in order to seize upon distant echoes and resonances in his own text itself. Consider, for example, the few very brief notes touching on the question of the *infinite*. When Leonardo refers to continuous quantity, that is to say geometry (*la quantità continua, cioè la scientia di Geometria*), and declares that it is infinite,[113] he does so with a view very different from that of Aristotle and scholastic philosophy and not so much in order to refute atomism by alleging the infinite divisibility of size,[114] but rather so as to use the proposition at the beginning of his *Treatise on Painting*: "Se la pittura è scientia," if painting is science, a possible object of mathematical demonstration, what can be said of this practice in so far as it relates to continuous quantity, that is to say the infinite?

If every quantity is infinitely divisible (and extendable), that is not to say that the infinite can be *given*: "What is this thing which is not given and which, if it were given, would cease to exist? It is the infinite which, if it could be given, would be limited and finite, for that which cannot be given is that which has no limits."[115] In Alberti's terms, a painter has no need to know about the question of the infinite, since he must only "fake" that which is visible (*quello si vede*), that is to say that which can be circumscribed by lines and reduced to an interplay of surfaces on a plane: the infinite has no extremities (no "surfaces"), no limits, and it cannot (even as Leonardo sees it) be "given" except as a concept. Now, perspective, considered as a *optical* schema, presents a contradiction that is at once theoretical and formal and of which Leonardo was perfectly well aware.[116] A perspective construction, being based, as it is, on the convergence of vanishing lines, makes visible that which cannot be seen: namely the meeting of two parallel lines in a given, visible point ("The eye between two parallel lines will never see them at so great a distance that they meet in a point"),[117] the *sign, marked by a hole* in Brunelleschi's experiment where, in the specular coincidence of the vanishing point and the point of view, the closure of the system is sealed.

It was a closure that the Quattrocento painters may well have tried to mask (by concealing the vanishing point by a screen, a stage wall, or a back cloth), or else, on the contrary, to get round (by opening up, within the perspective cube, a *veduta* [view] giving on to a landscape or a sky of indefinite depth).[118] But whatever they did, they could not prevent the geometrization of space, upon which the perspective order is founded, from the very outset secretly undermining the representational system that depended upon it. Whereas a perspective construction implies that the discontinuous and qualitatively differentiated space of pre-Brunelleschian art is replaced by the continuous, open, homogeneous and abstract, infinitely divisible and extendable space of Euclidean geometry, in practice it is defined, precisely as a *construction*, by its closure. The addition of atmospheric perspective (or even its substitution) will make no difference to this theoretical situation. For painters who resort to such a ploy, it can never be any more than a matter of escaping from the closure of the system by pictorial means that, given that they owe nothing to geometry and consequently do not lend themselves to demonstration, only *appear* to resolve the contradiction that lies at the heart of the paradigmatic code and that constitutes the productive mainspring of the system upon which the historicity of artistic practice depends. This contradiction—which Renaissance art stubbornly strove to overcome, after having itself laid down the terms for it—may be tackled from a conceptual point of view (in terms of the pairs closed/open, discontinuous/continuous, finite/infinite, etc.), as well as from a figurative one (in terms of the opposition between the demands of the "code," which imply reducing the representation to its earthly parameters, and those of a representation that does not forgo referring to heaven and the dimension of transcendence): from either point of view the definition of the paradigmatic code—if not of the system itself, for which it is both a "brake" and a "rudder"—has implications, as a contradictory structure, that reach beyond the strictly pictorial field.

PHYSICS AND METAPHYSICS

The use of Euclidean geometry to construct the setting of the *istoria* involved consequences that can be better addressed at a philosophical or epistemological level. Alberti himself no doubt expected that drawing a paved floor in perspective would *demonstrate* to him the way in which the "transversal" quantities seem gradually to diminish as they become more distant, continuing to do so as it were ad infinitum (*quasi per sino in infinito*).[119] But it is most remarkable that the (metaphorical) infinite here refers both to what is infinitely small (the diminution) and at the same time to what is infinitely large (the distance): the fact is that although Greek and, later, medieval thought had no difficulty accepting a *potential* infinite (obtained by indefinitely continuous division or addition), both adamantly rejected an *actual* infinite, beginning with the idea of an infinite, limitless universe. The Aristotelian (and the medieval) cosmos is a world with a finite structure, hierarchically organized and qualitatively differentiated. It is a universe divided into several regions of being, each governed by different laws, in which the worlds of the Earth and Heaven are placed in opposition. This idea is not without a solid basis in perceptible experience. As Koyré writes, "the concept of the necessarily finite nature of the stellar universe, the visible universe, is perfectly natural: we *see* a celestial vault; we can think of it as very far away, but it is extremely difficult to acknowledge that there is no such thing and that the stars are distributed in space without order, without rhyme or reason, at improbable distances that are all different. That implies a veritable intellectual revolution."[120] This was the revolution — the principles and consequences of which Koyré managed to disentangle — that was to be achieved by seventeenth-century science, when it replaced the qualitatively differentiated and concrete cosmic space of pre-Galilean physics by the homogeneous space of Euclidean geometry, and the idea of a finite, hierarchically ordered and qualitatively differentiated world by that of an open, indefinite universe, a world of *geometry made real*, subjected through and through to uniform laws, and in which physics and astronomy related to a single domain of being.[121]

The first person to assert "this geometrization of space and the infinite expanse of the universe, which is the premise of the scientific revolution of the seventeenth century and of the foundation of classical science,"[122] was not a scientist but a philosopher, and he paid with his life for his audacity. Giordano Bruno was excommunicated and publicly burnt at the stake in Rome, on February 17, 1600, for having — among other things and basing his argument on the omnipotence of God — defended the thesis that the universe is infinite.[123] Now, as Koyré stresses, Bruno was by no means a modern thinker: he was an execrable mathematician who defended a vitalist and magical concept of the world. Nevertheless, in his infinitism he certainly was very much ahead of his day. On this point his contemporaries simply *could not* accept what he said. Even the universe of a man such as Nicholas of Cusa, who professed some remarkably heretical ideas in this domain, was not an infinite but simply an indefinite one, the limits of which could not be determined.[124] As for Palingenius, the author of *Zodiacus vitae* (published in Venice in 1534), it is true that he maintained more forcefully than anyone else before Bruno that the limits of the creative actions of God were impossible to trace ("What are the Earth and the Sea compared to the immense and admirable space of the world?"). But all the same, he adamantly insisted upon the opposition between the terrestrial and the celestial regions:

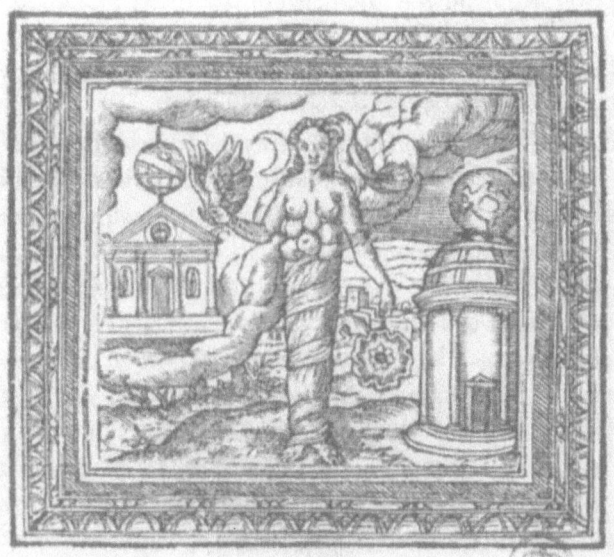

FIGURE 5. *Physicae ac metaphysicae differentia.* From Johannes Sambucus, *Emblemata et aliquot Nummi antiqui operis* (Antwerp, 1564).

The Earth is the last of the inhabited regions, and even if it is too good for Men and Beasts, . . . these are the dark Regions, because below the Clouds perpetual night reigns compared to the brilliant Light and eternal Splendor above. . . . But the air above the Clouds is a happy and serene Heaven. It is there that eternal Peace reigns: there lies the Royal dwelling of the Gods that are invisible to our corporeal eyes.[125]

The universe of Palingenius broke with the Aristotelian cosmos in one respect only: the infinity of the realm that, beyond the sky, sparkles with a wonderful brilliance. For the rest, he went along with the idea of a finite *material* world, surrounded by the nine celestial spheres.[126] It was a concept of the universe that, apart from that one correction, was in accordance with the pre-Copernican image of the world still being proposed in Peter Appianus's *Cosmography* (1539). In this, the earth occupies the center of a series of concentric circles from which it is separated by a double ring of fire and clouds that resemble the collars of undulating clouds to be found in medieval miniatures and also, in a completely different cultural context, surrounding the circular roof that represents the sky in the

Ming Tang, the "house of the calendar," of the ancient Chinese.[127] But the cloud that separates the realm of darkness from the realm of light is also to be found in a slightly later emblem found in the *Emblemata* by Johannes Sambucus (Antwerp, 1564).

Here, positioned diagonally, it marks the frontier between the physical and the metaphysical domains. Nature, in the guise of Diana of Ephesus, stands at the join of the two worlds, pointing with her winged right hand at a temple topped by an astrolabe, close to which is a crescent moon, occupying the upper left of the composition. With her left hand, which holds a heraldic rose (a symbol of organic matter), she indicates a small circular building topped by a globe of the earth, which occupies the lower left-hand corner of the picture. *Physicae ac metaphysicae differentia*: at the join between the two worlds, the band of cloud plays its twofold figurative *and* symbolic role; but while the physical domain is assimilated to the sublunary world, the field that stretches away beyond the clouds is here reserved, without further distinctions, for the metaphysical.[128]

Sambucus's emblem assigns no place to astronomy, either ancient or modern. It recognizes only the opposition between the physical and the metaphysical, the things of the earth and those of the heavens. And the concept of the infinity of the universe was, in fact, defended by philosophers, metaphysicians, before it ever was by astronomers. The men of science who worked most actively to do away with the opposition between physics and astronomy, between the laws of the earth and those of the heavens were, for reasons no doubt more metaphysical than scientific, not in so great a hurry to rally to the idea. Galileo himself, despite the fact that the notion was implicit in the geometrization of space for which he was one of the principal agents, never raised the question of the infinite.[129] As for Kepler, in whose opinion the idea carried with it "I know not what secret horror," he sought for the remedy in astronomy itself. The science of astronomy works only on data that are observable. It can accept nothing that runs counter to optics. And the optical world is a finite world: "Everything that is seen is seen by its extremities. . . . Indeed, to be infinite and to be limited is incompatible . . . consequently, nothing that is visible is separated from us by an infinite distance."[130] Kepler's astronomy does not break with perceptible experience: on the basis of the *appearance* of the sky, it rejects the idea of a uniform distribution of the stars, which is the corollary to the hypothesis of infinity, and concludes that "on the inside, toward the sun and the planets, the world is finite and, so to say, excavated. *What remains belongs to metaphysics.*"[131] Even the sky's appearance of a vaulted ceiling corresponds to the structure of *our* world (the world as an astronomer sees it from the point where he is placed): "There is therefore an immense cavity in the midst of the region of the fixed stars, a visible conglomeration of fixed stars around it, in which enclosure we are."[132] "Astronomy teaches only this: as far as the stars, even the least ones, are seen, space is finite."[133] And there is no sense in claiming, even metaphysically speaking, that beyond the visible world space may extend ad infinitum. Space as such is nothing. The void is nothingness, space only exists because bodies do: where there are no bodies, there is no space.[134]

COSMIC CUPOLAS AND CELESTIAL CUPOLAS

However, the world of the new astronomy was no longer that of medieval science. But that was more on account of dimensions than structure. For even if it is considerably larger,[135] the world of Copernicus is still a closed, finite one. And furthermore, it is still an ordered world, in which the earth and the heavens occupy the places that are theirs by right, in accordance with their relative importance and perfection.[136] The same can be said of the space that perspective makes it possible to construct: a painter deals only with visible things, things that—as Kepler, again, puts it—can be seen thanks to their extremities, and each of which must find the place that belongs to it on account of its importance and its role in the *istoria*; and even if, in the new depth instituted by trompe l'oeil, quantities regularly diminish *quasi per sino in infinito*, the vanishing lines still converge at a visible point, the point that—once again—seals the closure of the system.

Whatever the resistance that it provoked even in men of science—as the quarrel over the infinite, among other things, shows—the fact remains that the Copernican break marked the end of the period that covers both the Middle Ages and classical antiquity. Along with the earth that he inhabits and that Copernicus (following Aristarchus) tore free from its foundations and hurled into the heavens,[137] man was dispossessed of the central position assigned to him by traditional cosmology. But the *De revolutionibus orbium coelestium* was published in 1543. And when one notices that the makers of marquetry of the late Quattrocento were incorporating in their decorative compositions the instruments of the quadrivium, arithmetic, geometry, astronomy, and music, one immediately wonders about the conceptual influence that dictated their choice and arrangement.[138] Such decorative pieces affirm the unity of the arts and the sciences. But which arts and which sciences? And what arithmetic, what geometry, what music, and — above all — what astronomy? A full half-century before Copernicus, it can hardly be claimed that these iconographic designs implied the idea of a possible geometrization of space, let alone that the traditional cosmos was being challenged. On the contrary, the association of music and astronomy — itself traditional — indicates the persistent predominance of considerations of order and harmony in science itself (the "music of the spheres").[139] In the last analysis, the innovation that produced the most important consequences was the application of the procedures of descriptive geometry to the representation of the instruments of the quadrivium. The use of perspective favored painting's claims to be considered one of the liberal arts. But there were also epistemological consequences: by establishing geometry as the basis of representation, painters, in their own particular ways and with their own particular means, encouraged the geometrization of space in which Koyré rightly detected the first stirrings of the scientific revolution of the seventeenth century.

It is not hard to see the possible importance that the question of the status of perspective in its relation to representation might hold in the history of Thought (*Thought*, which, as Hegel remarked, is the basis for all thought*s*): Can perspective be reduced to a practical technique that makes it possible to draw a checkerboard floor or a *mazzocchio* (ducal cap)? Or does it, on the contrary, support a *theory*, as seems to be suggested by Brunelleschi's experiments and even the critique to which Alberti subjected the technical diminishing of transverse quantities?[140] If perspective can be reduced to a collection of formulas, painting's claims to be regarded as one of the liberal arts does not seem justified. But if, on the contrary, as Leonardo suggested, painting is based on theory, in other words it can be demonstrated and so counts as a science, why should painting not have figured, as philosophy did, in the great argument that was brought to a provisional and authoritarian conclusion with the two condemnations, thirty-three years apart, the first passed on Giordano Bruno, the second on Galileo? Within its own domain, it may even have posed a question that science refused to address. If the history of the sciences is not entirely logical, as Koyré has shown, and if scientific revolutions themselves have a *history*, the same is obviously true of art and its revolutions (even if art and science operate in different *times*). The problems raised by the geometrization of pictorial space (and the kind of contradiction that it involves) and the implications of perspective construction are such that the evolution of the system was bound to be chancy and affected by many deviations and blockages. There were too many interests at stake, of many kinds—not only aesthetic, but also philosophical, scientific, and theoretical—for the system to be able immediately to exploit all its consequences and possibilities. But that did not prevent art, considered as a semiotic practice, and insofar as its revolutions were primarily *theoretical* ones, from anticipating scientific symbolization and philosophical reflection on certain points.

In the artistic domain, Brunelleschi certainly seems to have been the man who made the *break*, not only through the experiments and demonstrations of perspective mentioned above, but also through the architectural prowess that was to establish his authority: the construction of the cupola of Santa Maria del Fiore. For the first time employing mathematical calculations to resolve a technical problem, Brunelleschi replaced medieval procedures of trial and error by a rational method founded upon the geometrization of space and the assimilation of the architectural space to Euclidian space. The technical problem was how to cover a space so vast that the use of scaffolding was ruled out.[141] But more needs to be said: in tackling the problem of the cupola, he brought his theoretical skills and his efforts at geometrization to bear upon one of the most sensitively symbolic points of a Christian church. The assimilation of the church's cupola to the cosmic sky (and vice versa) seems to have become one of the most fundamental features of both Christian art and Christian cosmology, as early as late antiquity.[142] Leaving aside the Byzantine and medieval cupolas, it is worth noting that it was in the sixteenth century that this kind of construction achieved its finest form as a centrally planned edifice crowned by a cupola that Palladio required to be built to resemble "that great temple produced in all its perfection by a single word prompted by the immense goodness of God."[143] The inscription that runs around the cupola of the little church of San Eligio degli Orefici, built in Rome later than 1509 by Peruzzi from drawings by Raphael, is perfectly explicit: *Astra deus nos templa damus tu sidera pende* ("God made the sky, we the temple; it is up to you to deploy the stars").[144] The cupola of the church of San Eligio was given no painted decor, but the Chigi Chapel in Santa Maria del Popolo (after

1515) was. Raphael designed a mosaic for it in which the Creator, from the lunette, directs the movements of the planetary gods, who are depicted in a series of medallions arranged in a ring. Far from anticipating the developments of the new astronomy, these designs conform to the most traditional of descriptions of the cosmos, just as Botticelli had portrayed it in his illustrations for the *Divina commedia*.[145]

An edifice crowned by a cosmic cupola—rather than a "celestial one," as we shall see—which is a structure closed upon itself, presents a suitable representation of mankind's region, "an immense cavity in the midst of the region of the fixed stars, a visible conglomeration of fixed stars around it, in which enclosure we are" (Kepler). In the art of late antiquity and the earliest Christian art, and also in earlier oriental art, the heavens—in the divine, theological sense of the term—were believed to extend beyond the vault that corresponded to the firmament through whose chinks glimpses of the celestial expanses could be caught.[146] But a Renaissance vault closes the cosmic space. As late as the sixteenth century, in the Arena Chapel of Padua, in the upper part of *The Last Judgment*, two angels were depicted completing the unrolling of the firmament (or beginning to roll it up), at the gates of heaven. The intense blue background that dominates this painting planned by Giotto seems to be peeled away *like a skin* (skin the reverse side of which is red) from the divine space to which it was fixed. Cosmic cupolas display nothing of the kind; in them there is no sign of any of those indiscreet hands emerging from the clouds, which seem to pierce through the Byzantine or medieval walls. Whatever would happen if someone stuck a hand right through the sky? was the question asked in one old argument against the possibility of the world being infinite (an argument probably taken over from Lucretius, and one that Giordano Bruno was also to appropriate). From the point of view of the Aristotelian tradition, the question makes no sense at all: beyond the surface of the heavens, there is no place and therefore no possibility of the existence of any thing or any body of any kind.[147]

It was this all too perfect arrangement, this assured closure, that, twenty years before the Copernican revolution, was upset by the celestial— or at least aerial, if not atmospheric—cupolas of Correggio (and we should also bear in mind the little cupola with its perspective *dal sotto in su* of the Spouses' Chamber where, as early as 1471–74, Mantegna had broken away from the traditional cosmic and astrological schemata. It was certainly not by chance that the first strictly speaking vertical piercing of the perspective cube took place in a secular decor rather than inside a Christian building). Riegl perceived the qualitative leap that was implied by the switch from a cupola constructed in separate quarters (Brunelleschi, Bramante) to a cupola conceived as a single unit (Raphael, Michelangelo):[148] it was upon just such surfaces that Correggio was to set up his "machines," most of whose prestige seems to stem from the force of the continuum, the suppression of all compartmentalization, and the removal of all material and figurative delimitation. Perspective with a single vanishing point had made possible not only the unification of the representational field, but also its considerable extension. As early as the turn of the sixteenth century, Perugino and Raphael began to open up their compositions onto indefinitely expansive skies.[149] But representation still remained subjected to a *horizon*. The contradiction that lies at the heart of linear perspective *vanishes*, is lost in a construction of horizontal perspective in which the linear horizon is confused with the line of the horizon.[150] However, that constraint, that constitutive limitation of the system, is removed with perspective *dal sotto in su*, which opens on to a *horizonless* depth, a space that—to borrow Philippe Sollers's fine expression—is *beyond lines*.[151] Giordano Bruno was to write as follows:

Thus not in vain the power of the intellect which ever seeketh, yea and achieveth the addition of space to space, mass to mass, unity to unity, number to number, by the science that dischargeth us from the fetters of a most narrow kingdom and promoteth us to the freedom of a truly august realm, which freeth us from an imagined poverty and straineth to the possession of the myriad riches of so vast a space, of so worthy a field of so many cultivated worlds. This science does not permit that the arch of the horizon that our deluded vision imagineth over the Earth and that by our phantasy is feigned in the spacious ether, shall imprison our spirit under the custody of a Pluto or at the mercy of a Jove.[152]

It remains to be seen whether Correggio's "machines" really did borrow something from that science, that power, or whether, on the contrary, they in truth constituted a kind of lure, designed to deceive.

There are two distinct aspects to this question, one of which relates to movement, the other to the function that falls to cloud in representation. To say that "in Correggio, everything is in movement" involves no theoretical characterization of the dynamics in Correggio's painting. The movement to which the figures of the cupolas of Parma (particularly that of the cathedral) are subjected is not neutral, not indifferent, and cannot be reduced to the coordinates of Euclidean geometry. The movement that carries the Virgin and the angels *upward*, toward their place of origin or final seat, is a movement that has something *natural* about it, in the Aristotelian sense of the word. For Aristotle, movement was natural if it tended to restore something to its proper, *natural* place, from which it had been removed as a result of some disorder or imbalance. But the movement by which a momentarily upset equilibrium is restored is by definition a finite movement, and one can see how Aristotle was able to exploit natural movement *upward* to prove that the world is finite.[153] Now, in a construction of horizontal perspective, the notions of down and up, left and right, and so on arise naturally;[154] but perspective *dal sotto in su* is itself governed, if not by a center, at least by one particular direction: the ascensional dynamism of Correggio's cupolas can be seen in the Virgin's rise toward the place destined for her, while the Apostles never leave the ground, which is their assigned place. And even if clouds take part in that movement, assisting it by providing a solid support for the figures that they convey or that pass through them, they too are where they belong, in their *place*, positioned in between the earth and the heavens.

But it is precisely that intermediate position of the clouds and their strongly contrasted relief that raise problems. Correggio is believed to have been the first in Italian art to conceive a unified decor for a vast cupola, a decor that no longer owed anything to either fake or real architectonic structures. We should not underestimate the novelty of an undertaking that challenged, at its most sensitive point, the closure of an edifice that was assimilated to the cosmos and known, metonymically, as the "dome" (*duomo*). But whereas Correggio's predecessors and contemporaries, for instance Bramante, deliberately aimed for the effect of an opening, by piercing their cupolas by oculi and lighting them through a lantern, the Parma cupolas are more or less blind, and the lighting of them posed considerable problems.[155] In truth, these cupolas that are claimed to be "celestial," far from opening onto limitless horizons—as Riegl thought they did[156]—on the contrary clearly fulfill a *covering* function. The solid-looking clouds with their heavily shadowed undersides, with which the cupolas are filled, do not operate as an intermediate term between the earth and the sky: rather, they mark the boundary of the Realm of Darkness, obscuring most of the splendor and light that reign beyond them. Even the way that they are arranged in concentric circles repeats a structure that is traditional. And, above all, in a kind of preventive reaction, these clouds *hide* the sky: not the astrological or cosmic sky, but the visible sky, the observable sky, the sky of the astronomers, in which modern physics originated and was to find its perfection and fulfillment.[157] It is as if the painter had been trying to anticipate an imminent revolution in thought by figuratively reaffirming the opposition of the two worlds, the one below the clouds and the one beyond them, in order to reduce it to an opposition between physics and metaphysics. In this *difference*, there was no place for astronomy, short of sweeping away the rampart of clouds that art, which saw itself as serving faith, was still trying to keep in place in between the earth and the heavens.

GALILEO'S CLOUDS

Cloud, banks of cloud hide something; but what? Movement, for example, that of celestial bodies the relative positions of which are changing —movement that traditional cosmology, attached to the theory of "fixed bodies," could not accommodate. The account in the *Sidereus nuncius* (1610) that Galileo gives of the discovery of four wandering stars, which he proves to be satellites of Jupiter, is most instructive in this respect.

Faced with this phenomenon, and unable to conceive in any way how stars could move in relation to one another, I began to hesitate, and asked myself how Jupiter came to find itself to the east of these fixed stars, whereas the previous evening it had still been to the west of two of them. Is it, by any chance, because the movement of Jupiter is not direct, contrary to astronomical calculations, and might it not, by its movement, have passed beyond those stars? I therefore waited impatiently for night to fall again, but my hopes were dashed: *on every side, the sky was covered by clouds.*[158]

We should not be misled by the tone, that of a narrator of *The Thousand and One Nights*, adopted by Galileo here: cloud did play a part in the new physics, and it was a part that was not solely figurative or allegorical. What Aristotelians asked was how the earth could possibly move and the clouds remain immobile? (According to Aristotle, clouds belonged to the first region surrounding the earth, the region that was common to water and air and was the seat of phenomena—exhalation and condensation—that related to the formation of water in the upper region, which produced clouds among other effects.)[159] Galileo's retort to this was that the argument was no more valid than for birds in the air, trees, or mountains. All these elements belong to the earth and are displaced along with it, by the same movement.[160] They are a part of the earth (as had already been stressed by Leonardo who, incidentally, believed that the earth is a star);[161] in other words, those things by no means constitute a limit, a threshold between one world and another, a frontier between two domains of being. With Galileo, cloud is definitely stripped of its metaphysical and allegorical functions. The astronomer was not one of those who thought that the truth only ever presented itself in disguise. That is why he did not consider himself to be bound by the Holy Scripture, the words of which are determined by constraints as rigorous as the effects of nature, and which, in their concern to cater for the capacities of uncouth and uncultured peoples, did not hesitate to veil their most essential dogmas.[162] The *Discourse on Comets* ends with the declaration that understanding the truth dissipates the clouds in which the mind is stuck (*che puote disnebbiar vostro intelleto*).[163] And what is true of demonstrative rationality is equally true of experimental proof: to discover "accidents" in the incorruptible heavens, spots on the sun, mountains on the moon—all of which are impious things—all you need is a telescope . . . and a sky free of clouds.[164]

The paradoxical fortunes of Correggio's cupolas indicate that they posed a problem that the thinking of the time was not ready to face, a problem that was, indeed, the subject of a veritable taboo. Far from heralding a profound formal mutation, the clouds of Parma, organized as they were to designate a space, helped to safeguard a system that no doubt had less to fear from attacks against linearity launched in the name of "style" than from the revelation of the prohibitions that safeguarded its equilibrium and "normal" functioning. The fact remains though that by concentrating most of his skills on the covering of the building and by imposing upon the decor as a whole a perspective *dal sotto in su*, Correggio drew attention to the point in the system that was both the most strategic and the most exposed. It is not hard to imagine the resistance provoked by this "frog soup" (*guazzetto di rane*) set in the sky of a Christian edifice. Burckhardt made no bones about it: how could divine Glory tolerate that the parts of the human body revealed by the upward perspective be pushed to the fore? The situation was such that it is said that only the peremptory opinion of Titian, who was passing through Parma with the court of Charles V, saved Correggio's cupola from destruction: "Turn it upside down and fill it with gold," he said, "and it will still be worth more than it cost":[165] a rather blunt way of putting it, perhaps, but it probably expressed the secret fantasies of the canons, who would have liked either to knock the cupola down or else to line it with gold, in the ancient manner, so as to thwart the desires that it aroused, first and foremost the longing to take a closer look, once the clouds had dispersed.

But painters and specialists were not deceived. We have noted the reaction of Annibale Carracci, and also the verdict of Titian. In 1587, Giovanni-Battista Armerini was to write that there was no way to decorate *tribunes* other than that chosen by Correggio.[166] But it was nevertheless some time before that solution was adopted. Now, there is at least one particularly striking indication that suggests that the history of ceiling painting in Rome and even the historical blockage that delayed its development were partly linked to the great ideological debate on the new astronomy and, in the last analysis, the question of the infinite. It was the Florentine Ludovico Cardi, known as Cigoli (1559–1613), a friend of Galileo, who corresponded with the latter from Rome and kept him informed of what was going on there,[167] and who presented Rome with its first celestial cupola, that of the Pauline Chapel in Santa Maria Maggiore. It took the form of an *Assumption* that, from a formal point of view, was certainly not very revolutionary as it shows the Virgin, seen horizontally, against a background of clouds, with her feet resting on a moon, in a perfectly conventional manner. But the moon in question is no longer the traditional, smooth, shining crescent: this is a moon with all its callouses, asperities, irregularities, and unevennesses, the moon whose true appearance had been revealed two years earlier in the *Sidereus nuncius*. The presence, beneath the Virgin's feet, in the papal chapel, of Galileo's moon, a moon subject, like the earth, to both generation and corruption, had the force of a manifesto, if not a provocation. Cigoli was perfectly well aware of this. He had himself engaged in important observations of sun spots, and in the course of warning Galileo of the intrigues directed against him, had compared the reactions provoked by his discoveries to the critics who accused Michelangelo of ruining architecture by departing from the rules of Vitruvius. But Cigoli died too soon to witness the triumph of great ceiling painting in Rome. In 1625, Lanfranco, at the request of the Theatine monks of San Andrea della Valle, reworked the schema of Correggio's *Assumption*, painted a whole century earlier, introducing one important correction: by giving the cupola a lantern that admitted a flood of light, this painter, also a native of Parma, opened up a gap that seemed to draw both the figures and the clouds out, rather like a drain in a basin. And in the very years in which Galileo was condemned, Urban VIII commissioned from Pietro di Cortona a ceiling for one of the new salons in the Barberini palace, to commemorate his reign. (This was the same Urban VIII who, while still a cardinal, had constantly supported Galileo, albeit counseling prudence, up until the day when, now pope, and overtaken by events, he had him brought to trial.) The ceiling in the Barberini palace shows the Virgin, ensconced on a pyramid of clouds, facing a group of angels carrying emblems of the papacy. But the most interesting thing about it is that the composition is organized according to a trompe l'oeil architectonic compartmentalization onto which the clouds overflow and the lines of which are governed by celestial perspective.

'PERSPECTIVAE PICTORUM ATQUE ARCHITECTORUM'

The Church had tolerated the new theories as long as they interested only mathematicians (*mathematica mathematicis scribuntur*: mathematics is written for mathematicians). But it found itself forced to censure them once the full extent of their metaphysical, or even theological, implications became apparent. Galileo was condemned for having moved on from a "hypothetical" position and for having claimed to base the mobility of the earth and the stability of the sun upon arguments that were demonstrative and *necessary*.[168] In this debate, art may have played an anecdotal role (the moon of Galileo in Cigoli's painting); but in the field of plastic arts as in that of science, the implications of the system were not clearly perceived.[169] Indeed, how could it have been otherwise if it was true, as Bruno claimed, that "he who demandeth to obtain this knowledge [of the infinite] through the senses is like unto one who would desire to see with his eyes both substance and essence."[170] *The infinite cannot be an object of the senses*; but the senses can serve to "stimulate reason." The destruction of the finite cosmos implied a break from the perceptive order that makes a celestial vault *visible*. But even if painting cannot *show* the infinite, it can suggest it, foster a presentiment of it, create a desire for it. And in the skies of his cupolas that is, objectively, exactly what Correggio tried preventively to remedy, by having a great gathering of clouds conceal the opening onto an indefinite perspective. As the great pictures of Venice show, sixteenth- and seventeenth-century painting accommodated the most distant perspectives perfectly well, just so long as they remained *horizontal.* That was not to be the case with perspectives *dal sotto in su,* liberated from the constraints of a horizon, where clouds would make it possible to drown outlines by substituting for a linear construction a "pictorial" kind of designation of space. But that was just a temporary, compromise solution, as was that of the *quadro riportato,* the fixing of an easel painting to the ceiling. In 1613, Guido Reni painted on the ceiling of the Casino Rospigliosi, in Rome, an *Aurora* with a horizontal perspective. Ten years later, Guercini completed a painting, on the ceiling of the Casino Ludovisi, of another *Aurora,* in which the foreshortened chariot was positioned in an opening giving on to a trompe l'oeil architectural composition seen in a *dal sotto in su* perspective: a linear perspective, but one in which the vanishing lines were no longer governed by a horizon and in which the contradiction that was the basis (and the mainspring) of the system seemed to dissolve into the colored depths.

The history of the seventeenth-century ceiling painting in Rome provides a good illustration for this debate.[171] The architectonic solution, according to Burckhardt the only one that arouses feelings in conformity with the dignity of the subject and that does not bring about a *diminishing* of space (see the beginning of Chapter 1), was only to become established in Christian churches at the end of the century, when Father Andrea Pozzo painted *The Triumph of the Jesuits* on the vaulted ceiling of Sant' Ignazio (1691–94). In the Gesù, *The Triumph of the Name of Jesus* (1672–83) by Baciccia (Giovanni Battista Gaulli) still features a strictly nebulous designation, with one innovation, which is that stucco clouds spill over from the frame of the fresco and a movement from the top downward, corresponding to the qualitative direction of a *fall,* counterbalances the upward movement of the elect: it is as if faith could cope with an opening that was limited, controlled, and above all *qualitatively* defined, with cloud functioning both as an operator of ascent and as an agent of descent.[172] But in his *Triumph of the Order of Saint Francis* (Rome, Santi Apostoli, 1707), the same Baciccia renounced perspective *dal sotto in su* in favor of a horizontal axis.[173] And it was only with Pozzo's fresco for Sant' Ignazio that the Church seems finally to have come to terms with the geometrization of pictorial space and the adoption of a horizonless perspective, which confers upon the opening up of space a violence that no construction could ever encompass. Here, cloud no longer serves as anything but a symbolic or pictorial accessory: the effect of spatial depth depends entirely upon the trompe l'oeil architectural perspective. As Pozzo himself stressed, "the structure which gathers within it so many varied figures is a false architecture in perspective, which serves as the field of the entire work." [174]

This same Andrea Pozzo was the author of a treatise with a telling title: the *Perspectivae pictorum atque architectorum*, which was to exert a considerable influence on artists of the eighteenth century. It reveals the extent to which the taste for movement-filled scenes and celestial perspectives was linked with a science of linear construction taken to the limit. The vault of Sant' Ignazio opens onto the infinity of the heavens thanks to a highly elaborate perspective arrangement. The painter has superimposed upon the walls of the real building a fake portico with pillars to which garlands of figures and clouds are attached. But these, far from being designed to "muddle" or unsettle the construction, on the contrary serve to strengthen it, both figuratively and allegorically. Even the iconographic programme that underpins *The Glory of Saint Ignatius* is extremely revealing as to the intentions of Father Pozzo. "Ignem veni mittere in terram, et quid volo nisi ut accendatur? [I am come to send fire on the earth; and what will I if it be already kindled?]": This quotation from the Bible (Luke 12:49) summarizes his argument. In Saint Ignatius's vision, Christ appears in the sky, carrying his cross; a ray of light passes from him into the heart of the saint, while another ray, streaming from his side, strikes a coin imprinted with a monogram of Christ (which is also that of the Company of Jesus), held by an angel; four rays that shine out from this coin strike four women, who symbolize the four parts of the world and are surrounded by various attributes, all of which are enumerated by Ripa. [175] This luminous triangle plays a role analogous to the triangulation of glances in Giotto's compositions: it confers upon space that is empty and undefined a truly representational coherence, meaning, and volume.

Pozzo's fresco corresponds to a decisive turning point in matters of faith. Pascal was soon to remark that nobody but an atheist freethinker had any reason to be bothered by the silence of infinite space. Far from continuing to censure the contributions of the new sciences, faith would henceforward seek to respond to them and to plumb the infinite, making it serve its own ends. The moment when the infinite began to be *written* into mathematics was also the moment when it began to be spoken about in the terms of faith. [176]

Let man therefore contemplate the whole of Nature in her complete and lofty majesty; let him avert his eyes from the common objects that surround him. Let him look upon that dazzling light, hung aloft like an eternal lamp to give light to the universe! Let the earth appear to him as a mere dot, compared with the vast orbit which our planet describes, and let him stand amazed as he considers that that mighty circuit is itself but a tiny point when compared with that traced by some of the other stars as they revolve in the firmament.

But if our vision halts there, let the imagination pass beyond. It will be more likely to weary of forming ideas than will Nature of supplying the material for them. The whole of our visible world is no more than an imperceptible speck on the ample bosom of Nature. No idea that we can form will come anywhere near it. In vain we seek to extend our thought beyond imaginable space: in comparison with the reality, the human mind gives birth to mere atoms. Nature is an infinite sphere, whereof the center is everywhere, the circumference nowhere.[177] The greatest perceptible characteristic of the omnipotence of God is that our imagination becomes lost in that thought.[178]

What is man in infinity? And what better preparation for faith than that could there possibly be? (Pascal, definitely breaking with tradition, then goes on to remark that if the infinitely small is less visible, the infinitely great is certainly more perceptible. But if that is so, how can anyone agree that painting is nothing but vanity, when it has the power to make man sensible of his own nothingness, his dependence, his *void*?)

Our Sheet's White Care

Une ivresse belle m'engage
Sans craindre même son tangage
De porter debout ce salut

Solitude, récif, étoile
A n'importe ce qui valut
Le blanc souci de notre toile

A lovely drunkenness enlists
Me to raise, though the vessel lists,
This toast on high and without fear

Solitude, rocky shoal, bright star
To whatsoever may be worth
Our sheet's white care in setting forth
—Stéphane Mallarmé, "Salut"
    (trans. Henry Weinfield)

In the Service of Clouds

The present work has so far concentrated on the relation between signs and representation in the pictorial context inherited from the Renaissance. The word *sign* has been used in its twofold (and possibly contradictory) sense of (a) a signifying unit, detectable as such on a figurative two-dimensional surface, and (b) a symptomatic feature through which the representational process, the *theater* of which painting at that time appointed itself to be, revealed itself in the very way that it was organized. We have seen what implicit structure these relations manifested, given that representation was in principle linked to the order of signs, and signs in their turn acquire their representativeness from the representation itself. However historically verifiable it may be, the fact that visual representation, in painting, is prior to the discursive variety is first of all a logical matter. Representation only exists as such if it presents itself as the representation of a representation, by means of a characteristic duplication that operates at the level of signs as well as at that of a "code" that ensures, thanks to its demonstrative nature, that the representational process is governed by certain rules. As for signs, while it seems that what they forfeit in semiotic meaning they gain in representational meaning (since imitation appears to win out over convention, and illusionism over "arbitrariness"), the symbolic shift caused by the introduction of the perspective regime, and the consequent opening up of the field of substitutions, definitively restored them to a discursive role. At the limit, a sign was denied all transitivity. Brunelleschi's experiment reduced /cloud/ to an effect of reflection, a mirror image engineered within the figurative field by dint of a material artifice. But the significance of that reduction is twofold, both rational and symptomatic: it was at once a consequence and evidence of the institution of perspective space as the *theoretical* space of (re)presentation. Far from illusionism (the limitations of which are clearly detectable here) being a principle of the system, it now appears as an effect of it, which is dependent upon the representational structure. This means that, even within the system of representation, the use of icons borrows not so much from mechanisms of illusion as from the lateral relations established between an iconic element and any other elements with which it may be associated on the figurative plane; and if that element can be seen as a symptom, that is not so much on account of its referent as because of the position and functions allotted to it in the system, insofar as the latter defines a specific field of production, historically dated and geographically localized.

The same logic that led Brunelleschi on principle to exclude "real" cloud from the domain of what is "depictable" dictated the use of the graph in order to get around the constitutive closure of the system, and to loosen its constraints, if not to mask the formal contradiction that was its mainspring. That was the situation up until the day when, ideology having finally come to terms with the theoretical implications of the geometrization of space, representation became able to accommodate the introduction of clouds, as a pictorial accessory, within the figurative framework regulated by the perspective regime. The same /cloud/ that, for Correggio, had served to "designate" a necessarily closed space would be used by the Venetian artists and landscape painters of the seventeenth and eighteenth centuries to give space the "quality" of indefinite openness. And as for the storm-laden skies that Ruysdael was to paint above the Haarlem plain and the dawn clouds that adorn the mythological landscapes of Claude Lorrain, they would no more undermine the consecrated representational order than would the massed clouds among which the figures of Tiepolo play, or the misty distances and evanescent perspectives of a *fête galante*. By taking in the element that used to constitute the negative connotation of its closed nature, representation was to demonstrate both the extent of the system's possibilities of adaptation and also the ambivalence of its formal potentialities.

If /cloud/ thus marks the closure of the system, it does so in opposition to the formal principle by which signs are governed, through its lack of any strict delimitation, as a "surfaceless body." But, as has repeatedly been pointed out, nothing could be more mistaken than to attempt to justify its pictorial fortunes on the grounds of a so-called thematic and/or plastic *nature*. In as much as it features in the figurative tissue, both because of its appearance and because of what it signifies, /cloud/ has no reality apart from that assigned to it by the representation. Does that mean that it simply has a use value, so can be classified as a *tool*, in an instrumentalist perspective?[1] Thus posed, in the terms and from the point of view that are those of thought about signs, the question cannot be given a simple, unequivocal answer: in a figurative situation regulated by the perspective model, the cloud element fulfills a number of easily identified functions. But the functional nature of a *sign* does not suffice to justify its value as a theoretical index any more than it exhausts the efficacy of the *figure* at the level of a signifier (on the sign/figure opposition, see Chapter 1, pp. 14 ff.). If /cloud/ assumes a strategic function in the pictorial order, it is because it operates alternately (or even simultaneously, if one takes into account the difference between the levels where it might come into play: its integration may skip a level), now as an integrator, now as a disintegrator, now as a sign, now as a nonsign (the emphasis here being placed on the potential negativity of a figure, on whatever in it contradicts the order of the sign, the effect of which is to loosen the hold of the latter). It may operate as an integrator to the extent that in a given situation it takes on a transitive or commutative force, guaranteeing the unity of the representation by the means peculiar to a sign and in conformity with the system's norm. Alternatively, it may operate as a disintegrator, insofar as, ceasing to operate as a sign and affirming itself as a figure (in the sense already explained), it seems to call into question, thanks to its absence of limits and through the solvent effects to which it lends itself, the coherence and consistency of a syntactical ordering that is based on a clear delineation of units.

This ambivalence of the sign—which reflects and duplicates that of the system—illuminates its fortunes: /cloud/ served a wide variety of purposes, ranging from those of a purely signaling and descriptive nature to constructive, or even destructive, ones. At the turn of the eighteenth and nineteenth centuries, and within the framework of a method explicitly defined as productive, it was to serve new ones. *A New Method of Assisting Invention in Drawing Original Compositions of Landscape*, published in 1785 by the Englishman Alexander Cozens,[2] set out to be strictly mechanical in principle, borrowing nothing from imitation of the masters or of nature, since it was founded solely upon the potential *information* to be derived from stains and blotting: "To sketch . . . is to transfer ideas from the mind to the paper . . . to blot is to make varied spots . . . producing accidental forms . . . from which ideas are presented to the mind. . . . To sketch is to delineate ideas; blotting suggests them."[3] Understandably, the repertory of cloud formations has a special place in this collection. Suitably enough, Cozens's method *caused much ink to flow*. Yet, however intolerable it may have seemed to academic minds, on account of its manifestation of an intrusion of chance automatism, or even shapelessness, into the pictorial field, the method involved no real theoretical break with the past. In principle, it was no more scandalous than Leonardo's or Piero di Cosimo's avowed interest in traces of humidity or sputum left on walls. Gombrich certainly realized that Cozens's schemata and the interpretations that he proposed derived, at least in part, from a well-established tradition according to which Claude Lorrain was the unchallenged master of the genre of landscape.[4] In the mark that a sponge leaves on a wall, just as in ever changing cloud formations, people see whatever they wish to see: configurations of their desires, images from their theater of life, signs of their culture.

"THE MARVELOUS CLOUDS"

"Et qu'aimes-tu donc, extraordinaire étranger?"
"J'aime les nuages, les nuages qui passent là-bas,
les merveilleux nuages . . ."

"And what do you like, extraordinary stranger?"
"I like the clouds, the clouds passing over there,
the marvelous clouds . . ."
—Baudelaire, *L'Etranger*

By getting clouds to serve ever more uses, art was risking having the situation turn to the advantage of the very element that it had so far seen fit to use as a tool, an instrument at the service of representation. And that is indeed what happened in the nineteenth century, according to the formula to which Ruskin reduced the pictorial production of his time: "the service of clouds."[5] It is an astonishing expression that, for the first time, explicitly recognized the precedence of the symbolic order and the nonfunctional character of the signifier that, in truth, is here still presented beneath an emblematic mask: the mask being, precisely, *cloud*, which painters, who had exploited it for so long, now set themselves to serve. Was not "cloudiness" the distinctive characteristic of "modern" landscape painting?[6] At the time when Ruskin set about writing his *Modern Painters* (around 1840), the interest in the sky manifested by English landscape painters such as Prout, Fielding, Harding, and, above all, Turner seemed to be unequivocally justified. According to him, modern painters surpassed their predecessors in the art of landscape painting, breaking with the academic tradition that recognized as its masters Poussin and Claude Lorrain; and to a large extent, the reason for their superiority was the "truth" of their skies, truth in which cloud certainly played a part. But the justifications that he produced in support of his thesis were still borrowed from the order of representation, if not from the register of what was signified: each individual can enjoy and profit from the spectacle or "scenery" of the sky,[7] a spectacle that is there for all to see; and nothing testifies better to the power of God than the *system of the firmament*, a system conceived, it should be emphasized, no longer in astronomical terms but in meteorological ones, and in which cloud was presented to the sight of men in the same way as air is presented to their lungs.[8] The painters of the past were able to express the *quality* of the sky, but did not seize upon its truth, for they did not perceive the calculated connection between the blueness of the atmosphere and the whiteness of the clouds, the regulated setting in which clouds were distributed in three different regions, *three scenic systems*, each of which corresponded to a specific formal datum: a central region, the only one to which earlier painters (in particular the Flemish) had paid attention; an upper region that Turner made his favorite domain, opening "to the world another apocalypse of Heaven";[9] and finally a lower region, that of rain clouds and mists without form or consistency, where modern painters excelled.[10]

But, as Ruskin later acknowledged (in a curious pseudo-scientific digression, which, as we shall see, set the seal on his critical remarks), there were negative as well as positive aspects to the service of the clouds. The very formula connoted the moderns' taste for open expanses, without limits or frontiers, their desire for liberty and a nature free from the rule of man (as was also testified, at another level and in another context, by their attraction to *ruins*, where the built-up constructed order had abdicated, just as compositional order appears to disintegrate amid the clouds). Similarly, their love of mountains was closely linked to their love of cloud, since the sublimity of the former was enhanced by the presence of the latter.[11] In short, the formula "in the service of the clouds" indicated a secular vision of the spectacles of nature ("Whereas the medieval never painted a cloud but with the purpose of placing an angel upon it . . . , we have no belief that the clouds contain more than so many inches of rain or hail").[12] Modern painters were interested in the perceptible aspect of clouds, their objective configurations, the effects of mists, the appearance of things seen through the screen of atmospheric formations. But there was another side to this interest. Whereas the painters of the past had sought for stability, permanence, clarity, the modern spectator was invited to take pleasure in obscurity, the ephemeral, change, and to derive the greatest satisfaction and instruction from that which was the hardest to fix and understand: wind, light, cloud shadows, and so on. As Ruskin was to write in 1853, the service of clouds was a formula that could unfortunately be used to characterize modern art in its most negative aspect. Did not much of the mystery cultivated by his contemporaries proceed from a desire to "speak ingeniously of smoke" that Aristophanes had long since denounced in his play entitled *Clouds*? According to Ruskin, Aristophanes was the only one of all the Greeks to have spoken ill of clouds, and also to have made a careful study of them (but where does that leave Aristotle and Epicurus?). Subservience to whatever was transient, uncertain, and unintelligible certainly also made itself felt at the level of composition (as was proved, according to Ruskin, by the importance acquired in landscape painting by the sky, to which the foreground was now subordinated, for the latter would be made darker in order to emphasize the whiteness of the clouds). But it also affected drawing: medieval painters drew with the greatest care and in great detail, whereas for moderns painting "is all concerning smoke": only this was really *drawn*, so everything remained vague, insubstantial, imperfect.[13]

Under the heading "ancient" Ruskin continued to lump together the painting of the Middle Ages, that of the Renaissance, and also the tradition linked with the academic interpretation of the works of Poussin and Claude Lorrain. At the same time, it was without reservation that he admired the art of Turner, which seemed to him to demonstrate the superiority of the moderns in landscape painting. For he classified *the truth of the heavens* among the most important "truths" that art had to understand, even more important than the truth of color, which he judged to be secondary to the truth of form. And if it is difficult to establish in what "the truth of clouds" consists, that is precisely because they can assume the most diverse forms; but it is also the reason why a study of clouds can be so profitable:

If artists were more in the habit of sketching clouds rapidly, and as accurately as possible *in the outline*, from nature, instead of daubing down what they call "effects" with the brush, they would soon find there is more beauty about their forms than can be arrived at by any random felicity of invention, however brilliant. (my emphasis)[14]

This may be regarded as an implicit criticism of Cozens's technique: for why, after all, seek from chance what nature dispenses, indubitably with a design, with such generosity and in ways that so wonderfully combine both mystery and beauty?[15]

Mystery:[16] this appears to be the specific characteristic of Turner's art, which is differentiated from all other art by its seemingly uncertain execution, to such an extent that this painter often appears to be the primordial representative of "cloudiness" and of the "vagueness" so characteristic of a whole section of the painting of the nineteenth century.[17] "Every one of his compositions [is] evidently dictated by a delight in seeing only part of things rather than the whole, and in casting clouds and mist around them rather than unveiling them."[18] Now, Ruskin must have perceived that, from the point of view of the masterpieces of Italian art and the works of the Pre-Raphaelites, this was something of a paradox: for is not "great drawing" clear and precise, and does it not aim for a clear delineation of forms? In contrast to Turner, whom Ruskin, despite everything, continued to regard as the strongest personality of the age, the Pre-Raphaelites—the greatest men, *as a class*, that modern Europe had produced in the arts— celebrated the veracity of light and were unanimous in condemning fog and all illusions based on "haziness." However, the debate that divided the partisans of cloud (Copley-Fielding and others) from their opponents (Stanfield, Harding), a debate that Ruskin could not ignore, was by no means new. It was, to repeat, in a different context, the debate that in their own day set Correggio and the Venetians, Rubens and Rembrandt, in opposition to painters of the Raphaelesque tradition and, in general, to all those who held indistinctness in horror and saw themselves as the champions of clear vision and "an endless perspicuity of space."[19]

"Truly, the clouds seem to be getting much the worst of it; and, I feel, for the moment, as if nothing could be said for them. However, having been myself long a cloud-worshipper, and passed many hours of life in the pursuit of them from crag to crag, I must consider what can possibly be submitted in their defence, and in Turner's."[20] These lines convey Ruskin's perplexity following his discovery of Italian art and the art of the Pre-Raphaelites. But his defense of Turner proves, even in the contradictions into which it leads him, that the author of *Modern Painters*, at the end of the day, jettisoned none of the ideas that he had nurtured before his travels in Italy. In the first place, clouds exist; whether one likes it or not, nature has decided that it be so, and a landscape painter ignores them at the risk of falling into mannerism. In the second place—and this was a fundamental rule of Ruskin's aesthetics—"there is no excellence without obscurity."[21] "Mystery includes not only the partial and variable kind that clouds and mists serve so well, but also the kind that is continuous, permanent, and that corresponds, *in all spaces*, to the infinity of things." The Pre-Raphaelites, like Turner (who, Ruskin insists, is "their true head"), are themselves full of mystery and suggest more than what can be seen. For the fact is that there is no absolutely clear and distinct perception: what matters is knowing where the mystery begins, "with the point of intelligibility [varying] in distance."[22]

Distance, visibility (and the effects of atmospheric perspective, and the notion of a "point of intelligibility"): there is nothing new about Ruskin's field of thought if you compare it with, say, that of Leonardo. The pictorial problem (not to mention the "aesthetics") with which he countered the academic tradition, far from introducing new developments, on the contrary marked out most precisely, with its paradoxical economy, the point at which theory was exhausted, and at the same time revealed the dead ends into which it led. The problem was really that of linearity; and here too cloud operates as an indicator. In absolute terms, no "cloud" can be drawn with a pointed instrument. Only a brush, used as delicately as possible, is capable of *expressing* a cloud's "edges" and textures in all their variety. Does this mean that Ruskin, at the conclusion of his enquiry, ended up accepting the very "effects" that he had condemned at the start? Not at all, for he is careful to explain that, faced with the task of *engraving* a cloud, an artist whose execution is delicate and careful will be able to obtain a good approximation to a *painting*; and as for drawing, a thick lead pencil is all that is needed, for this lends itself, when used at a slant, to rendering shadows, just as, when used as a point, it can be used for drawing lines.[23] The reason why cloud does not encourage drawing is not so much its shape, but rather its instability and evanescence. It is possible, at a pinch, for an artist to draw *a* cloud before it is dissipated or transformed; but if he sets out to reproduce the ordering of a whole sky filled with clouds, he must be content with a few hasty lines, and later finish the sketch from memory. And it was precisely Turner's prodigious memory that gave him his unequaled mastery in the rendering of skies. "Other great men coloured clouds beautifully; none but he ever drew them truly: this power coming from his constant habit of drawing skies, like everything else, with the pencil point."[24]

As used in Ruskin's text, the term "cloud" has an ambiguous status. (Sometimes it designates the meteorological phenomenon that modern painting established as its favorite referent; at other times it is used as both a distinctive image and an effective emblem, at points where symbols and imaginary representations overlap.) Bearing this in mind, it is not hard to see how cloud serves as a touchstone (to use a paradoxical expression) by which to judge the truth of a painting. For Ruskin, "truth" implies veracity in the restitution of a natural fact. This, unlike imitation, may be achieved by means of signs and symbols that take on a predetermined meaning even when they are not *images* and are not created in order to resemble particular phenomena.[25] This covert play—in which Ruskin is not the only critic to indulge—with the idea of substance and form, things and signs, imitation and representation, is characteristic of a particular critical method and accounts for both its detours and its reversals. This too can be verified, once again very simply, by considering the example of *cloud*. Where clouds in paintings are concerned, we have seen how Ruskin can be led into difficulties through having failed to elaborate an explicit theory of how iconic signs relate to their referents. When it comes to "real" cloud, he pretends that he is incapable of explaining it. Why is fog so heavy, and why are masses of even the most colossal clouds so light? What makes clouds float in the air? What is the explanation for the fact that vapor distributed in the air becomes *visible*? Above all, what can be said of the shape of clouds ("How is a cloud outlined?").[26] When the question is formulated like this, it proceeds from a certain confusion between the *signified* and the *referent* and between sign and substance. According to Ruskin, there are no answers to these questions. Nevertheless they increase the pleasure that is derived from "mystery." Similarly with the movement of clouds:[27] it cannot be explained, except by theories about electricity and the infinite, which nobody understands at all.[28]

The reference to the infinite that is introduced by the subject of skies and cloud is by no means fortuitous: it shows the extent to which Ruskin (if not Turner himself) remained a prisoner of the tradition of thought that stubbornly strove to set in opposition the scientific geometrical and abstract infinite (usually interpreted in a *privative* sense) and a pictorial and expressive infinite that religious and/or metaphysical ideology could accommodate. Is not the "expression of the infinite" the first thing to consider when judging a painting, even in its details? Is that not the proof of Turner's excellence, Turner, whose *blue*, even a tiny portion of it, is always "infinite," always of incommensurable depth and expanse?[29] But one has only to analyze the means by which he achieved this to discover that Turner's art is still governed by the same rules as the art of his predecessors: even clouds have to be put into perspective; and if one pretends that they present a flat basis, it is easy to set in place a grid (or checkerboard) that will show how they should be disposed. The wind may introduce an apparent confusion (as it did in Leonardo's painting), but nonetheless, perspective provides the rule for any sky arrangement, whether it be a rectilinear system or, as is more frequently the case, a curved one (Figure 6b). It is because painters are ignorant of perspective and the rules of proportion that stem from it that they fail in "the expression of buoyancy and space in the sky."[30]

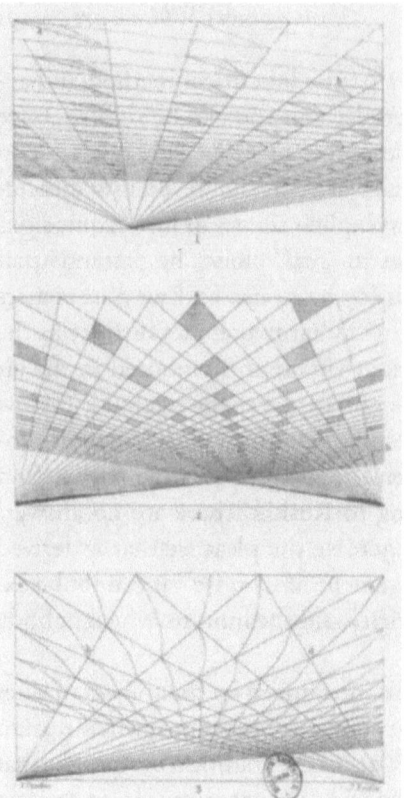

FIGURE 6A. Ruskin, *The Perspective of Clouds* (rectilinear), from *Modern Painters* (London, 1855), vol. 5.

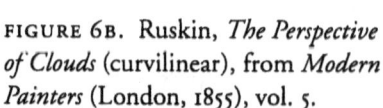

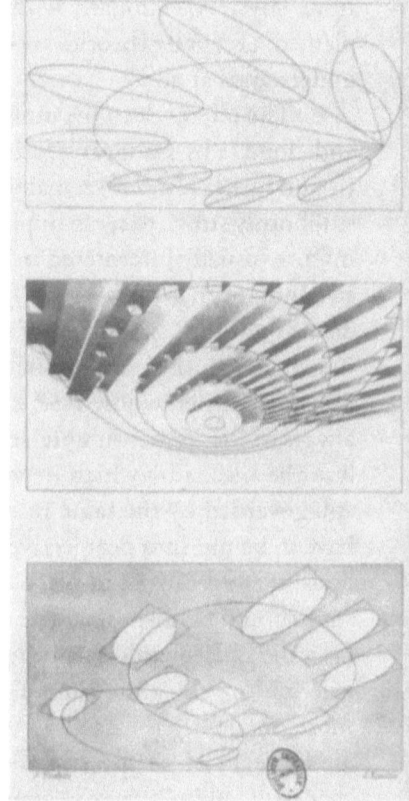

FIGURE 6B. Ruskin, *The Perspective of Clouds* (curvilinear), from *Modern Painters* (London, 1855), vol. 5.

The desire to include in his admiration both the art of Turner and that of the Pre-Raphaelites thus forced the ideologist to show his hand: behind the screen of mists and clouds, the paramount excellence of linearity remains assured, even in the functions allotted to the code that constitutes the best guarantee of its privileged position. "In the service of the clouds": for Ruskin the formula has the force of, not so much a program, rather a statement of fact. Far from introducing new theories, it reiterates the institution of perspective, with a warning note: this "service" must not be carried to such extremes that the system would destroy itself by renouncing its regulatory principle.[31] On the contrary, what it must do is accommodate itself to extending to the sky, until then treated as a back cloth, all the principles of organization that apply for the scene itself (although historically, as we have seen, the sky or "roofing" was converted into a checkerboard before the ground ever was). The schemata of celestial perspectives sketched by Ruskin are still governed by the traditional *point of view*, and the diminishing effect that they impose upon the clouds according to their position on the checkerboard is in keeping with traditional practice. After all, academicians were taught to exploit the shadows cast upon the ground by *clouds that themselves remain outside the framework of the representation*, in order to reinforce the effect of depth that is obtained, in a landscape, by means of perspective of one kind or another.[32] Initially, cloud had only found a place within the system thanks to a forcing of its principles and a slackening—in truth more apparent than real—of the formal constraints that regulated the functioning of the system as such. But such was the flexibility of that system and so extensive were its powers of adaptation that, at the very moment when it was about to be fundamentally threatened, it appealed to that very element, as both a syntactical tool and a factor of illusion, in order to preserve the coherence of representation that was governed ever more strictly by the regime of perspective.

METEOROLOGICA

But we need to return to the reasons that led Ruskin, on the pretext of raising those questions, to confess, somewhat complacently, his claimed inability to assimilate the progress of modern science.[33] Since the beginning of the century, the observation of the sky and atmospheric phenomena had been attracting renewed interest. As early as 1817–22, Goethe, in a series of studies and even poems, was paying homage to the Englishman

Luke Howard, who had written an essay on the scientific classification of clouds,[34] in which he allocated them to three zones or layers, just as Ruskin was to later. At the time, this work was regarded both as a first attempt at explanation, prediction, and *deduction* and also as the source of a number of poems, sketches, and paintings that took cloud as their referent but to which Ruskin, curiously enough, paid no attention. They included Goethe, Constable (see Plate 7a), and the Norwegian Johan Christian Dahl, to say nothing of Shelley's admirable poem.[35] Kurt Badt seems to have overestimated the direct impact of Howard's studies (of which Constable, for his part, appears to have been totally unaware),[36] and posed the problem of the relations between art and science purely in terms of "sources" and "influences." Nevertheless, he correctly perceived the reasons for the objective convergence that is noticeable, at the beginning of the nineteenth century, between the works of meteorologists and certain pictorial and poetic works, which, although they do not strictly speaking constitute a "genre," nevertheless certainly represent, among the general and individual works of the time, a group with an autonomy and specificity of its own.

Both the studies of Constable and the observations that Luke Howard's work prompted Goethe to make emphasize the temporal circumstances affecting phenomena. Constable himself was always careful to note on the back of his studies not only the date, but also the time, the place, the temperature, the direction of the winds, and so on.[37] The synchronic view of the Renaissance was now replaced by a diachronic if not historical view (see Howard's idea of "a history of climate"). It was all a matter of understanding the temporal development of the natural phenomena that occur in space. (Constable, who wanted landscape painting to be *scientific as well as poetic*, congratulated Ruysdael for having managed to produce a history—in the new sense of the term *natural history*—and for having understood that what he was painting and the landscape itself were the product not of the imagination but of a *deduction*.)[38] But the interest in clouds was not solely a response to new scientific or theoretical preoccupations. It also fueled romantic reveries on the part of Baudelaire as well as Goethe, on the infinite, the far distance, the indeterminate, and the powers of that which has no form. Kurt Badt appositely notes that in nineteenth-century painting /cloud/ occupies a place comparable to that held by draperies in fifteenth-century Flemish painting. His remarks prepared the way for the "exercises in shapelessness" that Valéry attributed to Dégas, considering the idea to be rather "da Vincian": by shapeless he did not mean that which had no shape but things whose shapes elicited in us "nothing that made it possible to replace them by an act of drawing or clear recognition."[39] A cloud is not a reducible form; and that was precisely why Goethe, for his part, considered it a special sign of the *Unvergangliches* (the imperishable), the relationship of reciprocal engendering that incorporates all things — including the I that knows — within the infinite cosmos, which was the subject of *Naturphilosophie*: "As I have perhaps paid too much attention to the study of geology, I am now tackling the domain of the atmosphere. *If it were only to discover how one thinks or may think, that in itself would be a considerable profit.*"[40]

Insofar as meteorology as conceived by Luke Howard and his contemporaries was claimed to be a primarily descriptive science, founded upon the observation and classification of phenomena, it is understandable that it overlapped, even in some of its graphic productions, with the work of landscape painting, painting that Goethe wished to see learning from rational knowledge: he even had the idea of producing a new art founded on science (*die Hervorbringung von neuerer Kunst aus Wissenschaft*).[41] But it is the strictly phenomenological, *nontheoretical* nature of this "science" that explains its attraction for a thinker such as Goethe, who was eager to condemn the materialist bases of Newtonian science and, following Schelling, defend the rights of a "speculative physics" (see his *Farbenlehre*). "Man must seize upon everything with his eyes [Alles muss der Mensch mit Augen fassen]."[42] The reason why Goethe, like Ruskin, pays attention only to the external aspect of phenomena is, quite apart from the fact that the rights of "mystery" are thereby preserved, that the "meaning" of those phenomena is more important to him than their *structure*. The science that both men were opposed to was that of the physicists who, following Newton, claimed, for example, to reduce light to a material process, whereas colors, like sounds, are *sensations* and as such are linked to the "animal frame," sensations that cannot be explained and can only be studied by a *moral science*.[43] Meteorology, conceived as a strictly phenomenological, nonmaterialist discipline, thus comes to the aid of a symbolic, if not mystical, interpretation of natural phenomena (as we shall see in Carl Gustav Carus and even Constable). That it does so is really not particularly surprising when one remembers the place given to these same phenomena in the great materialistic texts of antiquity. The fact is that *explanations* for how "clouds may form and gather either because the air is condensed under the pressure of winds, or because atoms which hold together and are suitable to produce this result become mutually entangled, or because currents collect from the earth and the waters,"[44] or for many other reasons—such explanations put a stop to all mythological digressions. Lucretius's admirable pages on the system of the sky (*ratio caeli*) were designed solely to deliver mortals from fear of the gods, which stemmed from their ignorance of causes. Neither the lightning that flashes when clouds produce many fiery atoms,[45] nor the thunder that is produced by thick clouds massed at a great height[46] are created by the gods. "Exclusion of myth is the sole condition necessary; and it will be excluded if one properly attends to the facts and hence draws inferences to interpret what is [invisible]."[47] However, both Goethe and Ruskin, each in his own way, strove to restore, in opposition to materialist science, the rights of myth and ideas, if not those of religion and of the "invisible," conceived in a mystical fashion.[48]

In Lucretius's materialist text, the formation of clouds provides a rough illustration of the process by which the atoms that move about in the void come together. "Clouds mass together when in the space of the sky above a number of flying bodies have come together, which are rougher and though they are entangled with very small catches are yet able to hold together in mutual attachment."[49] In this context, the distinction between the visible and the invisible implied no ontological difference. It was simply a matter of a *threshold*: "When the clouds first take their being, before the eye can see them, all thin, the winds drive and carry them together to the mountain tops. Now at length gathered together in greater mass and packed together, they are able to show themselves."[50] But Ruskin could not bring himself to accept such a notion, attached as he was to "mystery," to such a degree that, even when he himself had suggested the idea that cloud might be composed of spherical molecules held together by attraction,[51] he rejected it immediately, declaring that there was no solution to the problem of its *visibility, color,* and *outline.* What these problems allow us to glimpse, even as they mask it, is the very question of the signifier in painting, a question that Ruskin, for his part, systematically strove to conceal *beneath a mass of clouds.* Far from pondering upon the historical *and* theoretical reasons for the invasion of the pictorial field by cloud, and for painters entering into the service of cloud, he was to be found trying to explain this state of affairs indirectly and rather late in the day by a radical change in the aspect of the weather, a change indicated by the appearance in the European sky of a type of cloud formerly unknown. The two lectures that he gave in 1848 on the "storm cloud" in the nineteenth century marked a break with the traditional equilibrium of the "system of the firmament." In contrast to the text of *Modern Painters*, those two lectures manage to displace the question of the relation between the signifier and the signified in painting, and instead refer it to what is claimed to be a natural reality, itself defined in terms that are nothing if not scientific: in the face of encroaching darkness and increasing nebulousness, what consolation does materialism offer? Once again, the ideological function assigned in this context to *cloud* is unmistakable: we have seen how, far from invariably encouraging a systematic deconstruction of the perspective order, the invasion of the canvas by clouds in many cases on the contrary sanctioned a series of effects that, despite a superficial muddling of the geometric coordinates, strengthened the hold of that order, at the same time continuing, in conformity with the fundamental purpose of the classical system, to mask the *real* substratum of the painting, namely the surface upon which the images were inscribed. The resolutely antimaterialist tenor of Ruskin's text lends emphasis to the contradictory nature of the "service of the clouds": the very /cloud/ that, in the figurative context governed by the perspective model, seemed linked with the sensible component of the painting, its materiality, or even with color as opposed to delineation, turns out to function as a screen designed to mask the reality of the signifying process for which, in the symbolist period, it would even present itself as a fantastical substitute.

THE WAGNERIAN "IDEAL"

In truth, through the contradictions as much as through the pessimism evinced by many pages of *Modern Painters* and the aging self-taught expert's digression on the "storm cloud," and through his admiration for Turner and, conversely, his criticism of the facile nature of "haziness" and the attractions of "smoke," objectively Ruskin played a role in the debate that constituted the culmination of the quarrel that dominated the end of the century. This was the quarrel over Wagnerism. When Nietzsche, after a long "convalescence," decided to break with Wagner, his recovery took the form of a choice: a choice that rejected the dampness of the North and the mists of the Wagnerian ideal and opted instead for the *limpidezza* of the air of hotter countries and *gaya scienza*; that is to say, not only lightness, wit, fire, and grace, but also the *grand logic*[52]—and above all opted for a *different* art, "an ironic, light, subtle, divinely relaxed, divinely artificial art which, like a clear flame, spurts forth *in a cloudless sky!*"[53] This justification of *an art without clouds*, an art for artists (but not an art for art's sake) takes on its full significance when compared with the Wagnerian endeavor, which, in contrast, was committed to a search for effects, *expression* at all costs:[54] this put music at the service of the *stage*, of "total" theater, and of the public common to all the arts. In the sense that it has been said that the perspective regime involved the "neutralization" of the substratum of painting, this project implied as its precondition the neutralization of "the technical seat of the music," the organism that produces the sounds (i.e., the orchestra), which, in Bayreuth, was positioned below the stage, in a "mystical abyss" the function of which was to separate "reality" from the "ideal," and to make the action on stage seem to take place in an indefinite faraway place.[55] Now, the scenic effects sanctioned by this neutralization of the material productive source of the music were not unrelated to the effects aimed at by those who specialized in a certain kind of pictorial "haziness" or "cloudiness." For the fact that the orchestra was thus hidden from sight transformed the floor of the real stage into "a mobile, flexible, malleable, ethereal surface, the fathomless basis of which was the sea of feeling itself."[56] Nothing intervenes to block the view of the stage that the spectator enjoys from his seat, and his gaze encounters nothing but

a space that somehow, thanks to an architectural artifice, floats between the two prosceniums, presenting what seems to be a distant image with the inaccessible air of an apparition in a dream, while mysterious music, like vapours rising from the sacred bosom of Gaia, beneath the Pythia's throne, rises like a spirit from the "mystical abyss."[57]

It is possible to glimpse how the specifically musical features of Wagnerian opera—the "infinite" melody, devoid of "fixed points," in conjunction with the chromatism into which the system of tonality dissolves[58]—could be worked into what Wagner hoped would be a total representation, which incorporated all the various arts that were called upon to combine as the *means* to effect its realization. In a spectacle designed as much for the eye as for the ear,[59] similar means produced similar effects and these reinforced one another: the tonal disarray engendered by a succession of "vague" chords was increased by the collapse of the ground on which the story was unfolding, a collapse engineered by all kinds of optical procedures.[60] The important thing was that all specificity, all reality, all *materiality* should be denied to the means that combined to produce a work of total art—denied to the music, which was so spiritualized that it seemed like a "vapor," and likewise to the very painting of the scenery, which Wagner nevertheless regarded as "the ultimate and perfect conclusion of all the plastic arts" and which he thought should provide a model for the scenery for the dramatic art of the future, in which it would represent "the background of nature."[61] The neutralization of the technical instruments that produce sounds is simply the most striking indicator of a process of idealization that culminates, in *Tannhaüser*, with the Venusberg scene in which the fairylike nature of the apparition is underlined by means of transparent curtains painted with clouds, which are lowered one behind the other so as to obscure the outlines of the painted scenery of Wartberg that is set up at the back of the stage. In such a stage set, in which the painter-decorator is amalgamated with the operator of mechanical apparatuses, clouds—by seeming to dissolve all outlines—procure "the mysterious effect of distance."[62] Chromatism—which René Leibowitz identified as the element that destroyed the system of tonality, and which he never ceased to exploit from within,[63] in the same way that *color* works within the system of linearity—implies a considerable extension of the universe of musical sound. (Wagner himself compares this to the extension of the world at the time of the great discoveries and to the replacement of the closed sea of the ancients by the limitless ocean).[64] In the same way, "cloudiness," in which the contradictions of a culture still, in the last analysis, governed by the system of linearity are at once perceptible yet cease to be operative or productive, sanctions a series of effects that suggest that the framework of the representation is indefinitely extendable. But whereas chromatism (from the Greek *khrōma*, color, musical tonality) raised a series of questions of *form* that the musicians of the twentieth century, following Schönberg and Webern, were to work at and endeavor to resolve,[65] "cloudiness" in painting on the contrary represented a negation of the whole formal problem, if not a flight into the "ideal" similar to that which Nietzsche detected in Wagner. In a reversal that is in no way dialectical, the very element that in the classical context was an indicator of the materiality of the pictorial signifier, now served to reinforce the reigning idealism and functioned as the operator that transformed the scenic reality into an *image*.[66] Now we must find out whether, in a completely different context, that of the painting of the Far

East, the "service of clouds" did not take on a quite different meaning and at the same time have a far more radical effect. As we shall see, the digression should make it possible to introduce the question of modern painting in terms that are truly dialectical and materialist.

The Hieroglyph of Breath

> And with the hieroglyph of a breath, I wish to recover
> an idea of sacred theater.
> —Antonin Artaud, *Le Théâtre de séraphin*

> Huang Dachi had a face that was as fresh as that of a young boy.
> Mi Youren, at over eighty years of age, had preserved the clarity
> of his mind, with no sign at all of degeneration. He died after
> suffering no illness. That is because the clouds and mists
> in their paintings sustained them.
> —Dong Qichang (1555–1636), *Hua yan*

CHINA AND ITS CLOUDS

Except for the anachronism, and disregarding the prophetic utterances of Ruskin, it is tempting to assemble beneath the banner of an identical taste for cloudiness not only the "volatile" landscapes of Turner and the vaporous stage sets of Bayreuth, but also the monochrome landscapes of Song and Ming China and the Japanese paintings of "Zen" inspiration, with their monochrome mountains and forests drowned in mist and consumed by clouds. But an anachronism it would certainly be, since Chinese and even Japanese painting remained virtually unknown in the West even when European culture, tearing itself away from the "mists of the Wagnerian ideal" and the "lies of the grand style" (Nietzsche) was beginning to look toward other skies, in the first instance those of the Far East.[67] The prints that then appeared on the market provided the Impressionists and their contemporaries with many *hints* regarding the arrangement of a page, the way of rendering forms and the framing of figures. But, as Henri Focillon has stressed, Impressionist painting differed radically *in its matter*, which stemmed from the juxtaposition of dabs of color, from Japanese engravings, where flatness reigned (to the point where Japanese art came to be used as an argument by those who, breaking with Impressionism, were to advocate, following Gauguin, the separation of forms and the disjunction of lines and colors).[68] Whistler was perhaps to be the sole artist, working with color that had nothing heavy about it, to translate some of the elements of Japanism in his nocturnes and in the terms of a *taste* (but with the theoretical connotation that Mallarmé's version of "Ten O'clock" attaches to his name): he worked with an interplay of full spaces and empty ones, high horizons, compositions on several rising levels, separated by banks of mist, like those of, for instance, Hiroshige's *Stations of the Tokaido*.

But Whistler's art, all a matter of suggestion, "cloudiness" that does not speak its name,[69] was to have no follow-up. And despite the many pervasive resurgences, even today, of the moment of European taste to which the nocturnes discreetly testify, there is no historical justification for pressing on any further with a comparison with the pictorial tradition of the Far East. Does that mean that the comparison might be shifted to a theoretical level, that of the tasks imposed upon Western art by the exhaustion of the classical system of representation? It is indeed very tempting to draw a parallel between the reversal illustrated by Ruskin's formula, which placed modern landscape "at the service of the clouds," and what at first sight appears to be one of the most constant features of Far Eastern art and even of the Chinese painting that the West was now beginning to discover and in which "views of mists and clouds" seem to have occupied a privileged position (see Plate 7b).[70] Certainly, as early as the eleventh century, Mi Fu, in his *History of Painting*, underlined the great antiquity of the "concept" of clouds, when writing about the views of mists painted by Dong Yuan, which were then very much in vogue and in which the hard forms of mountains were now veiled, now unveiled, while the tops of trees now appeared, now disappeared, as the mists now lightened, now darkened. All this Dong Yuan painted without resorting to any artificial ploys.[71] In the eighteenth century, the *Jie zi yuan huazhuan* (Teachings on painting from the garden the size of a mustard seed) described clouds as "the adornment of the sky and the earth, the embroidery of the mountains and the waters," and on that account compared them to the quotations of poetry that writers used to make their style more powerful. In fact clouds constitute a *recapitulation* of the landscape: "For in their elusive emptiness, one sees many of the features of mountains and water courses hiding there. That is why we speak of mountains of clouds and seas of clouds."[72] As for the *Comments on Painting*, by Shitao (between 1710 and 1717), they seem to provide the key to the use of /cloud/ in landscape painting: in a landscape painting that succeeds "in getting at the principle of the universe," the form and impulse of which it expresses, "rivers and clouds, thanks to the way that they are gathered together or dispersed, *constitute what binds it together*"; "the Sky embraces the landscape by means of winds and clouds, and the Earth animates it by means of rivers and rocks."[73]

To judge by the letter of those texts, in the paintings for which they propose a theory /cloud/ does not seem to figure as a recurrent unit that can be precisely identified, circumscribed, or even localized. Buddhist painting, whether Chinese or Japanese, regularly made use of *cloud* as a support or vehicle, drawn with a precise outline the curve of which in many cases evokes the silhouette of a dragon or a fish with a long tail, upon which a bodhisattva is seated, Amida descends to earth together with his escort, or an *aspara* is borne aloft. And some scholars have detected in the *yun jian*, the "collar of clouds" embroidered upon the robes of the Chinese emperor, a vestige of an extremely ancient cosmic symbol that marked the threshold of the heavens, at the extremity of the axis of the world, in accordance with a schema that also cropped up in the medieval Western world.[74] But that iconography (and all the coincidences and affiliations that are sometimes claimed to be established, at this level, between European art and Chinese art) are of no account at all compared to the extremely original theoretical status that Chinese treatises assign to cloud: that of an element or principle which, depending on whether it is gathering or dispersing, constitutes in its "elusive emptiness" the *bond* that ties together the landscape where it interconnects, even as it hides, "the lines of the mountains" and the "water courses." In the representational system of the European Renaissance, cloud at the very outset presented a problem, as is shown by Brunelleschi's demonstration. But all the indications are that the extent and, even more, the nature of the specifically pictorial functions imparted to mists and cloud and—as we shall see—their cosmological connotations suggest that Far Eastern painting, on the contrary, regarded this element both as a particularly prized motif and as a particularly prized principle. Even where it is introduced into a picture by mechanical or conventional means, in the West cloud marks the limitations of a representation that is governed by the finite nature of linearity. Beyond a certain point, a proliferation of clouds, more or less deliberate and controlled, seems to be a symptom: it signals the beginning of the dissolution of an order (but not its *deconstruction*). In other words, at first sight the Chinese system appears to function, practically, in a fashion that is quite the opposite to that of the Renaissance system, for it seems to begin and find its way forward at the very point where the latter meets its limit, its closure. That is perhaps a superficial impression that may mask questions thrown up by even a rapid and superficial examination of Chinese works, both practical and theoretical, in the pictorial domain. (But the purpose of this digression through China and its "views of mists and clouds" — which cannot be avoided at this point — is simply to make it possible to pose the question of the conditions for a possible theory of /cloud/ that is *general* [and not just *local*, geographically and historically speaking] and through that to a general theory of painting itself.)

THOSE WHO DRAW CLOUDS AND
THOSE WHO BLOW THEM

Yet, *in its treatment*, Chinese /cloud/ at first sight appears to relate to a problem that is not all that different from that which the present work has attempted to define in the Western context.

To paint clouds, one uses only pure colour. When one looks at them, they seem to mass together. It is preferable not to draw any lines in ink. When, in *qing lü* landscape paintings,[75] one draws fine lines, the [clouds] must be in harmony. So one uses dilute ink to trace [their outlines: but perhaps that addition is not really necessary] and one tints them with blue.[76]

The advice of *Mustard Seed* seems perfectly clear. Cloud, as Aristotle established, has more to do with color than with line, insofar as the latter creates *figures*. If the method for suggesting depth dictates the use of lines, however fine, then /cloud/ must be outlined in dilute ink, and even those faint lines may subsequently be sprinkled with powder, as the ancients taught:

The Tang painters had two ways of painting clouds. The first is known as *chui yun* (literally: the blowing of clouds). It consists in painting silk with a light coat of white. This way of doing it corresponds to layers of clouds floating in the wake of the wind; it is light and pure, people like it; it is extremely graceful. The second method is called *goufen* (literally: tracing with powder). In these *jin bi* landscape paintings,[77] the ink strokes are covered with powder. Marshal Li the younger often used this method; it produced a very powerful effect. This increased the richness of the painting.[78]

Does this mean that in China, as in the West, /cloud/ defies the order of linearity, and that the way in which it operates in landscape painting can be defined and appreciated in terms of the drawing/color (if not the "linear"/"pictorial") opposition mentioned at the beginning of the present work? Does it mean that Chinese painting (and here we should bear in mind the *resistance* that European landscape art encountered in its early days) proceeded from the very element that was supposed to be excluded in the West, and—as seems to be indicated by the privileged place assigned to "views of mists and clouds" in the Chinese hierarchy of pictorial forms— exploited that which was suppressed in Western painting, of which it would thus seem to constitute as it were a *negative* and also a necessary theoreti- cal complement? Expressed in that form, the proposition is not tenable given that, even in China, the traditional classification of schools of paint- ing was massively dependent upon the opposition of drawing and color: "The *Huainanzi* says: 'The people of Song excelled at drawing, and those of Wu at painting with colors.' Is that not the truth?"[79] As early as the Tang dynasty (seventh to tenth centuries), during which pure landscape painting became definitively established, a distinction that was to provide a frame- work for all "histories of painting" was introduced between the "Northern school," that of professional painters, who were inevitably traditionalist and whose "blue and green" mode was, right down to the nineteenth century, to represent a kind of ideal, and the "Southern school," that of amateur painters, scholars who, using ink in a spontaneous and liberated way, were a few centuries later to become the apostles of monochrome landscapes, in opposition to the academicians. Now, despite its borrowings from tra- dition, the theory of painting expounded not only by Mi Fu's *Huashi* but also by Shitao's *Comments*, and to a lesser degree by *Mustard Seed* was pro- duced solely by scholars.[80] It tells us nothing about paintings in the official style, which are characterized, *even in their drawing of clouds*, by clear and precise outlines and a technique in which color is used—seldom in an in- discreet fashion—to supplement the drawing. Of course, set out in this categorically geographic form, the opposition is only useful as an indica- tion: it does not explain the real, historical evolution of Chinese painting. However, the theoretical debate about /cloud/ is illuminated by the rivalry between, on the one hand, an orthodox if not academic art, committed to the strictest linearity and, on the other, painting that explicitly sets out to be transgressive. (There are countless anecdotes about the practitioners of *po mo*, "spattered ink," and *yi pin*, "painting with no constraints": one artist would spread out his scrolls upon the floor and bespatter them with blots that he would then turn into landscapes by adding the odd stroke here and there; another would use his pigtail as a paintbrush or would paint with his back turned on his work, or in a state of intoxication or trance, and so on.)[81]

THE BRUSH STROKE

In reply to a painter who boasted in Zhang Yanyuan's company that he could paint "blown clouds" (ninth century), the latter retorted:

The ancients have not yet achieved perfect skill in the painting of clouds: scattering on a piece of silk light [specks of] powder blown there by one's mouth is called "blowing clouds." Clearly, it seizes upon a natural principle. But although one may call it a wonderful solution, nevertheless there is no sign there of any brush stroke. So it cannot be called painting either. It is impossible to copy.[82]

That last remark reveals Zhang Yanyuan to be a believer in an eclectic academicism that insisted upon scrupulous imitation of the models and masters of the past, thereby certainly complicating the task of experts and "connoisseurs."[83] The scholars resolutely opposed such a conformist attitude.[84] It was not that they defended originality at all costs, only the rights of *transformation*, freedom, the need for a painter who studied the ancients to make changes, *conversions* ("If there are rules, there must be change. . . . The minute one knows the rule, one must endeavor to transform it").[85] To be sure, individualistic professions of faith abound among the painters and calligraphers,[86] and Shitao himself insisted that "it is by oneself that the Rule must be established." However, the desire for originality, and the wish to owe nothing to anyone but oneself were affirmed within a predetermined, already established field: that of *painting*, in the sense of a specific practice the principle and basis of which lay in the *brush stroke*. It was a principle, a basis that was discovered in a breakaway from knowledge that was a slave to imitation; and it led to a return to sources in a quest for the supreme rule that was born from an absence of rules and that encompassed a multiplicity of rules.[87]

The primacy of the brush stroke was, after all, enshrined, right from the start, in the definition of painting imposed by the very use of the character *hua* 畫. According to the *Shuo wen jie zi*, a dictionary that went back to the first century (c. 100) and that provides the etymology of characters, the character *hua* is a complex one, a compound of two graphic elements that, taken separately, constitute the characters *yu* 聿, derived from a hand writing with a stylus, which took on the meaning of "brush," and *tian* 田, an image of a plot of land crossed by furrows. The meaning of "to paint" is thus to trace with brush lines that give the outlines of forms just as paths mark the edges of fields and fix their shape. That certainly seems to be the meaning understood by *Mustard Seed*: "The *Shuo Wen* says: painting is made up of limits; it resembles the paths that delimit the fields."[88] From there, it is but a step to declaring that Chinese painting only exists thanks to its graphism,[89] and that lines or, to be more exact, *outlines* dominate Chinese painting from start to finish of its history;[90] and it was a step that Western art criticism took without bothering much about the pertinence of the categories that were imposed by the culture, if not the ideology, from which it stemmed. It is true that theory—and not just academic, traditionalist theory, but even that of the "amateurs," the scholars, which culminated in the concept of the "unique brush stroke," *yi hua* 一畫 that governs Shitao's system—denies the name of "painting" to any practice where the "brush stroke" plays no part, starting—as we have seen—with the technique of "blown clouds" and that of "spattered" ink. It is furthermore attested that Chinese art criticism considered the first Western oil paintings imported into China to be heterodox. But the absence of precise details as to the nature of those paintings is telling: should we take it that the Chinese connoisseurs were united in their condemnation of not only Turner but also the Pre-Raphaelites, Ingres, and Delacroix? When posed in those terms, the question reveals the *limits* of the system of thought according to which Western criticism operates. For clearly, the connoisseurs of Ming China would no more have accepted as painting a panel of figures delineated as precisely as those of Uccello than they would a fresco by Masaccio in which forms are defined by more strictly pictorial means. Outlines, surfaces, even limits (in the sense in which, in Alberti's theory, outline is limited to the surface that it *denotes*)—those categories can guarantee no grip upon a practice that, in truth, is regarded from the point of view of *productivity* rather than from that of its products, a practice dominated not by lines—as an overhasty analysis would have it—but (as the texts make abundantly clear) by *brushstrokes*.

FLESH AND BONES

"The ancients say 'having a brush, having ink.' Many people do not understand those two characters. If one only has outline, without a method of brush strokes, that is called 'lacking a brush.' If one has a method of brush strokes but no nuances [literally, no light and no dark], one can indicate neither what is exposed to the light, nor what is the opposite, nor the shadow of cloud, nor what is brilliant, nor what is dark: this is called 'lacking ink.'"[91] The *brush* and the *ink* are the two notions, the two categories to which Chinese criticism most often resorts in order to assess the value of a painter: "Once you have the brush and the ink, you are a master; to have the brush but not the ink is a mistake; to have the ink but not the brush is also a mistake."[92] In other words, neither the ink nor the brush can be reduced to elements, formal components that are identifiable as such: rather, at this stage in the analysis, they correspond to complementary productive principles whose relative usefulness is measured by the extent as well as the nature of the effects that they engender in practice. "The brush serves to position forms, the ink serves to differentiate shadow from light."[93] Despite appearances, that opposition does not duplicate the opposition between line and color, as used ideologically in the West. For whereas in the figurative tradition of Europe, line, frequently in covert ways, assumes most of the semiotic functions conferred upon the painted image (outline *denotes* figures, while color is just an extra), Far Eastern painting imposes a radical theoretical shift that involves, first and foremost, relinquishing the notion of *outline* and at the same time any idea of classifying pictorial aspects either under the heading of line (or figures) or under that of color ("graphism" or "tonality").

As Henri Focillon has observed,[94] in the West painting has always fashioned etching in its own image (Ruskin, after all, thought that where clouds were concerned, an engraving could produce a good approximation to a painting). The truth is that, in the last analysis, Western painting has always been conscious of its relation to engraving, a relation that involved not merely reproduction and popularization, but in the first place a translation or *conversion*, for the engraving was expected to convey by strictly graphic methods the essential "message" of the painted image: to return to Rousseau's terminology, the lines of a painting still affect us in an engraving. Indeed, classical European painting was not averse on occasion to borrowing something of the character of an engraving (even though as Alberti insisted, it was better not to overemphasize outlines). In China, by contrast, the manner of engravers (*ke hua*) was considered by most theorists to be a fundamental mistake where painting was concerned.[95] Indeed, engraving, even Japanese engraving, could never serve as an introduction to or even a term of comparison for pictorial analysis (whereas Western engraving did find ways of reproducing relief, lighting, etc.). The fact is that Far Eastern pictoriality eludes any *reduction* and cannot be classified in terms of the opposition between line and color. "If one only has outline, without the method of brush strokes, that is called 'lacking a brush.'" There is a dispute about the translation of *lun kuo* 輪廓 (from *lun* 輪, "wheel," and *kuo* 廓, "wide," "to widen": the circumference of the wheel as its rotation carried it away of its own accord). Petrucci favors "outline," Rymans "broad lines."[96] But the meaning of the passage from *Mustard Seed*, cited above, remains enigmatic from the point of view of Western categories: for it seems that it is not the *brush* but the *ink* that provides the "outline," the "broad lines," the work of the brush being assimilated, on the contrary, to the method of "brush strokes" or "wrinkles" (that is to say, according to the classical definition, that which is obtained by means of a pointed brush wielded at a slant, or the "overpainting added with a dry brush": "They [the broad lines] are added later, so as to break up the volumes").[97] The ink, not the brush (but nevertheless the ink applied by the brush), makes the forms of the Mountains and Rivers (that is to say the landscape) "expand" (we shall be considering what that means, in a moment). The ink provides the "broad lines," if not the "outline," while the task of the brush (but a brush full of ink) is to determine their lines of force[98] and, by means of "strokes" or "wrinkles," to suggest the "living depth of things." While the various kinds of "wrinkles" or "strokes" listed in the Treatises ("raindrops," "bushes in disorder," "the peaks of clouds," "whirling water"; and also "the cranium of a skeleton," a "shaving block," "the hairs of oxen," etc.) refer to natural structures and each corresponds to a particular characteristic aspect of a "real" mountain, at the same time those wrinkles and brush strokes have a primarily diacritical, differential, function. "Because the shapes of mountains may take on a thousand different aspects, it follows that the expression of their relief cannot be reduced to a single formula."[99] The brush derives its expressivity from the differences that distinguish the various types of wrinkles (*hua*, meaning one particular "stroke" among others, the outline obtained by the brush moving to and fro, used for buildings and for pine trees).[100] But by doing so it frees itself from being in any way subordinated to the order of delineation. As Pierre Ryckmans forcefully puts it (but without drawing all the possible theoretical conclusions),

Once the broad lines have placed the outline of a given object (a stone, a mountain, a tree trunk, etc.), the "wrinkles" are drawn inside the broad lines or lean against them, to describe the relief, texture, grain, luminosity, and unevenness of surface and the volume of the object; *that is to say that, in Chinese painting, they combine the various functions which in the West belong now to line, now to colour, now to shadows, now to perspective, since they describe the shape, matter, lighting, and mass of things, all at once.*[101]

The function of the brush is thus not to make the outline stand out or to delimit the forms whose structure, or even texture, it is required to express by its own particular means. "Whatever the plastic form, it can always be reduced to the elementary principles that belong to the different kinds of lines and wrinkles."[102] From this follows a notion of form—if that is the right word and if it is acceptable here—that owes nothing to delineation in the Western sense, nor to the Aristotelian distinction between color and figure (since the painting of the "scholars," which is the subject of most of the Chinese treatises on landscape painting, first and foremost among them Shitao's *Comments*, is monochrome painting that uses only tonal differences). Ink and brush are like *flesh and bone*. If the ink makes the forms of the landscape expand, it is insofar as it confers its flesh upon the skeleton that the brush must provide. It fills in shapes, gives them a contour, just as flesh gives a body its "figure." But just as flesh cannot do without bone and bone cannot do without flesh, so ink cannot do without a brush and a brush cannot do without ink.[103] And it is this that makes the technique known as "blowing clouds" and that of "spattered ink" so aberrant; it is also the reason why such techniques do not deserve to be called "painting." Taken literally, such a technique surely leads to separating what should not be separated and isolating one of the terms of an opposition that makes no sense unless it is presented as *dialectical*, unlike purely formal or analytical oppositions of the line/color, linear/pictorial, or even form/matter type. Ink is not the same to the brush as color is to line, nor as matter is to form; for (and this is the point that needs to be stressed in order to avoid any interpretation of an idealist nature, of the kind that abounds in Western literature devoted to Far Eastern art) bones are no less material than flesh.

'HUA'/DELIMITATION

Painting, which is "made up of limits" and "resembles paths" that limit fields, is nevertheless not a matter of delineation and, above all, is not governed by any predetermined closure (a "path" is not a closure). That, at any rate, is what is suggested by the character *tian* 田, "field," which is one component of the character *hua* 畫 (or, in its simplified form, 画), "painting," if one pays attention to the order in which the strokes that make it up should be traced: according to this order, in the pictorial field "delimitation" assumes a meaning radically different from that which was explicitly assigned to it, from the time of Alberti onward, by European tradition. "Here alone, leaving aside other things, I will tell what I do when I paint. First of all, where I draw I inscribe (*scrivo*) a quadrangle, as large as I wish, which is considered to be an open window through which I see what I want to paint."[104] Alberti's window is just one interpretation, among other possible ones, of a pictorial field of the kind that a Western painter begins by circumscribing. The field's opening thus coincides with the marking out of its closure (see Paul Klee: "The scene is the surface or, to be more precise, the delimited plane").[105] Now, *the order of the strokes* that governs Chinese writing on the contrary requires that the delimitation or closure be established only at the last moment (any failure to do this constitutes the equivalent of a spelling mistake): when the whole or a part of a character is confined within a closed space, that space must not be closed until all the strokes within it have been traced.[106] In the case of the *tian* 田 character, one should thus begin by tracing the "left" side of the "field" l, then proceed to draw the upper edge and the right-hand side ⅂ in a single movement; next, one traces within the space thus marked out two strokes forming a cross, beginning with the horizontal 闩, 用; lastly, one "closes" the character by drawing the base of the rectangle 田.

But such a "closure" does not do away with the discontinuous aspect of a delimitation that operates with simple strokes that are *separate*,[107] and the openness of which is maintained, preserved, right up until the last stroke has been completed. In this connection, we may compare the character *hua*, in its simplified form 画, in which a kind of *open* frame duplicates the outline of the field, with the very similar-looking schema that Klee, starting from the position *P*, uses to explain the various possibilities of orientation on a previously delimited flat surface. His schema confers upon

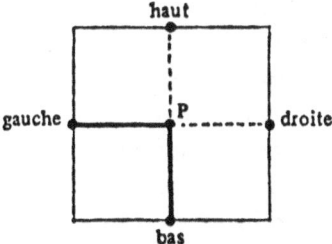

FIGURE 7. Klee, *Figure*, from *Das bildnerische Denken* (Basel, 1956), p. 39.

the surface (*die Fläche*) a material identity in which "left," "right," "top," and "bottom" all appear as attributes. In contrast, the Chinese character forces one to think of the orientation (the definition of the "directions") as primary: within the space, indicated rather than delimited on three sides, and *with the bottom left open*,[108] an initial horizontal stroke divides the top from the bottom, at the same time connecting the left and the right, while the vertical stroke assures the link between the top and the bottom, at the same time marking out the left/right articulation. The priority given to the position of the orients, and first and foremost the division between above and below, or "sky" and "earth," over delineation and the substantial production of a surface as such corresponds to a constant feature of Chinese thought. This is not based upon the principle of identity together with the notion of substance, which is its corollary, but takes as its point of departure *relative orientations* or rather (as the categories *ink* and *brush* clearly show) the opposition of contraries.[109] That is a point made forcefully, in relation to landscape painting, by a passage from *Mustard Seed*, which it is worth comparing to the text of Alberti cited above:

In general, before beginning to paint, it is necessary to reserve the place for Sky and the place for Earth. What do we call Sky and Earth? In a picture of one and a half *chi* [a *chi* is about 35 cm.], the upper part is kept for the Sky, the lower part for the Earth; only in the middle part does one set about determining the landscape.[110]

The operation by which the painter *sets aside* the place of the sky and that of the earth and inserts the landscape *between the two* does not, it must be stressed, proceed from a division, a partition of the surface conceived in its identity, its substantiality, and as the substratum of the painting. Shitao, too, criticizes the method of three successive planes (a foreground for the ground, a second plane for the trees, a third for the mountain) and that of two sections (with an intermediate zone of clouds),[111] for being too close to the system of *cutting the area into pieces*, which characterizes engravings.[112] If some kind of delimitation is introduced, its role is not so much to isolate, separate, *define, identify*, but rather to establish a correlation between two terms that it presents as separate only in order to open up a field in which they can interact and in which the dialectical process prompted by their opposition can take place.

OPENING CHAOS

Let us return to the ink/brush pair, which we can use as an example to confirm the aversion in Chinese thought (an aversion based in *language*) to any kind of logical definition following the Western model that attributes a predicate to a subject. Neither the ink nor the brush can be defined otherwise than in their "correlative duality" (which is why it is impossible to introduce or translate those concepts into a Western language except by reducing them to substantial terms that are identifiable as such).[113] As we shall see, the same applies to oppositions of a similar nature that obey the same laws even though they operate at other levels and far more extensively. The relations between the ink and the brush are by no means simple and linear. In fact, the priority of "structure" over "form" (insofar as such concepts are acceptable in the context of a system of thought in contact with which the "entries" that Western thought endeavors to keep open become confused) seems to have been explicitly affirmed at an extremely early date. In his *Guhua pinlu* (late fifth century), Xie He set out a number of principles that lie at the basis of all Chinese aesthetics. The second of these principles concerns the quest for a skeletal framework, using a brush;[114] the third concerns the determination of forms and the representation of objects;[115] the fourth relates to the application of color;[116] and the fifth to the distribution of lines, composition.[117] The principles thus lay down that setting in place the skeletal framework should precede the determination of forms. But that does not mean to say that the brush has the slightest *logical* precedence over the ink, even if, as Ryckmans points out, a certain hierarchy, in which the brush is of more account than the ink, may seem to be suggested by the fact that ink is more easy to have than the brush.[118] Insofar as it is provided by nature and represents a technical acquisition,[119] ink is opposed to a brush, which represents man's contribution (man who controls it) to pictorial production; and it is up to the ink to prepare the way for the *opening up of chaos* that will be accomplished by the brush:[120] the synthesis between the ink and the brush is realized in the brush stroke—the unique brush stroke, *yi hua*, as Shitao puts it—the variants of which constitute the simplest and most elementary method of handling ink and brush, which represents "the first elementary step in the apprenticeship of calligraphy and painting."[121]

"Ease with the ink is a matter of technical training; the spirit of the brush is a matter of life." "To have ink but no brush" means that you are invested with the ease bestowed by technical training, but remain incapable of giving free rein to the spirit of life. "To have the brush but no ink" means that you are receptive to the spirit of life, but without being able to introduce the metamorphoses produced by the ease that results from technical training.[122] Such is the shift made necessary by the reference to China here that the brush may be associated with receptivity, while ink is associated with invention or rather, as the Chinese text puts it, with *transformation*. But the fact is that painting is a matter not of imitation, but of reception, and is founded on a dialectic of *hospitality*, in which each term successively adopts the position of host (*zhu*) and guest (*bin*), a dialectic that governs the relations between the ink and the brush, and even more profoundly those between painting and the universe, since "the principle of painting and the technique of the brush are nothing other than the inner substance of the universe on the one hand, and its external beauty on the other."[123] (But "inner substance" is a particularly ill-chosen expression given that Chinese thought is not familiar with such an expression or even with the notion of "substance," being aware only of interaction, correlations, and implications between terms, between *signs*, and altogether unpreoccupied with the substance that underpins them,[124] the notion of which, as Granet has shown, merges with that of *rhythm*.) Nevertheless, that inner structure (provided one understands it in a strictly logical sense, and as a principle of the dialectical order) is in no way an intelligible Form, a Platonic Idea. This is a good point at which to condemn the idealist reduction to which virtually all Western criticism systematically subjects both the pictorial theory and the pictorial practice of the Far East.[125] For instance, Chinese theory does not recognize any separation such as the European "iconological" tradition makes between the "body" and the "soul" of an image (see Chapter 2, p. 51), and at the same time it rejects the copying or imitation of external appearances. "Resemblance seizes upon the figure (*xing*) and allows the animating breath (*qi*) to escape; truth (*zhen*) seizes upon both the breath (*qi*) and the matter (*zhi*)."[126] But this breath is not of a different order from that of a figure or matter. It is not a "spiritual" principle. Rather, it constitutes the primary, simple, fundamental element that is the Unique Brush Stroke, which animates what is originally un-differentiated and draws beings and things out of chaos.[127] Breath (*qi*) is precisely the movement of life (*sheng dong*), which the first of Xie He's principles associates with the harmony (or/and, through homophony, the revolutions) of breath (*qi yun*). And, through one of those reversals that are so common in Chinese thought, it appears to be an effect of the "ink" rather than the "brush"—the ink, the function of which is to "open up chaos" and which is assimilated to "water," the element in which painting finds its "movement."[128]

YIN/YANG

Although the character *yin* may be used by analogy to designate a painting, whereas its primary meaning is "shadow," that does not mean that it is through shadows that a painter sets out to *copy* objects: the ancient fable according to which painting originated in the delineation of a shadow that a body projected on to a wall[129] only makes sense in the context of an art that serves *mimesis*, that is to say *outward aspects* and, in the last analysis, *outlines*. There are shadows and shadows, as the West certainly recognizes when it uses the expression "*Chinese* shadow play" for a type of spectacle that implies no specular duplication. When a Chinese painter draws the shadow of a bamboo that falls upon a window screen, he does not do so in order to fix its outline, but because that device, which acts both as a mark and as a substratum, produces the equivalent of a painting in which, through abstraction or rather *abbreviation*, the principle of things is made manifest. "How could anyone learning how to paint a landscape act other-wise?"[130] "For even if, where animals, oxen and horses, are concerned, or people and objects, it is enough to copy what they look like, in landscapes copying does not succeed. In a landscape, the place where creative thought operates is on a higher plane."[131] Or, as Shitao puts it, "the substance of the landscape is realized by reaching the principle of the universe," and that is why, in the hierarchy of genres, landscape ranks very highly. The very term that in Chinese designates "landscape" is made up of two associated characters, *shanshui* 山水, "mountain/water," but once again it does not so much indicate "substance," but rather manifests the *law*: a landscape is governed by the same rhythm as the order of the world; like the universe itself, it is engendered by the interplay of two principles, two antithetical emblems, in accordance with the rhythmic concept that governs all Chinese thought, whether mythical or philosophical, the concept that is expressed by the formula *yi yin, yi yang* (literally "one yin, one yang"), which may be symbolized by any *image* that conveys two, contradictory aspects.[132] *Yin* and *yang* constitute an efficacious pair that corresponds to the classification of all aspects of reality and their universal alternation, for the order of the world results from interaction between the two sets of complementary aspects, according to the rule of bipartition that governs the notion of totality and that, as Granet has emphasized, is utterly dominated by the category of *sex*. Now, the distinction or opposition between the brush and the ink, respectively assimilated to *mountain* and *water* (Shitao: "You must make the ocean of the ink embrace and carry things, and the mountain of the brush set itself up and dominate"),[133] is itself determined by that rule. ("The union of the brush and the ink is that of Yin and Yang. The indistinct fusion of Yin and Yang constitutes the original chaos. And without the means of the Unique Brush Stroke, how could one clear the original chaos? . . . The metamorphosis of the One produces Yin and Yang, and that is how all the potentialities of the world come to be accomplished.")[134]

The above observations are by now common knowledge, albeit sometimes misunderstood, despite Granet's decisive analysis. The couple formed by the *yin* and the *yang*, which "evoke [all other emblems, grouped in couples] with such force that they seem to bring them into being, both them and their coupling," wields an authority that is founded upon sexuality. What most frequently prompts Western criticism is the fact that it is an authority that does not lend itself to an idealist interpretation since, as Granet stresses, the two terms "constitute the two antithetical aspects of what we should call matter or substance."[135] But that applies equally to the ink/brush couple (and "couple" should here be given its sexual connotation). But whereas in Western categories activity (assimilated to spirit or mind) tends to be valued more highly than passivity (linked to matter), in Chinese thought, if the brush is in any way privileged over the ink, it is certainly not the kind of privilege that is associated with the Western "idea." Even to speak in terms of passivity/activity is to go either too far or not far enough: the ceaseless exchange of oppositions and functions (between ink/brush, host/guest, etc.) is characteristic of a type of thought that systematically plays upon opposition as a means of expression, and in which every term needs its contrary if its meaning is to be revealed in all its complexity:[136]

Anyone who can seize upon the sea only to the detriment of the Mountain, or the Mountain to the detriment of the sea, in truth has only an obtuse form of perception! But as for me, I do perceive! The Mountain is the Sea, and the Sea is the Mountain. The Mountain and the Sea know the truth of my perception: everything resides within man, through the free impetus of just the brush and just the ink![137]

So it is not hard to see that, in landscape painting, copying is not enough. "Painting is not a copy of a preexistent universe, it is itself a universe. . . . Painting is not a description of the spectacle of creation, it is itself a creation in the literal sense of the word, a microcosm the essence and mechanism of which are identical and parallel to those of the macrocosm."[138] In other words (with the important reservation that Chinese thought is not familiar with the idea of "Creation" in the Judeo-Christian sense of Creation with a capital *C*), painting does not borrow from the representational order or structure: and it would appear that, in its highest form (landscape), Chinese painting, like Chinese poetry (but possibly unlike Japanese painting and poetry), owes nothing to spectacle, that is to say theaters.[139] In principle it may be closer to geomancy (*fengshui*), the object of which is to determine the propensities of sites by taking into consideration the water currents (*shui*) and the air currents (*feng*), which are always studied in relation to the mountains.[140] The "field" of painting is defined by a double orientation, the horizontal/vertical, as much as by its "limits" (and has nothing to do with a "scene," in the sense understood by Paul Klee). And the function of a painter is analogous, *within his own field*, to that of the Leader whose primary function is to establish a certain order in space, and whose forays to the four *limits* of the empire correspond to the need to reconstitute its expanse rhythmically, classify its various spaces virtually, and at the same time define the cycle of the seasons.[141]

You must make the ocean of ink embrace and carry things, and the mountain of the brush set it itself up and dominate; then, their use must be greatly extended so as to express the eight orientations, the various aspects of the Nine districts of the Earth, the majesty of the Fire Mountains, and the immensity of the Four Seas; and it must develop so as to include all that is infinitely great, and be focused minutely so as to accommodate all that is infinitely small.[142]

Everything, right down to the controlled to-ing and fro-ing of the brush, from left to right, and from top to bottom, and so on must obey the same cosmological determinations as the movements of the sovereign. Epic poetry celebrated the voyages of Qin Shi Huangdi, the founder of the Chinese empire, and those of Emperor Wu, the great sovereign of the Han dynasty. Both endeavored to set the empire in order by building *an immense network of roads* running from north to south and from east to west.[143]

The job of a Western painter, as defined by Alberti, is, by means of a kind of *repetitio rerum* that forms the basis of the representational structure, to establish the scene where a story will then take place (in the same way, as Georges Dumézil has shown, that priests of the Vedic hymns and the Roman *fetiales* opened up their perspectives). In contrast, at his own level and within his own field, a Chinese painter, just like the prince or the responsible regent, is the regulator of rhythm, the person who gets the yin and the yang to act in concert.[144] In order to animate space and time, the sovereign has to take up position in the middle of the *Ming tang* and provide the year with its *empty center*.[145] Similarly, the painter participates in the metamorphoses of the universe, plumbs the shapes of the mountains and rivers, measures the far-distant immensity of the earth, gauges the disposition of the peaks, and deciphers the dark secrets of the clouds and mists; but he must always return to "the fundamental rhythm of the Sky and the Earth,"[146] and, in order to contribute to their creative work, he must "make up the third": this expression (based on the etymology of *can* 參, "to participate") confers upon the work of the painter its full dialectical significance. *Before beginning to paint*, far from defining the space of the representation in advance, as a finite area, he "reserves" the place for the sky and the earth and, only then, in the middle left empty, decides upon the landscape, which is seen as a resolution of the contradiction of sky/earth and at the same time as a way of passing beyond it.[147] In Western art, the sky and the earth appear as two levels, the one above the other, conventionally defined and identified by a number of accessories (for sky, for instance, clouds, which also make communication between the two regions possible and, on occasion, serve as vehicles for the actors in the *istoria*). In contrast, in Chinese art, sky and earth constitute two antithetical terms that, through their very opposition, combine to produce the landscape, provided that the painter has been able, with his brush stroke, to seize upon the rhythm in which they interact.[148] This rhythm is not finite, but infinite, for the scope of painting should both "expand to include the infinitely great" and contract "to accommodate the infinitely small."[149] Similarly, the Unique Brush Stroke, which represents the basic unit of the painting, "confers the infinity of the brush strokes."[150] It is a formula that, at the same time, confirms that this is an essentially *productive* idea, the potential infinity of which is deliberately underlined. However, it should not be imagined that, to create a painting, all that is necessary are a few brush strokes, a few marks on a sheet of paper or a piece of silk: if the rhythm of sky and earth is not established *at the very start* (and we now understand the logical function of that "rhythm"), all that will be obtained is a meaningless daubing,[151] with no instructive value at all (whereas "among the paintings of the ancients, there is not one that fails to offer encouragement or advice").[152]

CITATIONS

The question of /cloud/ in Chinese painting should be posed in the context of this symbolical logic or *emblematic system* (in the efficacious and productive sense in which Granet understands an emblem) that governs the painting of a landscape and that is summed up in the signifying dialectic between ink and brush. Once we accept that "form" can be conveyed otherwise than in and by "outline" (or outline is understood altogether freely, without reference to delineation),[153] the opposition that Leonardo draws between solid bodies (reducible to a composition of surfaces) and "bodies without a surface," with no precise or definite limits, ceases to be pertinent in the pictorial order. If the painter's mission is not to capture the fleeting appearance of things, but to seize upon their organizing principle, then clearly, despite superficial analogies, one is a long way away from any kind of "impressionism." And, again similarly, where "space" is concerned, it cannot be claimed that cloud in Chinese painting serves to dissolve the geometric frame and open it up onto the infinite. Chinese painting does relate to a space, but in a very different sense from Western painting, which is always obsessed with the fantasy of trompe l'oeil and is constrained by illusion, whether it surrenders to it or, on the contrary, rejects it. "If one en-

gages in painting mindful of its ability to render distance, the picture will
not equal the (real) landscape. If, on the contrary, one takes as one's theme
the marvelous work of the brush and the ink, the landscape will certainly
not equal the picture."[154] All speculation on the status of "perspective" in
Chinese painting is pointless given that what the latter pays no attention
to is not "depth" (once again, it is quite clear that depth is not necessarily
linked to perspective), but submission to a geometric order or the authority
of one particular point of view. As Brecht remarks in a particularly illuminating note, "Chinese composition is not affected by the element of constraint with which we are so absolutely familiar. *Its order costs no violence.*"[155]
Faced with an unrolled scroll painting, the spectator does not remain in
the position of a passive observer. He or she leaps across the space to reach
the four frontiers of the world, always responding to the impulses that are
received from the antithetical elements that go to produce the landscape.
He or she echoes the call of solitude and "responds" to that of the peaks
of the majestic mountains and the forests swathed in clouds stretching far
into the distance.[156] Cloud may contribute to this "concert," but it should
not dominate the other voices or *violate* them as did, in Mi Fu's opinion,
the landscapes of the famous Li Cheng: "Painted in very dilute ink, these
landscapes look like dream [landscapes]. In the midst of the mist the rocks
seem to move like clouds. It is very skillfully done, but there is little truth
in the inspiration."[157]

Even if, as *Mustard Seed* states, clouds provide a *recapitulation* of
painting methods, the purely technical question of how they are "rendered"
is less crucial than the logical position assigned to this element in the landscape. According to the "two sections" method, which Shitao criticizes,
clouds should be *added* between the mountain *above* and the scene *below*,
so as to emphasize the separation of the two levels. This mechanical and
conventional division should not be confused with the separation into the
sky/earth levels from which Western painting derives many of its effects. It
conveys nothing of the interchanges and interaction between the above and
the below, the sky and the earth, the mountain and the water, since, quite
literally, the sky and the earth have no place in a Chinese landscape, which
is established in between the two. The landscape must express the (antithetical) structure and the (dialectical) impulse of the universe by means
of a whole series of effects, which Shitao lists.[158] The rivers and the clouds
assume a decisive function in such a work, for "in their gathering or their

dispersion they constitute the link," while the sky embraces the landscape with its winds and clouds and the earth animates it with its rivers and rocks, according to the rhythm that accounts for all the metamorphoses of the landscape, and all the reversals and inversions of signs of which it is both the product and the place. In other words, cloud, in these circumstances, does not signify any kind of transcendence. Its function is not solely to emphasize the height of the mountains and the depth of the forests; for it also constitutes one of the elements through which the mountain communicates with its contrary: in relation to the mountain the clouds assume the role that is assigned to water in the mountain (brush)/water (ink) dialectic.[159] Rocks are the bones of the mountains, but waterfalls are the bones of rocks: nothing is stronger than water, for water, in the last analysis, wears away rocks and shakes even the highest mountains (Laozi: "Nothing is more flexible and weaker than water, yet to remove what is hard and strong, there is nothing to surpass it"). Water is like the blood that engenders bones, and like the marrow that nourishes them; a dead bone, one with neither blood nor marrow, is no longer a bone. A mountain with neither water nor clouds is no longer anything but a dead skeleton.[160] And that is why rock is called "the root of cloud" (*yun ben*): the mountain gives birth to cloud; vapor is as it were its breath.[161]

The clouds are the ornament of the sky and the earth, the embroidery of the mountain and the waters. They are as swift as a galloping horse. They dash themselves against the rocks [of the mountains in such a way] that one hears the noise that they make. Such is the vigor of cloud! In general, when the Ancients painted clouds, they had two secrets.

Firstly, at the spot in the landscape where a thousand peaks and ten thousand precipices gather in great numbers, they would conceal them by means of clouds. The blue-tinged peaks penetrate the sky and all of a sudden scarves of white spread out horizontally, separating them in layers. When the upper part of the mountains pierce through the clouds, their blue tips reappear. As writers say, this is the way to seek calm in precipitation. Using the five colors, this is how to charm the eyes of spectators.

Secondly, at the spot in the landscape where the mountains and the precipices are too few and far between, by using clouds one adds movement. Where there is neither water nor mountain the layers [of clouds] begin. They undulate like [the waves] of the great sea and seem to be formed from mountain peaks. It is the same as what writers call "making citations from poetry so as to increase the power of one's style."[162]

Mustard Seed provides a precise summary of the plastic functions assigned to /cloud/ in landscape painting. Where the composition is too cluttered, cloud makes it possible to articulate it and abbreviate it (this is called seeking calm in precipitation). Where, on the contrary, there is mostly emptiness, cloud serves to introduce movement. And the notion of *citation* makes a timely intervention at this point, for cloud makes *intertextual* exchanges possible between the mountain and the water — thereby imparting to this concept its full material resonance. The water and the mountain exchange their texts (their "characters") and even their *texture*, with the clouds seeming to be both the product and the sign of that exchange.[163] *Mustard Seed* places the question of clouds right at the end of its list of methods of painting landscapes because, yet again, they provide a *recapitulation*, "for in [their] elusive emptiness one catches many glimpses of mountains and water courses hidden there. That is why one speaks of mountains of clouds and seas of cloud."[164]

THE INSCRIPTION OF EMPTINESS

However, "the winds and clouds do not surround all landscapes in the same way, and rivers and rocks do not animate all landscapes following a single method of wielding the brush."[165] This brings us back to the question that was at stake in the debate between on the one hand the "blowers of clouds" and other practitioners of "spattered" ink, on the other those who favored the *brush stroke*. Chinese painting, like Western painting, was familiar with clearly delineated clouds as well as with mists with indistinct outlines. As we have seen, it was a difference in treatment that overlapped with an opposition that made a massive impact on "the history of art" as China knew it. The opposition was between the "blue and green manner" with clear and precise outlines and the *yi pin* manner of the scholars, the "manner without constraints," and in the eighteenth century this was duplicated by another opposition between two styles of landscape painting. The *mi* style was essentially descriptive and punctilious over details, and was favored by Gu Kaizhi and Lu Tanwei, who worked with such care that "one could not see where their lines ended." Meanwhile, the "abbreviated" *shu* style was adopted by Zhang Sengyou and Wu Daozi who on the contrary spaced out their dots and lines so much that completely void spaces appeared ("in this their thought was fully expressed, even though the drawing remained incomplete").[166] The names that tradition links with the two styles are of artists who were famous for their talent not only as painters but also as calligraphers. The work of painting could not be dissociated from that of writing, which, using the brush and ink as it did, involved the same elements and proceeded from the same dialectic (to such a degree, indeed, that by the Yuan period scholars already preferred to designate painting by the *xie* character 寫, "to write," rather than by the *hua* character 畫).[167] But writing and calligraphy had, even as early as the Han period, been subject to a similar division between the regularity of "correct" (*zheng*) writing, used for official documents, and the cursive so-called grass writing (*cao shu*), the rough style, which was for a long time prohibited and possibly owed part of its success to that disapproval, which, however, became increasingly relative as the years passed.[168]

Of course both historical and theoretical consequences stemmed from the fact that, both in the Far East and in the West, painting was linguistically associated with writing: in both, the same word or character (graphein, xie, 寫) was used to designate the two practices. In the West, the paradigm, given that it operated phonetically, led to the pictorial process being analyzed in terms of representation, derivation, *projection* (see Chapter 3, p. 116), on the basis of a structure that was already verbalized and articulated in accordance with the ways of *logos*. In China, in contrast, writing did not lead to a phonetic analysis of language and was not seen as a more or less faithful transfer of speech, and that fact in its turn delivered painting from its dependence upon a preexistent "neutral" totality that claimed to be the totality of all that was signified. In other words, painting was definitively liberated from any dependence upon language, upon *logos* as *phonè* and—with the backing of the argument that a deep connivance running through the entire instituted system linked the lines traced in pictures with the lines traced by writing—it was established at the origin of the interaction from which language, alongside "logic" or "science," stemmed.

Suffice it to note at this point that, where both writing and painting were concerned, cloud was associated with the kind of transgression that is justified by a quest for *spontaneity*: "Calligraphy has the spontaneous [*ziran*]. Once the spontaneous is there, the Yin and the Yang are manifest. Once the Yin and the Yang are manifest, the drawing of forms appears."[169] It was while watching the summer clouds glide by on the wind that the monk Huaisu (eighth century) suddenly seized upon the idea of the brush and obtained *samadhi* (ecstasy) from ink;[170] and, similarly, it was the views of clouds painted by Mi Fu that revealed the *samadhi* of ink to Dong Qichang, the great critic of the Ming period.[171] In his *Hua yang*, the same Dong Qichang declared that a landscape painter "should apply his thought to the engendering of clouds. Colors cannot be used. [The effect] should spring from the purity of the ink, with a result resembling the vapours of breath, which always seem on the point of falling. Only then can one speak of the harmony of the vital movement."[172] One can see how it was that Zhang Yanyuan, while agreeing that through this "wonderful solution" one seizes upon the principle of nature, assimilated those who operated with *spattered* ink or (by virtue of the homophony of the characters) *broken* ink to the "blowers of clouds." And although, as Nicole Vandier-Nicolas observes, it is hard to tell quite what the first masters of water washes understood by this, later texts are perfectly explicit:

In the seventeenth and eighteenth centuries, painters who worked with spattered ink first constructed their subject by casting light (*zhao yin*) upon the mountains and the forests. They thus left empty spaces between the forms that they had suggested, so as to allow the breath to circulate throughout the painting. Next, with a few light touches, they made sure that "the empty spaces and the full ones" were connected, and then they animated the whole scene by setting in opposition the blacks made with dry ink and the pale tones treated with ink saturated with water. Once the painting had dried, what remained to be done was to "encage" the darkest points with a veil of pale ink: namely, the mountain summits, the peaks, and even the spaces left empty between the swathes of mist.[173]

So just as *Mustard Seed* declares, it truly is the ink — the ink that differentiates between the shadows and the light — that first gives the "outlines" by "shedding light" on the mountains and the forests. But, once again, that does not mean the ink on its own, without the brush; rather, it means that, in this particular operation, the brush must not be detectable.[174] The endeavor of many calligraphers-painters to allow no trace of their labor to be detectable in the finished work was matched by the painters of views of mists and clouds. However, it is also important to repeat that Chinese painting can by no means be reduced to a more or less pictorial "cloudiness." As we have seen, Mi Fu himself harbored reservations about the landscapes of Li Cheng, painted in very dilute ink. But it was left to an eighteenth-century critic, Tang Zhiqi, to rally to the traditional and *dialectical* concept of landscape painting, and condemn the reduction implied by procedures that gave precedence to the ink over the brush, to the water over the mountain, and—in the last analysis—to painting over reality:

The *qiyun shendong* (revolutions of breath, movements of life) and vaporous or unctuous effects are not [things] of the same [order]. Sophisticated people are much mistaken when, in connection with those vaporous and unctuous [effects], they speak of the movement of life (*shendong*). . . . Breath can [manifest itself by means of] lines, ink and color. One can also consider its organizational capacities, its strength, and its dynamism. All these have to do with its revolutions. As for the vital movement itself, revolutions of breath [as conveyed by the brush] cannot take its place.[175]

All credit to Tang Zhiqi for having made clear the fundamental principle of the Chinese theory of painting: painting should be recognized to be a specific signifying practice. It is on the basis of that specificity, of the *difference* upon which it is founded as a signifying practice, that painting should be considered in its relationship to reality—a relationship of understanding rather than expression, of *analogy* rather than duplication, of *working* rather than substitution. But *working* and *understanding* mean not just a working with and understanding of the ink and the brush. The text quoted above reveals another "trait" that has nothing to do with linearity and that is characteristic of Chinese painting, namely the role that falls to empty space and, through it, to the substratum (or material upon which the picture is painted).[176] "Chinese artists also have a great deal of room on their paper. Some parts of the surface seem unused. Yet they play an important role in the composition. Their dimensions and shape seem to have been planned with as much care as the outlines of objects. In these gaps, the paper or linen has a value of its own. The artist does not wish to deny the surface as a whole by covering it entirely. A mirror in which something here is reflected retains its value as a mirror. What this implies is a happy rejection of the complete submission of the spectator, for he can never be totally persuaded by the illusion."[177] In the West, the fact is that the brush is called upon to *cover* the canvas, in other words, to make it disappear under the applied layer of preparations, oils, pastes, and varnishes, thereby, at the level of the senses, magnifying the "annihilation" of the substratum that is implied by the perspective construction. Also, in the West, phonetic writing tends to efface its materiality and strive to make itself independent of any substratum in order to operate as closely as possible to *phonè*. In contrast, Chinese painting and Chinese writing always derive part of their radiance (*shen zai*) from the quality of the silk or paper that is used. Some calligraphers even thought that that radiance should take precedence over the *xing shi*, the material body of the character drawn. So it is even more remarkable to find Guo Xi declaring in his *Linquan gaozhi* that mists and clouds give the mountain its *shen zai*,[178] thereby clearly indicating the link between /cloud/ and the substratum as it is revealed in its specific radiance by the paradoxical *inscription of emptiness* that is authorized by the dialectic between the ink and the brush. (But by the same token, clearly /cloud/, once it assumes this eminently scriptural function, will not necessarily lend itself, on a flat, two-dimensional surface, to being detectable or delimited, as a denoted unit or an identifiable graph: "views of mist and clouds" may well occupy an enviable place in the hierarchy of pictorial forms; but quite possibly the whole of Chinese painting contains not one single *cloud* in the sense in which Ruskin *understood* the word.)

Fabric and Garment

> We see then that everything our analysis of the value of commodities previously told us is repeated by the linen itself, as soon as it enters into association with another commodity, the coat. Only it reveals its thoughts in a language with which it alone is familiar, the language of commodities. In order to tell us that labour creates its own value in its abstract quality of being human labour, it says that the coat, in so far as it counts as its equal, i.e., its value, consists in the same labour as it does itself. In order to inform us that its sublime objectivity as a value differs from its stiff and starchy existence as a body, it says that value has the appearance of a coat, and therefore that in so far as the linen itself is an object of value, it and the coat are as like as two peas.
> —Karl Marx, *Capital*

Throughout its entire history—a history that a pictorial text *describes* within its own order and at its own specific level—Western thought, from Aristotle down to Leonardo da Vinci and to Descartes, has stubbornly rejected the idea of *emptiness*. It has done so through an ideological impulse, which, in its turn, indicates a far more profound rejection, in which—as now, following Althusser's and Sollers's work on Lenin's text, we are beginning to see—materialism turns out to be what that thought has *suppressed*. In the pictorial field such a rejection or suppression finds expression in the "annihilation" of the material and technical substratum of the painted image. And that neutralization or annihilation was accomplished by the institution of the perspective space in the guise of an objective setting, a perceptible continuum that borrows its substance from light. East/West: as we have seen, cloud seems to fulfill perfectly symmetrical functions in the two systems. In the Western system it serves to conceal the very principle that, in the Far East, it is its function to produce (although one question that remains unresolved is whether it is justifiable to associate under the same rubric, purely on the grounds of their sharing the same *denotation*, elements that, in two such heterogeneous systems, assume values and functions whose symmetry may be no more than apparent). More immediately, however, the digression by way of China has manifested the fundamentally materialist articulation between emptiness and the substratum to which a painting is applied—emptiness acting as the substratum, the substratum acting as emptiness; and at the same time it has revealed the meaning of the work by means of which Western painting has systematically endeavored to obliterate the fact of that annihilation. To take but one particularly revealing example, let us consider *The Dog in the Arena*, one of Goya's bullring scenes in the series of "black paintings" in the Qinta del Sordo. Even when the painting seems to be opening up to admit emptiness, that opening up still obeys the rules of trompe l'oeil: the zone that is left empty only presents itself as such thanks to its figures being evacuated and rearranged round the edges of the composition. The impression of "emptiness" obtained in this way is reduced to the effect of a "lack" that simply emphasizes the fullness of the "background" against which the figures stand out. This is in conformity with the requirements of a system that could only produce the scene of the representation by replacing the medieval background of gold by an illusionist space constructed according to the rules of a more or less constraining perspective, *an illusionist space that itself constituted a figure.* The formal coherence of the system was definitely provided not so much by the rigor—a frequently relative rigor—of the figures portrayed but rather by the degree to which the substratum was foreclosed. (That is why it was not until our digression by way of China that we could introduce the word *dialectic*, at last giving it its full theoretical resonance.)

The decisive break came with the last works of Cézanne, in which, in the gaps, in what is lacking in the image, the canvas itself manifests its material nature, while the attention paid to the flat surface of the picture wins out, once and for all, over endeavors to create an illusion of depth. It is through this shift from an *image*, offered to the imagination, to a *picture*, offered as such to the spectator's perception, even more than through the deconstruction of the traditional space that made that shift possible, that Cézanne's work at the turn of the twentieth century marks a *break*. To be sure, Impressionism, by replacing the in-depth superposition of layers of paint and glaze by the juxtaposition of discrete dabs of color on the flat surface, had already restored a semblance of physical reality to the pictorial surface. (Huysmans even remarked, in connection with Whistler's paintings, that "the canvas is hardly full, here and there it even shows its grain.")[179] But for Monet, a dab of paint was still an instrument used to get forms to dissolve in the atmosphere. It was only with the advent of Cézanne that dabs of paint on the canvas assumed a constructive function.[180] Concerned as he was not to note his impressions, but to transcribe *sensations*, stubbornly pursuing "the realization of the part of nature which, falling before our eyes, gives us a picture,"[181] Cézanne finally enabled the letter of the picture to triumph over the cipher of representation. Between the interpretation of the model and its realization,[182] the picture offers itself as a *finite* place of a conversion in which the dab of paint, which repeats that place in its form and position,[183] presents itself as the operator. A dab of paint, not a line: for the method of construction cannot be reduced to drawing, which only conveys the configuration of what is visible but cannot capture the "reflection," the light that, through "general reflection," envelops things. (But a reservation of capital importance is called for at this point, namely that *light does not exist for the painter* and, from a theoretical point of view, it is necessary to distinguish between *optical* sensations, which are produced in the visual organ and which classify planes according to their degree of relative luminosity, and *coloring* [not "colored"] sensations, that is to say the signs that *represent* those same planes on the canvas, by means of the conversion that Cézanne denotes as a "realization," and that makes it necessary to raise the question of the signifier in painting in theoretical terms.)[184]

The radically new relations that Cézanne introduced between drawing and color are such that "the strokes which build up the objects in all their compactness are open forms," the outline of an apple being constituted by "innumerable touches that overlap each other but also slip into the surrounding objects," and sometimes, paradoxically, are to be found outside the apple.[185] But this partly double attention — attention paid to the text of nature, and attention paid to the picture as a text, a text, as Clement Greenberg points out, whose density in the last analysis predominates over that of the object—also explains how it is that the canvas finally breaks through in the intervals between the dabs of paint applied to it here and there. The *thesis to be developed* is: "whatever our temperament or our power in the presence of nature, to provide an image of what we see, forgetting all that appeared prior to ourselves." As he grew older, Cézanne was faced with problems that he had not expected and that increased the difficulty of his task: "Now, at about seventy years of age, my coloring sensations which give light are causing abstractions that prevent me covering my canvas or proceeding with the delimitation of objects when the points of contact are tenuous and delicate; *as a result, my image or picture is incomplete*." [186] Of course, the extraordinarily innovative aspect of Cézanne's oil paintings and also of the watercolors that he produced in his last years should not be ascribed to any waning in his visual acuity; on the contrary, what is important is that the painter found his diminishing "power in the presence of nature" good reason for allowing free rein to his coloring sensations, [187] in other words, again, to the signs of his optical sensations, thereby opening up the way for *abstraction*, an abstraction that, as he himself said, with considerable *prescience*, broke away both from the annihilation of the substratum of the painting and from delineation, and was the consequence of the dialectical conjunction of color and surface. For at this point there was no question of continuing to try to save figures by means of lines, as the Neo-Impressionists were to: on the one hand the image (but not the *picture*, as the painters who took over Cézanne's work where he left off were to be keen to show) was incomplete; "on the other hand the planes fall one upon another, and this gives rise to the Neo-Impressionism that circumscribes contours with a black line, *a mistake that should be opposed as strongly as possible*." [188]

It is remarkable that, even as he lamented the infirmities of old age, Cézanne perceived the scale of the break that his work heralded. The idea repeatedly crops up in the correspondence of his last years, sometimes with a prophetic tone ("I am doggedly working, and catch glimpses of the promised land. Shall I be like the great leaders of the Hebrews, or shall I be able to enter it?"). [189] So well aware was he of the altogether transgressive nature of his work in relation to the traditional canons of depiction that he had no hesitation in comparing himself to the painter of *The Unknown Masterpiece* ("Frenhofer, that's me"), the author of the "fictional picture."

This "wall of painting," as Balzac called it, owed nothing to the outward forms of representation and offered nothing to seize upon to any analysis that proceeded from denotation—unless, that is, amid "this chaos of colors, tones, and vague nuances, a kind of formless fog," the spectator could make out the tip of a naked foot, "a fragment that had escaped a slow, incredible destruction," as Pourbus did in Balzac's story and, feeling somehow reassured, cried out, "There's a woman underneath it all." The canvases of Cézanne's last period are certainly not palimpsests; but, as in that "wall of painting," the unknown masterpiece, the lateral relations between one dab and another, one tone and another, definitely win out over the vertical relation between figures and their referent, the only relation understood by any reading limited to the order of verbal denotation. It is laterality that in its turn establishes itself as the order of *literality*, in which even the "blanks" "assume an importance" (Mallarmé: "Un coup de dés jamais n'abolira le hasard"),[190] and in which the substratum of the painting becomes the equivalent of emptiness, the same void that constituted a substratum upon which, according to Epicurus, atoms clustered together in variable order and arrangement, "like letters which, although there are not many of them, nevertheless, when they are arranged in different ways, produce innumerable words."[191]

Cézanne, who was an assiduous reader of *De rerum natura*, declared that, in order to paint a landscape well, he first had to discover its geological foundations.[192] Accordingly, instead of the studies of skies and clouds of his Romantic and Impressionist predecessors, he produced a mass of studies of rocks, depicted in all their stratification and fragmentation and with all their faults—rocks that he analyses so thoroughly in terms of his coloring sensations that he makes them look like clusters of cloud, thereby reversing Lucretius's observation on clouds ("Often we seem to see high mountains advancing, trailing loosened rocks attached to their sides . . ."). But this reversal, unlike that implied by the "service of clouds," carries a truly dialectical impact in so far as Cézanne's deconstruction, far from following in the wake of a more or less fantastical "cloudiness," on the contrary produces, as a material component of the pictorial process, the very element that Romantic landscape painting was still endeavoring to obliterate, namely the surface, as the substratum for any inscription and any construction, the substratum as surface, the raw material that it is the job of painting to articulate. The production of that surface, that substratum,

confers meaning upon the interminable theory of /cloud/ in Western painting, and at the same time makes manifest the way that it has been used as a
screen to mask a suppression: the suppression of the signifier, the suppression of *painting* as a specific practice, a materialist process of production.
It was up to theory, understood not as a procession of works or texts but
as a production of concepts, to show how a certain "cloudiness" was used
to mask the return of something suppressed that was indicated in the very
contradictions of Ruskin's text, and to pinpoint the moment when, in the
work of Cézanne, that censorship was lifted. However, as the present text
draws to a close, it is important to dispel or prevent any confusion. The
substratum, the "canvas" as revealed by Cézanne and set up as a signifier,
is by no means a *given fact*, as positivist ideology would have it. (So much
is perfectly clear when you reflect, as Meyer Schapiro does, that the prepared, delimited, pictorial flat surface, let alone the free-standing panel, is
a relatively recent acquisition in the history of the human race, and that
for many thousands of years painting used as its substratum the extremely
irregular surfaces of cave walls, sometimes going so far as to exploit certain
accidental bumps in the rock, as for instance in Altamira.)[193] The canvas
is the product of a history, the history of Western painting, a history that
has yet to be written from a materialist point of view.

In particular, the emptiness that Cézanne resuscitates in his watercolors should not be confused with the emptiness exploited in Chinese
painting, which, for its part, is the product of a *different* history, also one
that has yet to be written. It is today possible to put forward the idea of such
a history and also that of a general theory that would interweave particular
histories within a dialectical perspective; and the reason why this is possible
is that contemporary painting has developed in such a way that it demands
that the task be undertaken. Hence the paradox in which contemporary
painting finds its point of departure, a paradox that El Lissitzky expressed
in his own way in his lecture "New Russian Art" (1922), when he declared at
one and the same time that, since art never evolved, the new pictorial order
owed nothing to the past, and also that the flat surface of the suprematist
canvas appeared as the ultimate expression of space, the last link in the long
sequence of "impressions of space" that constituted that evolution.[194] The
infinite white plane of Malevitch, the *finite*, square, or rectangular panels
of Mondrian (although now the notions of both the finite and the infinite
no longer mean what they did in the context of the Renaissance), appear at
once as the direct product of the break made by Cézanne and as the point

of departure for modern painting: painting at last liberated from the clouds that used to burden it. But of course we should also note the returns that cloud made in the Cubist period, as an element used both constructively and as a gap filler in Delaunay's *Windows* and *Towers* and in Léger's *Wedding* (see Plate 8), its recurrence, again in Léger, as a "demystified" object, surrounded by the same outline as a leaf or the handlebars of a bike, a tricky element, to be fixed, as the painter put it, by the metallic structures of his *Builders*; and later, in Liechtenstein, where it appears in the guise of a rent in the continuity of a regular fabric. Nor should we forget the role that it assumed, as a reflection repeated ad infinitum in Monet's series of *Water-lilies* (contemporary with the first compositions of Mondrian), the format of which prepared for the work of Pollock, extended to the dimensions of whatever canvas was able to sustain its impact. But those returns and detours only take on their dialectical meaning in relation to another history, improbable as yet, that of modern art, from which the present study both proceeds and at the same time claims to do no more than mark its threshold. In such a history, painting, to borrow Marx's language at last, would cease to pretend that the sublime reality that the garment confers upon the fabric/canvas can be separated from the latter's more or less stiffly woven body. This history would endeavor to express its thinking in a language other than that of *merchandise*, the language that made Pontormo consider a painting as a kind of woven fabric indistinguishable, apart from the relative value attached to it, from the woven fabric of the canvas itself. This would be in conformity with the principle set out by Marx, according to which all work, in as much as it produces value, can be reduced to one and the same abstract measurement that makes the garment the equivalent of the fabric and the mirror of its value. (In the classical system, the fabric/canvas has a value, a derisory value, only because it is related to the picture painted upon it, a reference that abolishes its materiality, for only the superficial crust that covers it, the clothing that conceals it, is worth something.) The history that I have in mind, because it would be materialist, but qualitative rather than quantitative, and dialectical rather than quantifiable, far from treating pictorial work as merchandise solely from the point of view of the value that it produces, would conceive of it as materially determined, as a specific practice the productivity of which would be measured by the extent of the effects to which it could lay claim in the symbolic order.

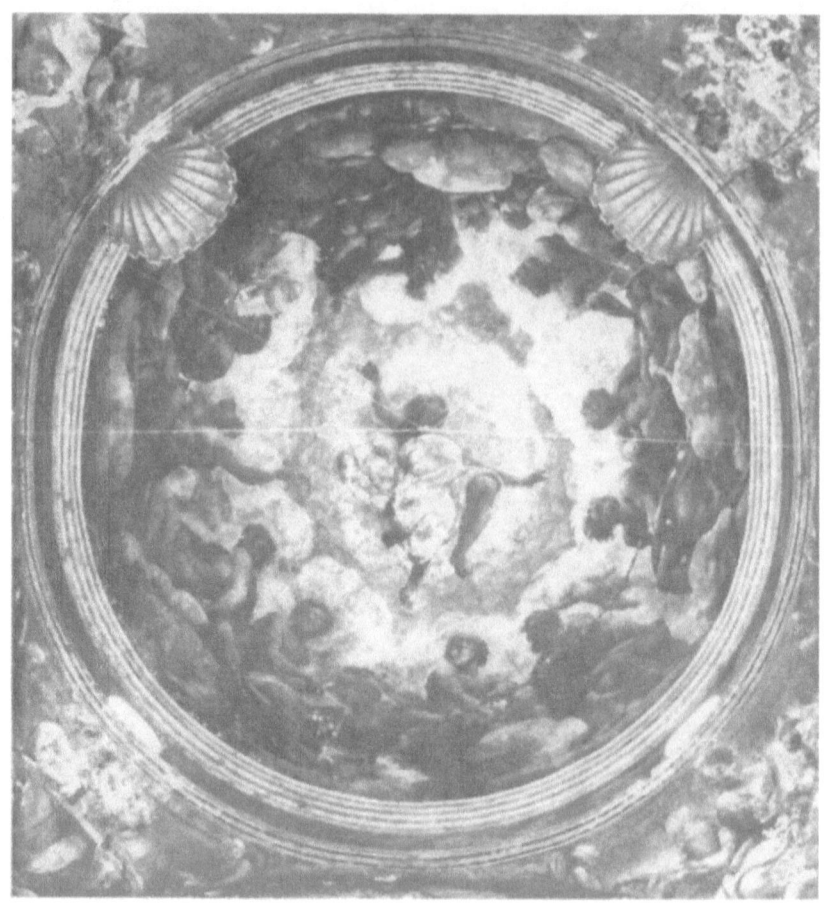

PLATE IA. Correggio, *The Vision of Saint John on Patmos*. The cupola of San Giovanni Evangelista, Parma (Alinari-Giraudon).

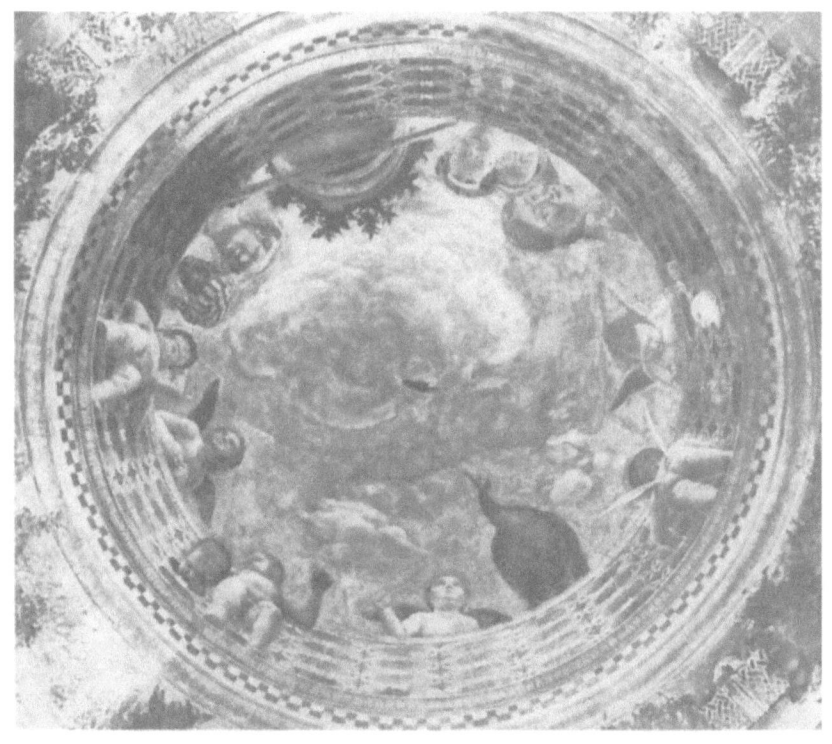

PLATE IB. Mantegna, lunette on the ceiling of the Spouses' Chamber, Ducal Palace, Mantua (Anderson-Giraudon).

PLATE 2. Zurbarán, *The Vision of the Blessed Alonso Rodriguez*. Academia San Fernando, Madrid (Anderson-Giraudon).

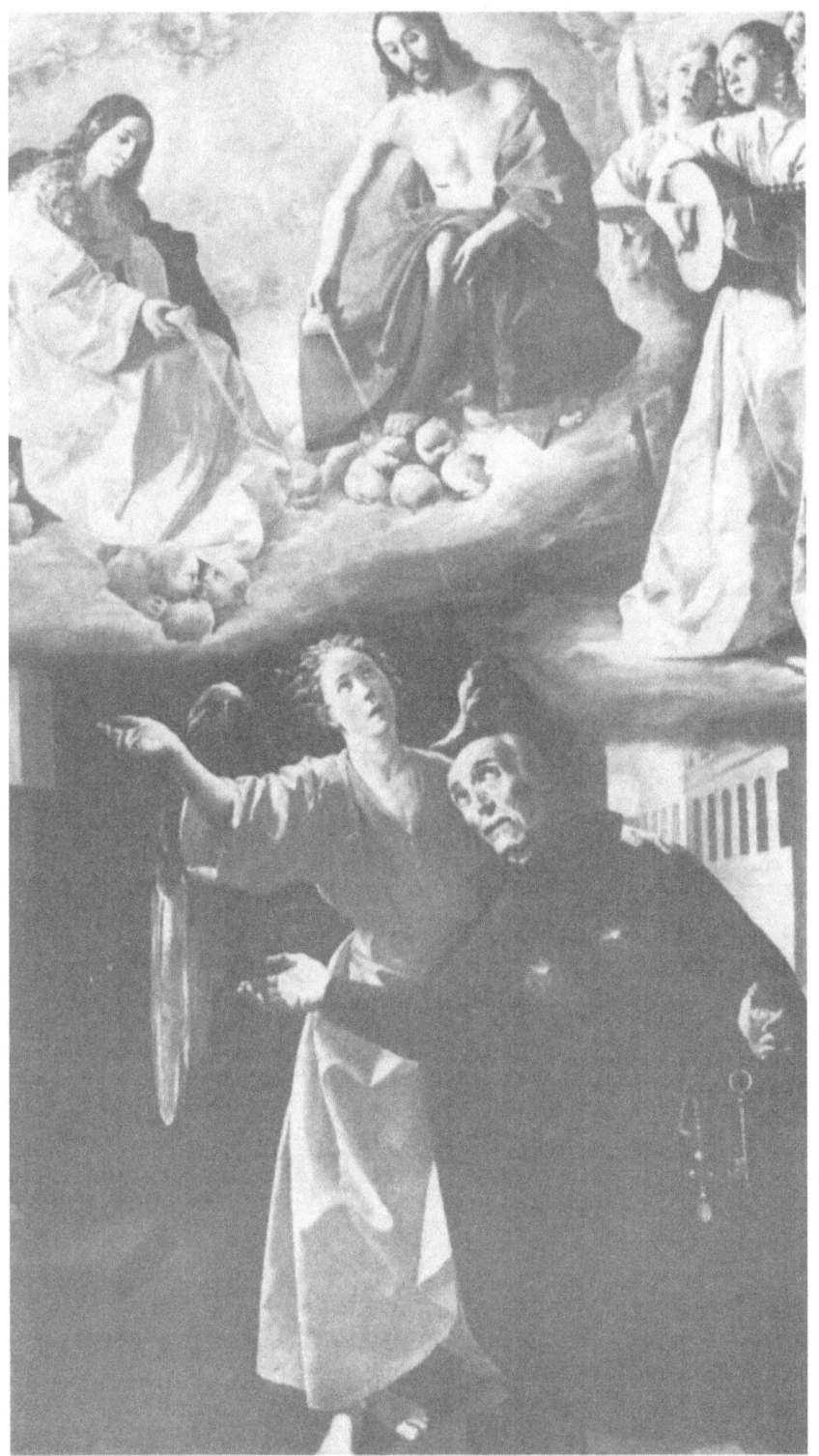

PLATE 3. Mantegna, *The Ascension of Christ*. Right-hand panel of the triptych *The Adoration of the Magi*. Uffizi Gallery, Florence (Anderson-Giraudon).

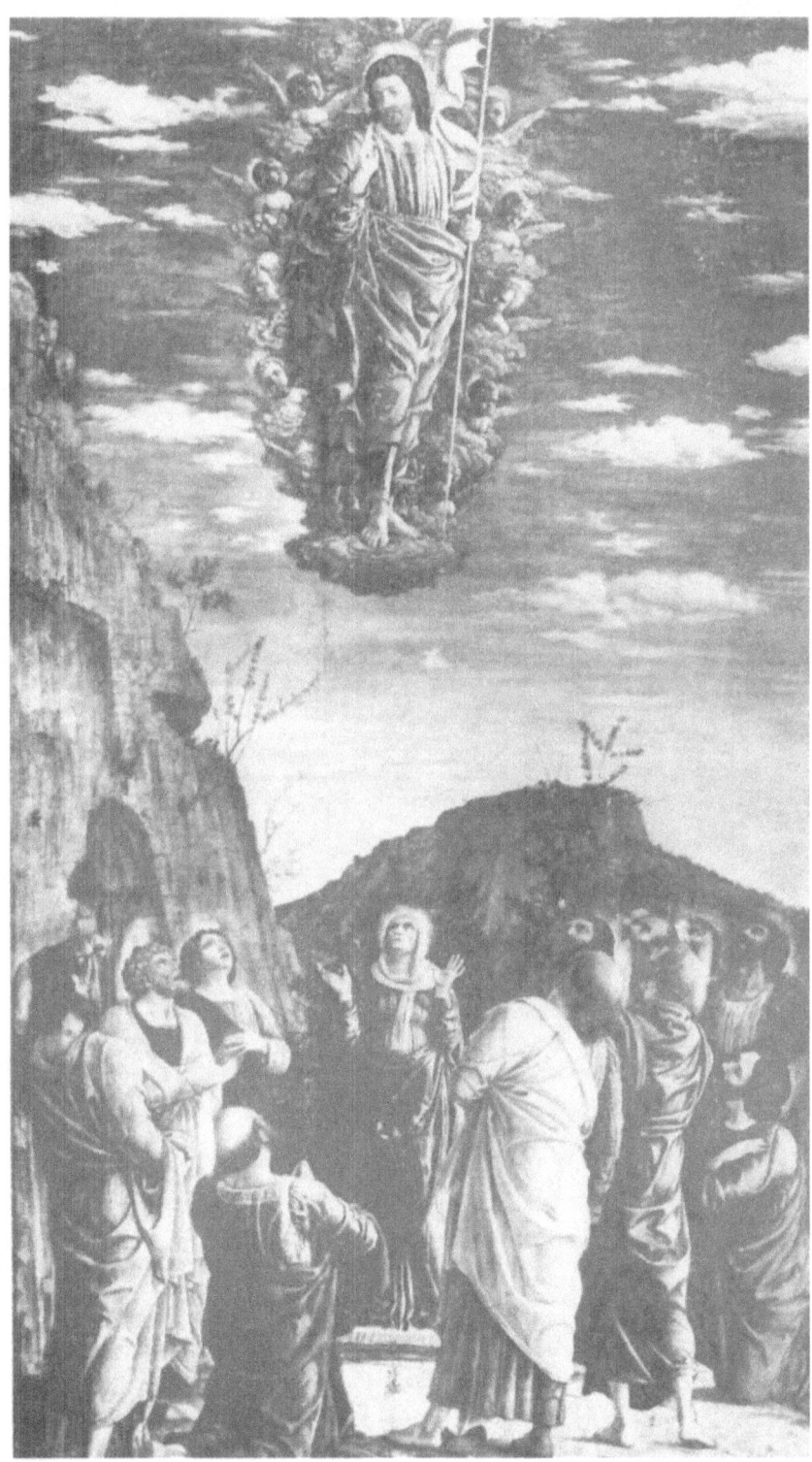

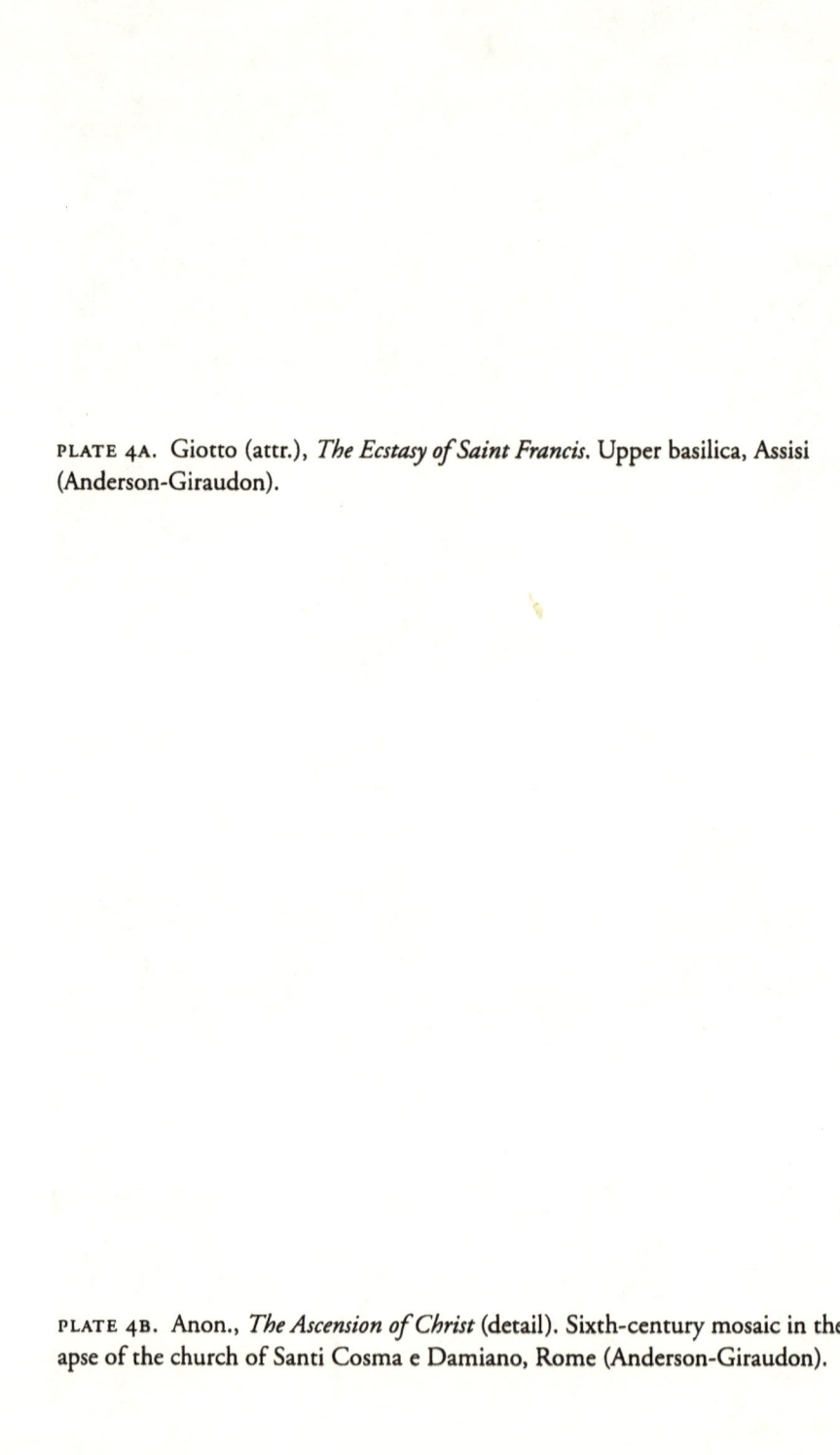

PLATE 4A. Giotto (attr.), *The Ecstasy of Saint Francis*. Upper basilica, Assisi (Anderson-Giraudon).

PLATE 4B. Anon., *The Ascension of Christ* (detail). Sixth-century mosaic in the apse of the church of Santi Cosma e Damiano, Rome (Anderson-Giraudon).

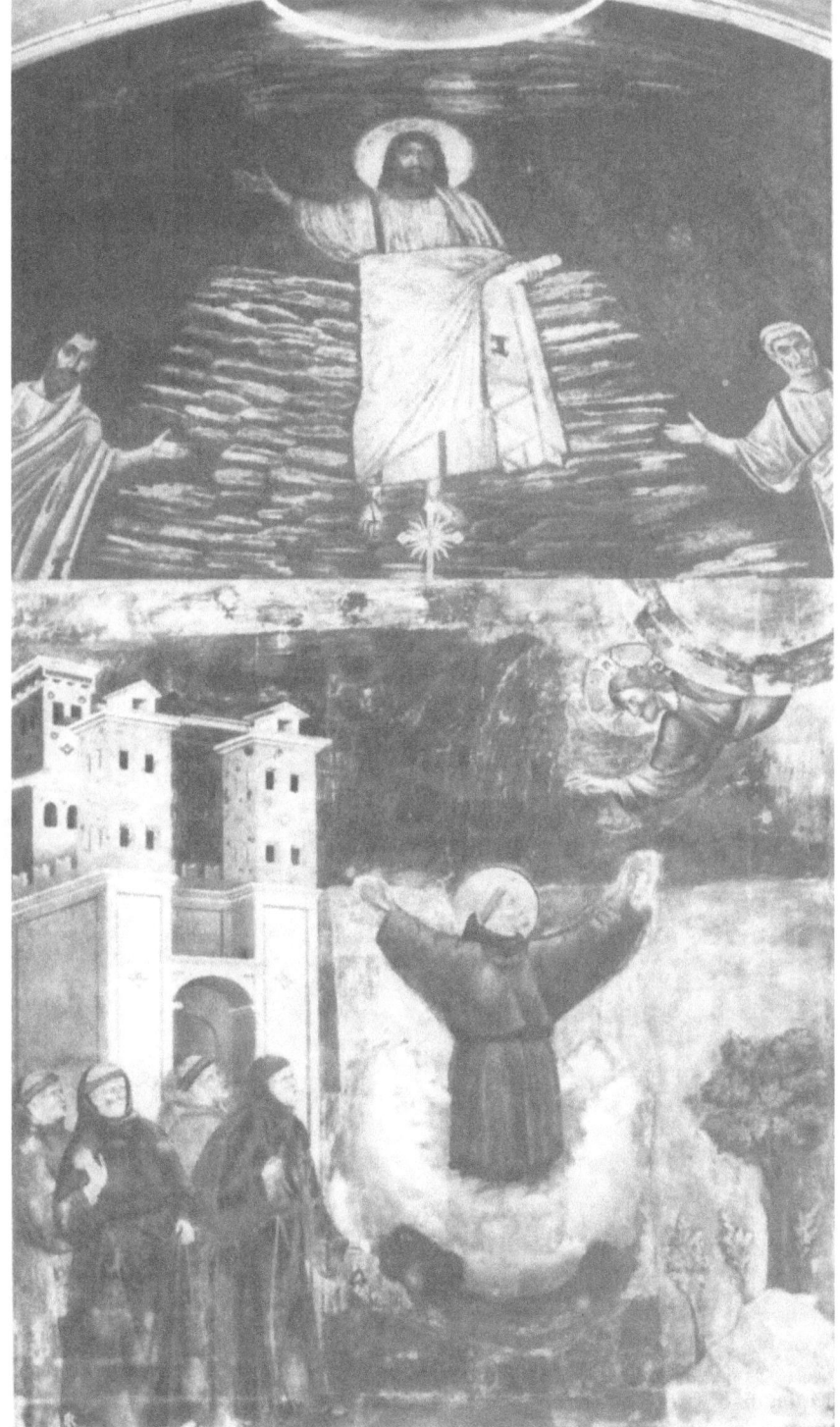

PLATE 5. Masolino, *The Foundation of Santa Maria Maggiore*. Capodimonte Museum, Naples (Anderson-Giraudon).

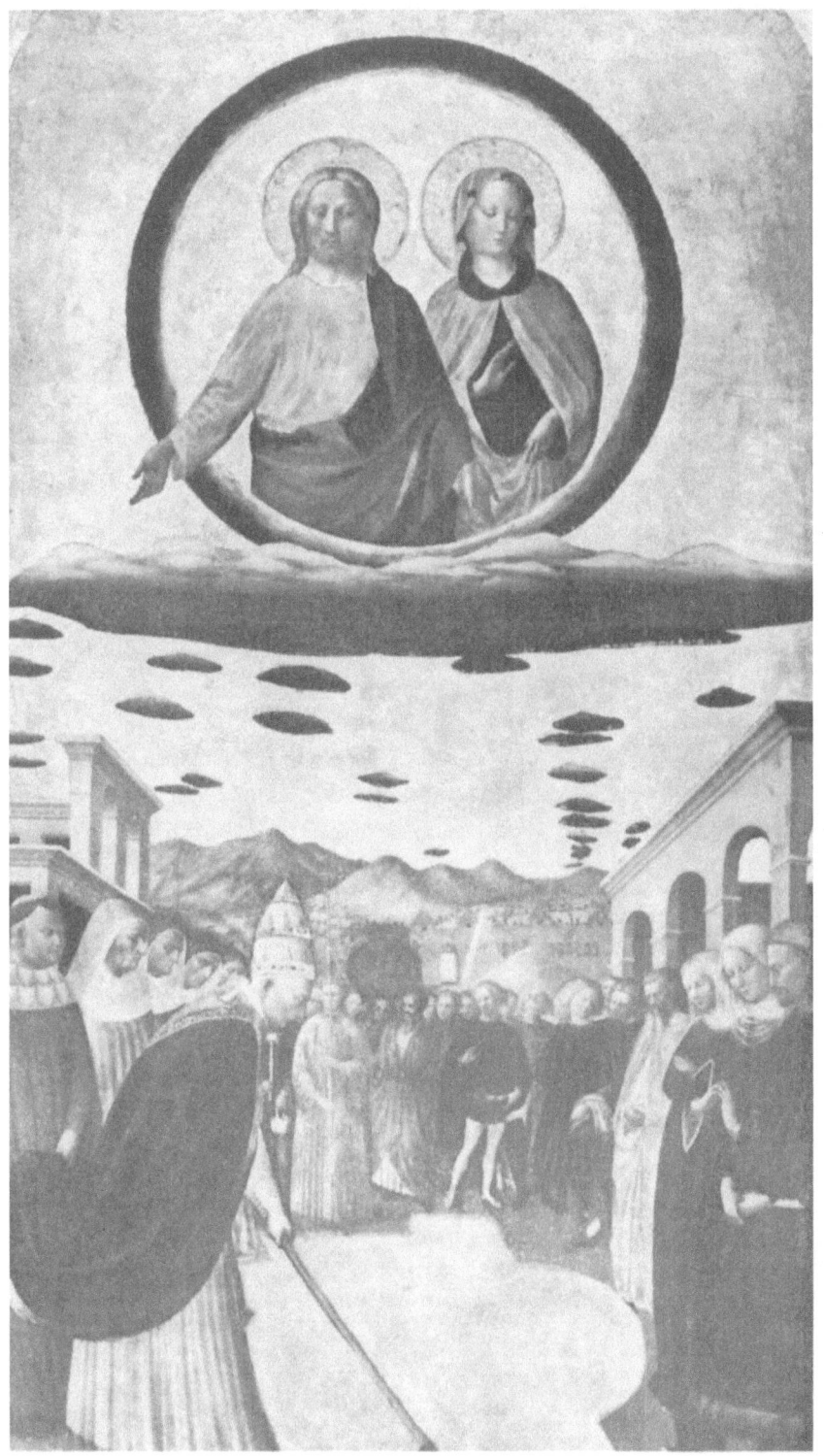

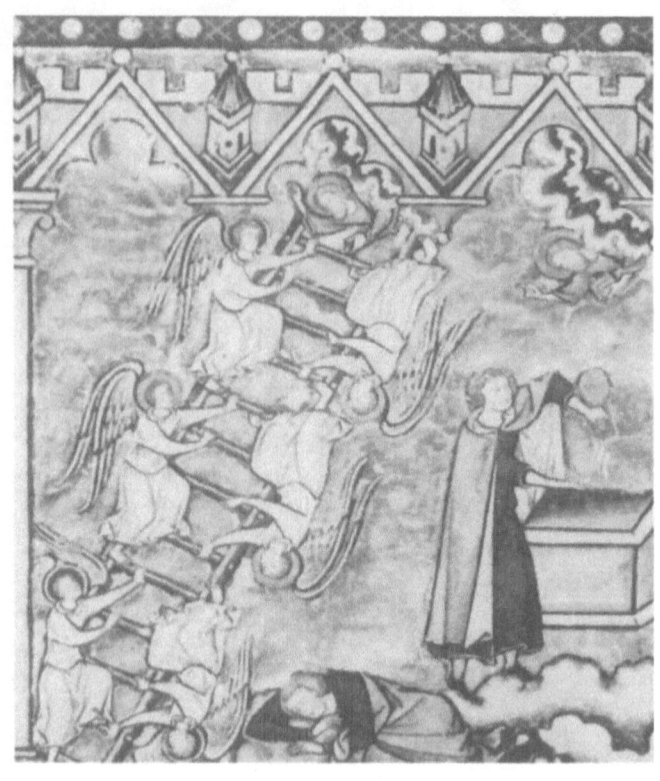

PLATE 6A. Anon., *Jacob's Dream* (Bible, c. 1300). MS 138, fol. 42r, Pierpont Morgan Collection, New York.

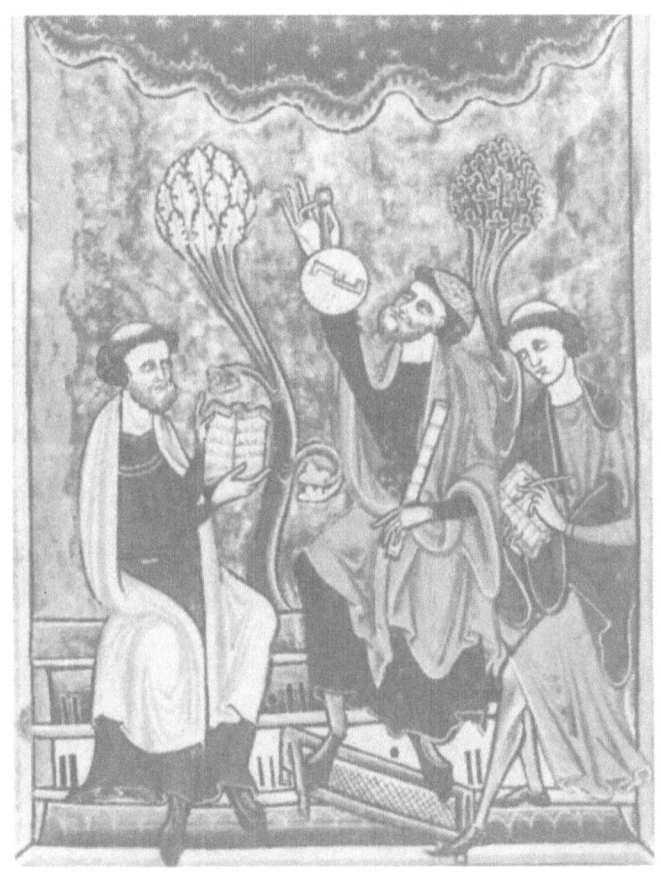

PLATE 6B. Astronomer, copyist, and calendar maker (Paris Psalter, thirteenth century). MS a.46185, fol. 1v, Bibliothèque de l'Arsenal, Bibliothèque Nationale de France, Paris.

PLATE 7A. Constable, *A Study of Clouds* (dated 6 September 1882). Victoria and Albert Museum, London.

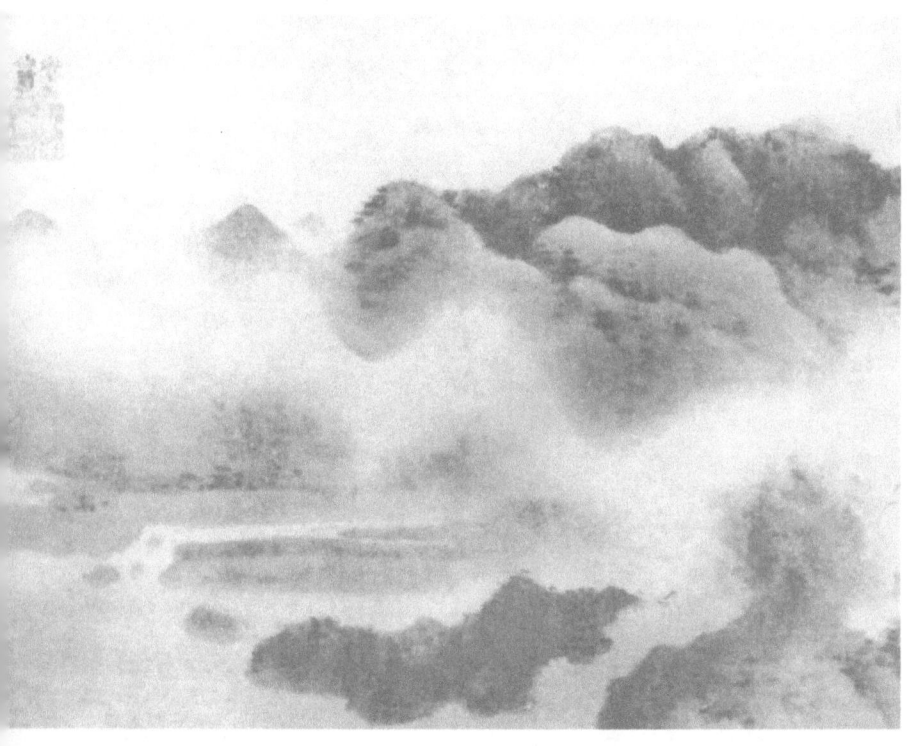

PLATE 7B. Gao Ranhui (attr.), *Summer Mist* (thirteenth or fourteenth century).
Traditionally attributed to Mi Fu. Private collection, Japan (Photo J.-R. Masson).

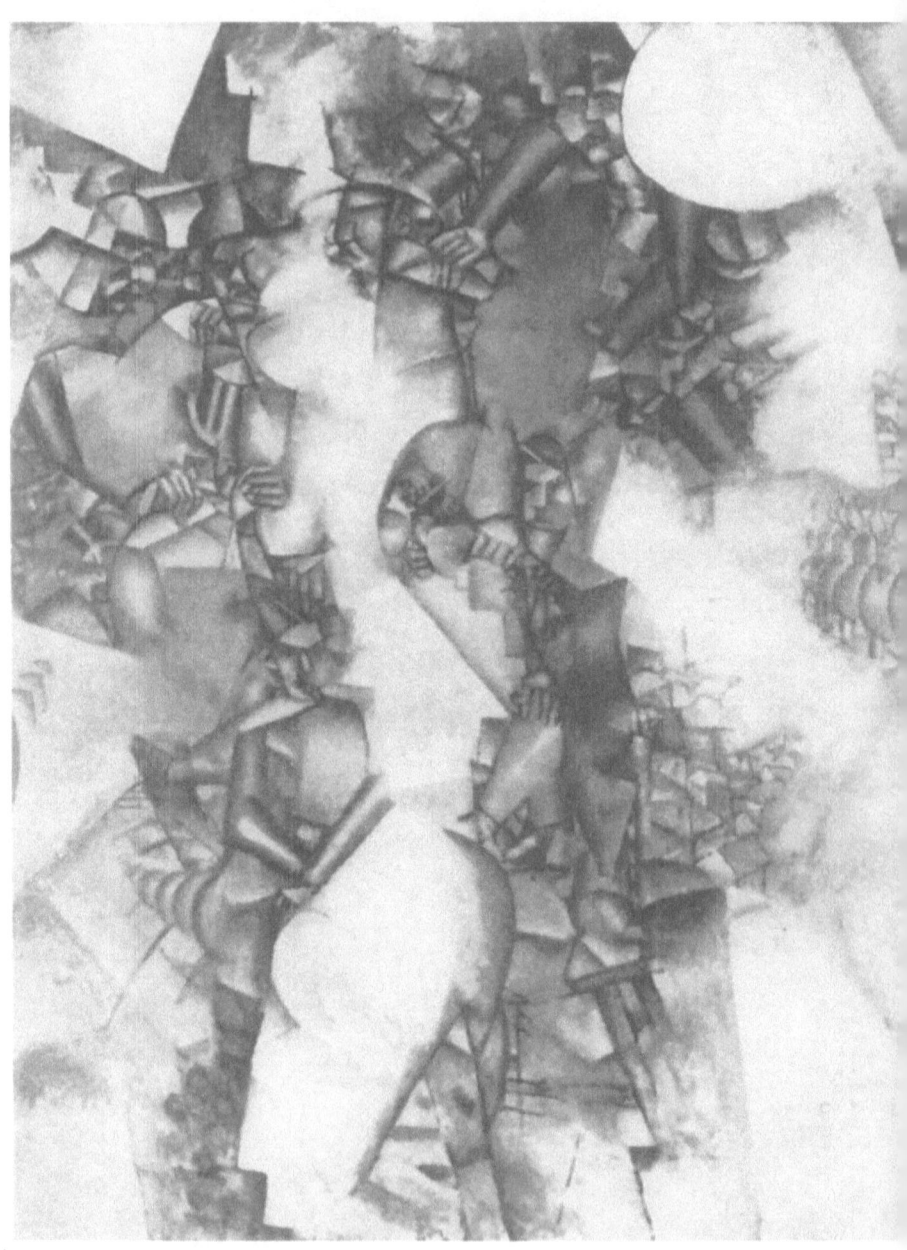

PLATE 8. Léger, *The Wedding* (1911–12). Museum of Modern Art, Paris (Giraudon).

REFERENCE MATTER

Notes

CHAPTER I *Sign and Symbol*

1. "The most beautiful of all the cupolas that have been painted before or after him." Anton Raphael Mengs, *Memoirs Concerning the Life and Works of Anthony Allegri, Denominated Correggio*, in *Works* (London, 1796), vol. 2, p. 23.

2. See Augusta Ghidiglia Quintavalle, *Gli affreschi del Correggio in San Giovanni Evangelista a Parma* (Milan, 1962), figs. 8–11.

3. Burckhardt is here alluding to the tradition according to which, on the occasion of the consecration of Correggio's cupola, one canon compared it to "frog soup." See Corrado Ricci, *Corrège*, French trans. (Paris, 1930), p. 103.

4. Jacob Burckhardt, *Der Cicerone*, 4th ed. (Leipzig, 1879), vol. 2, pp. 698–700; English trans.: *The Cicerone: An Art Guide to Painting in Italy* (London, n.d.), pp. 181–82.

5. Heinrich Wölfflin, *Renaissance et Baroque*, French trans. (Paris, 1967); English trans.: *Renaissance and Baroque*, trans. Kathrin Simon (London, 1964), p. 64.

6. Wölfflin, *Renaissance and Baroque*, trans. Simon, pp. 64–65.

7. See Giovanni Bottari, *Raccolta di lettere sulla pittura, scultura, ed architettura* (Rome, 1767), vol. 1, p. 86.

8. Probably the cathedral cupola. But this same Annibale was no doubt also able to study the whole of the interior decoration (then intact) of San Giovanni Evangelista. In 1586, having settled in Parma "to devote himself entirely to the study of Correggio," he even set to work, with his brother Agostino, to copy the fresco in the apse, which was about to be destroyed in order to make it possible to extend the choir. The fresco is now known only through the fragments that were collected and preserved in the National Gallery of Parma and the National Gallery of London, and through the copy by Cesare Aretusi, which took the place of the original in the reconstructed apse. See Giovanni Bellori, *Le vite dei pittori, scultori, et architetti moderni* (Rome, 1672), pp. 22–23.

9. Anton Raphael Mengs, *Reflections on Beauty and Taste*, in *Works*, vol. 2, p. 45.

10. In the above-mentioned sense of the term. For there is a great difference, according to Mengs, between taste, which stems from the choice of parts and which (according to the doctrine of the academicians) aims to "improve" nature, and what is generally called "manner," a kind of fiction or imposture that consists in omitting some elements or inventing some that defy the limits of nature (ibid., p. 50).

11. See Marie-Christine Gloton, *Trompe-l'oeil et décor plafonnant dans les églises romaines de l'âge baroque* (Rome, 1965).

12. Wölfflin, *Renaissance and Baroque*, trans. Simon, p. 64.

13. See Anton Raphael Mengs, *Reflections upon Raphael, Correggio, and Titian, and upon the Works of the Ancients*, in *Works*, vol. 1, p. 131.

14. Gloton, *Trompe-l'oeil*, p. 99.

15. "Accrescevasi in lui sempre più lo spirito, e l'habilità alla pittura; perche invaghitosi de' modi del Correggio, disegnava, e coloriva le sue opere, e s'invogliò tanto della cupola del duomo di Parma che ne formò di coloretti un picciolo modello, praticando l'unione, e lo stile delle figure vedute dal sotto in sù in scorto [His spirit and his skill at painting were growing all the time. He was so fascinated by Correggio's modes of painting that he sketched and colored his works, and was so impressed by the cupola of the cathedral of Parma that he made a tiny, colored model of it, trying to reproduce its unity and the style of the figures seen foreshortened from below].": Bellori, *Le vite*, p. 99. (Translator's note: Unless otherwise stated, all translations into English are mine.)

16. Cf. Gloton, *Trompe-l'oeil*.

17. "Als der modernste aller jener Renaissance-Italiener." Aloïs Riegl, *Die Entstehung der Barockkunst in Rom* (Vienna, 1908), p. 47.

18. Stendhal, *Rome, Naples, et Florence* (Paris, 1919).

19. Burckhardt, *Der Cicerone*, vol. 2, p. 701: "Correggio was the model for Rococo painting [*das Muster der Rococomalerei*]." See Riegl, *Die Entstehung der Barockkunst*, p. 54: "The whole development came from there [*Die ganze Entwicklung geht von hier aus*]"; from, that is to say, the cupola of the cathedral at Parma.

20. See Riegl, *Die Entstehung der Barockkunst*, pp. 51–53; and, for a more general position of the problem, *Die Spätrömische Kunstindustrie* (Vienna, 1901), chap. 1.

21. See Erwin Panofsky, "Der Begriff des Kunstwollens," in Panofsky, *Aufsätze zu Grundfragen der Kunstwissenschaft* (Berlin, 1964), pp. 33–47.

22. Mengs, *Reflections on the Talent of Correggio*, in *Works*, vol. 2, p. 45.

23. Ibid.

24. Mengs, *Reflections on Beauty*, p. 43.

25. Ibid.

26. See Burckhardt, "Correggio's durchhaus subjektive Kunst," in Burckhardt, *Der Cicerone*, vol. 2, p. 700.

27. It is important to underline the *theoretical* nature of the functions assigned

by Riegl, and also to a lesser degree by Wölfflin, to an opposition that classical doctrine had already used (as Mengs's texts show), but in an empirical fashion, without
claiming to confer any particular status or "scientific significance" upon it.

28. Mengs, *Reflections on Raphael*, in *Works*, vol. 1, p. 131.

29. Mengs, *Reflections on Beauty*, p. 44.

30. Ibid.

31. "Higher" is here placed in quotation marks to indicate that the idea only
makes sense in relation to the procedure of analysis. As we shall see, we are dealing
with a process that in fact implies a subversion of all hierarchies and a blurring of
all distinctions between levels.

32. See Roland Barthes, "Le Message photographique," in *Communications*
(Paris, 1961), vol. 1, pp. 127–38; reprinted in Barthes, *L'Obvie et l'obtus: Essais cri-
tiques* (Paris, 1982), vol. 3, pp. 9–24.

33. Heinrich Wölfflin, *Principes fondamentaux de l'art* (Paris, 1952), p. 13; English trans.: *The Principles of Art History*, trans. M. D. Hollonger (New York, 1950).

34. That denotation here owes less to the iconic relation or the relation of resemblance between the "sign" and the thing (or "reality") and more to the proposi-
tional order in which a pictorial graph is associated with the concept "cloud." The
relation between a /cloud/ in a painting and a real "cloud" is initially simply one
of *homonymy*, in the sense that Aristotle gave the word in his *Categories*:

> Things are equivocally named [homonyms] when they have the name only
> in common, the definition (or statement of essence) corresponding with
> the name being different. For instance, while a man and a portrait can prop-
> erly both be called "animals," these are equivocally named. For they have
> the name only in common, the definitions (or statements of essence) being
> different.

Aristotle, *Categories* 1, trans. Hugh Tredennick, Loeb Classical Library (Cambridge, Mass., and London, 1983); French trans. cited in Bernard Pautrat, *Versions
du soleil, figures et système de Nietzsche* (Paris, 1971), pp. 13–14, n. 1. In other words,
as will continue to be emphasized, occurrences of the /cloud/ theme should be considered primarily in semiological terms, rather than as any reference to the *painted*
reality, which is a physical phenomenon or material object.

35. Mengs, *Reflections on Raphael*, p. 129.

36. Ibid., p. 132.

37. See Louis Hjelmslev, *Prolégomènes à une théorie du language*, French trans.
(Paris, 1968), p. 83; English trans.: *Prolegomena to a Theory of Langage*, trans.
Francis J. Whitfield (Madison, Wisc., 1961).

38. See Jean-Paul Sartre, *L'Imaginaire, psychologie phénoménologique de l'imagination* (Paris, 1948); English trans.: *The Psychology of the Imagination*, trans. Mary
Warnock (London, 1972).

39. Hjelmslev, *Prolégomènes*, pp. 63–70.

40. Ricci, *Corrège*, p. 126.

41. Gaston Bachelard, *L'Air et les songes: Essai sur l'imagination du mouvement* (Paris, 1942), p. 30.

42. Ibid., pp. 212–14.

43. Ibid., p. 118.

44. See Jean-Pierre Richard, *L'Univers imaginaire de Mallarmé* (Paris, 1961), pp. 17 ff.

45. See Erwin Panofsky, *The Iconography of Correggio's Camera di San Paolo* (London, 1961).

46. See Egon Verheyen, "Correggio's *Amori di Giove*," *Journal of the Warburg and Courtauld Institute* 29 (1966): 160–92.

47. According to the reconstruction proposed by Verheyen, the decor planned by Giulio Romano for the Hall of Ovid was to have consisted of Correggio's eight panels illustrating the loves of Jupiter. It is known that after the painter's death the duke of Mantua tried in vain to gain possession of the canvases that he had commissioned, and of other works related to them (see Piero Bianconi, *Tutta la pittura del Correggio*, 2nd ed. [Milan, 1960], p. 32). Whatever hypothesis one adopts regarding the other planned episodes, on the basis of particular sketches that exist (see Arthur H. Popham, *Correggio's Drawings* [London, 1957], catal. no. 80, pl. Ca), it is significant that all the panels that were actually executed are governed by a single theme, in which the aerial connotation is strongly marked.

48. See Landino's *Commentaire sur Dante*, cited by Panofsky, *Studies in Iconology* (New York, 1972), p. 215: "Ganymede, then, would signify the *mens humana*, beloved by Jupiter, that is: the Supreme Being . . . ; being removed from the body . . . , and forgetting corporeal things, it concentrates entirely on contemplating the secrets of Heaven." This reading was adopted by the humanists, in particular the authors of collections of emblems, led by the initiator of the genre, Andrea Alciato (*Emblemata*, 1530). In his *Ovide moralisé*, Ganymede had already been interpreted as a prefiguration of Saint John the Evangelist (with the eagle representing Christ). As Panofsky realized, it is an interpretation that illuminates a curious passage in a letter dated July 7, 1533, from Sebastiano del Piombo to Michelangelo. It may be considered to be a joke, but that does not prevent it from shedding interesting light upon the problem of the celestial cupolas, and likewise upon certain slightly earlier works by Correggio, in particular the relation between Ganymede and the cupola of San Giovanni Evangelista: "As to the painting in the vault of the lantern [of the Medici Chapel], Our Lord [Pope Clement VII] leaves it to you to do what you like. I think the Ganymede would look nice there. You could give him a halo, *so that he would appear as Saint John of the Apocalypse being carried up to heaven*." Panofsky, *Studies in Iconology*, p. 213; my emphasis.

49. Verheyen, "Correggio's *Amori di Giove*," p. 189. The importance of the cloud is clearly demonstrated in a drawing on the same theme preserved in Wind-

sor and attributed to Correggio by Venturi and Corrado Ricci (*Corrège*, pl. 292), but not mentioned by Popham (*Correggio's Drawings*, p. 196, catal. no. A-132).

50. Verheyen, "Correggio's *Amori di Giove*," pp. 189–90.

51. See Giordana Canova, *Paris Bordon* (Venice, 1964), fig. 135.

52. Verheyen, "Correggio's *Amori di Giove*," pp. 185–86.

53. The significance of a cloud as a fantasy is confirmed, at the level of mythology, by the tale of Ixion, which forms a pair with that of Io: this king of Thessaly had tried to violate Hera, who had told Zeus about it. To ascertain that she was telling the truth, Zeus gave her appearance to a cloud that he then laid alongside Ixion. When the latter's reaction left Zeus in no doubt as to Ixion's desires, to punish him he attached him to a (flaming?) wheel, which whirled him away through the skies. Ixion's union with that cloud brought forth the centaurs (see Georges Dumézil, *Le Problème des centaures: Etude de mythologie comparée indo-européenne* [Paris, 1929], pp. 191–92). The Ouranian dimension to the myth of Io is further confirmed by the girl's ultimate transformation into a constellation.

54. A letter from Aretino to the duke of Mantua, dated August 6, 1527, and cited by Verheyen, testifies to Federigo II's partiality to "suggestive representations": "Credo che Mess. Jacopo Sansovina rarissimo vi ornerà la camera di una Venere si vera e si viva che empie di libidine il pensiero di ciascuno che la mira [I think that the most rare Sir Jacopo Sansovino will decorate your room with a Venus so true and lifelike that the thoughts of anyone who looks at her will be filled with lust]." See Bottari, *Raccolta di lettere* (Milan, 1822), vol. 1, p. 532.

55. Verheyen, "Correggio's *Amori di Giove*," p. 186.

56. Bachelard, *L'Air et les songes*, p. 297.

57. See Rensselaer W. Lee, "*Ut Pictura Poesis*: The Humanistic Theory of Painting," *Art Bulletin* 22 (1940): 197–269.

58. See Roman Jakobson, *Essais de linguistique générale*, French trans. (Paris, 1963), pp. 30–31 and 218; English trans.: *Fundamentals of Language* (The Hague, 1980).

59. Boris Eikhenbaum, "La Théorie de la méthode formelle," in *Théorie de la littérature* (the texts of the Russian Formalists), trans. Tzvetan Todorov (Paris, 1965), pp. 40–42.

60. "Those who have influenced me the most are artists, not scholars: Picasso, Braque, Khlebnikov, Joyce, Stravinsky. In 1913 and 1914 I lived among painters and was friendly with Malevitch. He wanted me to go to Paris with him" (Jakobson, cited by Jean-Pierre Faye, *Le Récit hunique* [Paris, 1967], p. 281).

61. Jean-Jacques Rousseau, *Essai sur l'origine des langues*, chap. 13, cited by Jacques Derrida, *Of Grammatology*, trans. G. C. Spivak (Baltimore, 1976), p. 206.

62. "Ma se una medesima superficie cominciando ombroso, a poco a poco venendo in chiaro continua, allora quello che fra loro sia il mezzo si noti con una sottillissima linea adcio che ivi la ragione del colore men dubbia [If the surface seen

proceeds from a dark colour gradually lightening to bright, then you should mark with a line the mid-point between the two parts, so that the way in which you colour the whole area is made less uncertain]." Leon Battista Alberti, *Della pittura* (Florence, 1436), bk. 2, p. 85; English trans.: *On Painting*, trans. Cecil Grayson (Harmondsworth, Eng., 1991), p. 67.

63. In his *Libro dell'arte* (Florence, 1859), chap. 29, p. 18, Cennino Cennini wrote that a painter should lead a life that is honest and temperate, avoiding *anything that might make his hand tremble* (throwing stones, being too frequently in the company of women, and so on). (The sexual connotation of the prohibition is clearly detectable here.)

64. "In ogni nostro favellare molto priegho si consideri me non come mathematico ma come pictore scrivere di queste cose. Quelli con solo ingegnio, separata ogni materia, misurano le forme delle cose. Noi perchè vogliamo le cose essere posta da vedere, per questo useremo quanto dicono più grassa Minerva [In everything we say I earnestly wish it to be borne in mind that I speak in these matters not as a mathematician but as a painter. Mathematicians measure the shapes and forms of things in the mind alone and divorced entirely from matter. We, on the other hand, who wish to talk of things that are visible, will express ourselves in cruder terms (*useremo . . . una più grassa Minerva*; literally: 'we will use a fatter Minerva')]." Alberti, *Della pittura*, bk. 1, p. 55; *On Painting*, p. 37.

65. "Pero che la circonscriptione è non altro che disegniamento del orlo quale, ove sia fatto con linea troppo apparente, non dimostrera ivi essere margine di superficie ma fessura e io desidererei nulla proseguirsi circonscrivendo che solo l'andare del orlo [Circumscription is simply the recording of the outlines, and if it is done with a very visible line, they will look in the painting, not like the margins of surfaces, but like cracks. I want only the external outlines to be set down in circumscription]." *Della pittura*, p. 82; *On Painting*, p. 65.

66. See Lev Yakoubinski, "Sur les sons de la langue poétique," cited by Eikhenbaum, "La Théorie de la méthode formelle," p. 39.

67. Henri Focillon, *Vie des formes* (Paris, 1947), p. 10.

68. See Julia Kristeva, "Pour une sémiologie des paragrammes," *Tel Quel* 29 (Spring 1967): p. 55; reprinted in *Recherches pour une sémanalyse* (Paris, 1969), p. 177.

69. Nazi cultural policy judged the waywardness that was the principle of modern art to be pathological and it was upon this that it claimed to found its theory of "sick" or "degenerate" art. But the point is—and this is what provokes such resistance—modern art defines itself, in all that is most radical about it, as refusing to accept that it is *wayward, deviant*, an *anomaly*, and sets itself up as, to borrow Kristeva's expression, an *analytical zone*, within which a specific knowledge is elaborated.

70. René Leriche, "Introduction générale," in vol. 6 of the *Encyclopédie française*, cited by Georges Canguilhem, *Le Normal et le pathologique* (Paris, 1966), p. 52.

71. Translator's note: "Tongue" in the sense of language and "tongue" in the sense of the physical organ are both rendered as *langue* in French.

72. Maurice Merleau-Ponty, *L'Oeil et l'esprit* (Paris, 1964), p. 13.

73. "Fermavasi talora a considerare un muro dove lungamente fusse stato sputato da persone malate, e ne cavava le battaglie de'cavagli e le più fantastiche città e più gran paesi che si vedesse mai; simil faceva de' nuvoli dell'aria [He would sometimes stop to gaze at a wall against which sick people had been for a long time discharging their spittle, and from this he would picture to himself battles of horsemen and the most fantastic cities and widest landscapes that were ever seen; and he did the same with the clouds in the sky]." Giorgio Vasari, "Vita di Piero di Cosimo," in *Le vite de' più eccellenti architetti, pittori, et scultori italiani* (Florence, 1550), vol. 4, p. 134; English trans.: "Life of Piero di Cosimo," in *Lives of the Painters, Sculptors, and Architects*, trans. Gaston du C. de Vere (London, 1912–14), vol. 4, p. 127.

74. See Meyer Schapiro, "Style," in *Anthropology Today*, ed. A. Kroeber (Chicago, 1953), p. 29; reprinted in Kroeber, ed., *Theory and Philosophy of Art: Style, Artist, and Society: Selected Papers* (New York, 1994), pp. 51–101. As another example of a *symptom*, it is worth citing the little pictures painted by Antonio Tempesta (Rome, Borghese Gallery) on stones bearing images or landscapes. By adding a few painted figures to the forms of clouds, plants, or buildings revealed by sawing through a block of marble, he obtained a *Temptation of Saint Anthony* (in which the image of the tempter appeared in a cloud drawn by "nature"), a *Crossing Through the Red Sea*, and so on (on the collaboration between art and "nature," see the next section).

75. Leonardo da Vinci, *Trattato*, cod. urb. lat. 1270, 35v, *Modo d'aumentare e destare l'ingegnio a varie inventioni* (How to expand the mind and conduct various inventions). See Philippe MacMahon, trans., *Treatise on Painting* (Princeton, N.J., 1956), vol. 1, pp. 50–51, and vol. 2, *passim*. Raphaël Petrucci (*La Philosophie de la nature dans l'art d'Extrême-Orient* [Paris, 1910], pp. 117–18) was the first to compare this passage from the *Treatise on Painting* to a text by Song Di, a Chinese painter of the eleventh century, published by H. Giles (*An Introduction to the History of Chinese Pictorial Art* [Shanghai, 1905], p. 100):

> Choose an old, ruined wall, spread over it a piece of white silk. Then, every morning and evening, look at it until at last you can see the ruin through the silk, its bumps, levels, zig-zags, and cracks, fixing them in your mind and your eyes. Make the bumps into your mountains, the deepest parts your rivers, the hollows your ravines, the cracks your streams, the lightest parts your closest points, the darkest parts your most distant points. Fix all that deeply within you and soon you will see men, birds, plants, and trees, and figures flying or moving between them. Then you can use your brush as you will. And the result will be a heavenly, not a man-made thing.

76. For Pliny, the sky (*caelum*, from *caelatum*, "chiseled," according to the etymology proposed in his *Natural History*, which again stresses the double meaning of *mundus*, in Greek *kosmos*, both "world" and "ornament") is not smooth, polished like an egg, but is engraved with countless figures of all the animals and things ("esse innumeras ei effigies animalium rerumque cunctarum impressas [stamped upon it are countless figures of animals and objects of all kinds]"; *Historia naturalis* 2.3, trans. H. Rackham, Loeb Classical Library [Cambridge, Mass., and London, 1967]), the seeds of which fall upon the earth or into the sea and give birth to beings: the eye can make out here the image of a bear, there one of a bull, elsewhere one of a letter.

77. See Jurgis Baltrušaitis, *Aberrations: Quatre essais sur la légende des formes* (Paris, 1957), pp. 47–72.

78. "Varietates colorum figurarum in nubilus cerni, prout admixtus ignis superet aut vincatur [Variations of colour and shape are seen in the clouds in proportion as the fire mingled with them gains the upper hand or is defeated]." Pliny, *Historia naturalis* 2.61.

79. See H. W. Janson, "The 'Image made by Chance' in Renaissance Thought," in *De Artibus Opuscula XL: Essays in Honour of Erwin Panofsky*, ed. Millard Meiss (New York, 1961), pp. 254–66.

80. Pliny, *Historia naturalis*, 35.36.102–3.

81. "He is not versatile who does not love equally all things that are contained in painting [Quello non sia universale che non ama equalmente tutte le cose che si contengono nella pittura]. For example, if one does not like landscape, he esteems it a matter of brief and simple investigation, as when our Botticelli said that such study was vain, because by merely throwing a sponge full of diverse colors at a wall [*col sol gittare d'una spanga piena di diversi colori in un muro*] it left a stain on that wall, where a fine landscape was seen. It is really true that various inventions are seen in such a stain. I say that a man should look into it, and find heads of men, diverse animals, battles, rocks, seas, clouds, woods, and similar things, and note how like it is to the sound of bells, in which you can hear what you like." Da Vinci, *Trattato*, 33v; MacMahon, trans., *Treatise on Painting*, vol. 1, p. 59. Clearly, here a study of stains can serve in the invention of clouds in a painting, just as cloud, in its turn, in another context, may also play a productive role.

82. Philostratus, *The Life of Apollonius of Tyana*, bk. 2, chap. 22; trans. F. C. Conybeare (Cambridge, Mass., and London, 1948–50), vol. 1, pp. 174–75.

83. *Mimētikēn men en physeōs tois anthrōpois hēkein, tēn graphikēn de ek tekhnēs*, ibid.

84. "The heritage of Antiquity, like nature itself, is a vast space requiring interpretation; in both cases there are signs to be discovered and then, little by little, made to speak." Michel Foucault, *Les Mots et les choses* (Paris, 1966), p. 48; English trans.: *The Order of Things* (London, 1970), pp. 33–34.

85. "Often giants' countenances appear to fly over and to draw their shadow

afar, sometimes great mountains and rocks torn from the mountains to go before and to pass by the sun, after them some monster pulling and dragging other clouds." Lucretius, *De natura rerum* 4.136–40, trans. W. H. D. Rouse, Loeb Classical Library (Cambridge, Mass., and London, 1953).

86. It is worth noting that, for Lucretius, the images to be seen in clouds are simulacra of a kind, which do not emanate from bodies, but are produced spontaneously in the atmosphere. This was a possibility already considered by Epicurus: "formed in many ways, they are carried aloft and melting incessantly change their shapes and turn themselves into the outlines of all manner of figures: as often we see clouds quickly massing together on high and marring the serene face of the firmament while they caress the air with their motion." *De natura rerum* 4.129–35.

87. From Pliny down to Alberti and Rousseau, Western tradition regarded the delineation of the shadow of a human being projected onto a wall as the first painting act, the founding act (in the phenomenological sense of the term) that foreshadowed the kind of historicity associated with the practice of Western painting (see Chapter 3, pp. 116 ff.).

88. Aristotle, *Meteorologica* 1.5.342a, trans. H. D. P. Lee, Loeb Classical Library (Cambridge, Mass., and London, 1962).

89. Ibid., 3.6.

90. Ibid., 3.2.72a–372b.

91. Ibid., 3.3.373a. I shall return to this point in a work devoted to a painting by Paul Klee; see Hubert Damisch, "Egale Infini," *Critique*, nos. 315–16 (August–September 1973); reprinted in Damisch, *Fenêtre jaune cadmium ou les dessous de la peinture* (Paris, 1984); trans. S. Bann, "Equals Infinity," in *Twentieth-Century Studies*, nos. 15–16 (December 1976): 56–81.

92. *Ei gar mē touto prattoi geloia doxei khrōmata poiousa euēthōs*, Philostratus, *The Life of Apollonius of Tyana*, bk. 2, chap. 22.

93. "Troppo ardito e volonteroso di imitare tutte le cose che ha fatto la natura, co' colori, perchè le paiano esse (e ancora migliorarle) perfare i suoi lavori ricchi e pieni di cose varie, faccendo dove accade, come dire, splendori, notte con fuochi e altri lumi simili, aria, nugoli, paesi lontani e dapresso, casamenti con tante varie osservanze di prospettiva, animali di tante sorti, di vari colori e tante altre cose [Being so bold and desirous of imitating all the things that nature has made with colours, because they seem (even better) to make (the painter's) works rich and full of a variety of things, making here and there, how shall I say, splendid things, nights with raging fires and other similar lights, air, clouds, landscapes far away and close to, windows with many different views and prospects, animals of many kinds, of various colours, and many many other things]"; Pontormo, letter to Benedetto Varchi, 18 February 1548. For a good edition of this text, which is often cited, see Luciano Berti, *Pontormo* (Florence, 1964), pp. 91–92.

94. "E la pittura panno acotonato dello inferno, che dura poco et è di manco spesa, perchè levato ce gl'ha quello riciolino, non se tiene più conto [Painting is

a hellishly woven material, ephemeral and of little worth, because if the superficial coating is removed, nobody any longer pays attention to it]"; ibid. *Panno dello inferno* refers to the material of an inferior quality used for painting, in contrast to the material used for clothing: the more it costs the longer it lasts: "Pensomi dunque, che sia come del vestire, che questa sia panno fine, perchè dura più che è di più spesa [I think then that it is as with clothing: the finer the material the longer it lasts and the more it costs]. . . . Ma dovendo ogni cosa aver fine, non sono eterne a un modo [But as all things come to an end, in no way can they be eternal]"; ibid. (a prophetic remark, considering the now irredeemably degraded condition of Pontormo's own frescoes).

CHAPTER 2 *Sign and Representation*

1. "The wonder is that an assembly of so many forms could, in fact, be made to produce a unified impression, and it would have been impossible *without the strongly accentuated framework.*" Heinrich Wölfflin, *L'Art classique* (Paris, 1911), p. 71; English trans.: *Classic Art*, trans. Linda Murray and Peter Murray (London, 1994), p. 53; my emphasis.

2. See Martin S. Soria, *The Paintings of Zurbaran* (London, 1953), p. 6.

3. Roman Jakobson, *Essais de linguistique générale*, French trans. (Paris, 1963), p. 37; English trans.: *Fundamentals of Language* (The Hague, 1980).

4. Soria, *Paintings of Zurbaran*, p. 139 (catal. no. 24, pl. 9 and fig. 16).

5. Saint Teresa of Jesus, *Oeuvres complètes*, French trans. (Paris, 1949), pp. 194–96.

6. It was, on the other hand, perfectly acceptable to reduce the space taken up by the supernatural, divine level, rather than that of the terrestrial world: in the *Premonition of Peter of Salamanca*, from the Guadalupe cycle (Seville museum, Soria, catal. no. 151, fig. 104), the composition is confined to the terrestrial level, and only the underside of the prodigy is shown, in the form of premonitory clouds.

7. "The curtains of her alcove billowed gently around her, like clouds."

8. Teresa of Jesus, *Oeuvres complètes*, p. 193: "We could well believe that the very cloud of infinite Majesty is with us in this exile."

9. On the symbolism of the sky and "the world above," see Mircea Eliade, *Traité d'histoire des religions* (Paris, 1953); English trans.: *Patterns in Comparative Religion*, trans. R. Sheed (London, 1971), chap. 2; also see Gerardus Van der Leeuw, *La Religion dans son essence et dans ses manifestations*, French trans. (Paris, 1948), pp. 54–65.

10. Mircea Eliade, *Traité d'histoire des religions*, pp. 103–4.

11. On the notion of the "figurative object," see Pierre Francastel, *La Figure et le lieu* (Paris, 1967), chap. 2; and *La Réalité figurative* (Paris, 1965), pt. 3.

12. See Mircea Eliade, *Images et symboles* (Paris, 1952), pp. 33–72; English trans.: *Images and Symbols* (London, 1961).

13. Seville museum, Soria, catal. no. 70, pl. 46.

14. One of the most famous ecstatic experiences of Saint Teresa, in which she was joined by the vicar of Saint John of the Cross, who had come to the Convent of the Incarnation to visit her, is noteworthy for the impact that it had on the domain of images. A painting was immediately commissioned to commemorate the scene, and was placed in the parlor of the convent along with an inscription recalling the event: "Siendo priora deste convento de la Encarnacion nuestra Santa Madre, y vicario de dicho convento San Juan de la Cruz, estando en este locutorio hablando en el misterio de la Santissima Trinidad, se arrobaron entrambos, y el santo subio elevando tras si la silla, como se vede en la pintura [When the prioress of the convent of the Incarnation, our Holy Mother, and the vicar of the convent of Saint John of the Cross were in this parlor speaking of the mystery of the most Holy Trinity, they were both entranced and the saint rose up levitating, *as can be seen in the painting*]." Cited by Olivier Leroy, *La Lévitation: Contribution historique et critique à l'étude du merveilleux* (Paris, 1928), p. 100; my emphasis.

15. Teresa of Jesus, *Oeuvres complètes*, p. 194.

16. See, in Saint John of the Cross, the theme of night used as a necessary plastic expression of the absorption of apparent being into real being: "Through a prodigy of mystical imagination, night is both the most intimate translation of the experience and also the experience itself." Jean Baruzi, *Saint Jean de la Croix et le problème de l'expérience mystique* (Paris, 1924), p. 330.

17. For other examples of the influence of pictorial representation on mystical visions, see Erwin Panofsky, *Early Netherlandish Painting* (Cambridge, Mass., 1958), vol. 1, pp. 469–70, n. 277/3.

18. Teresa of Jesus, *Oeuvres complètes*, p. 764.

19. Ibid., p. 715.

20. Saint John of the Cross, *Oeuvres spirituelles*, French trans. (Paris, 1964), p. 443.

21. See *Le Décret sur l'intercession des saints, l'invocation, la vénération des reliques, et l'emploi légitime des images*, promulgated by the Council during its last session, in 1563. As Pierre Francastel has shown, it was designed not so much to bring Christian art back within the limits of decency or orthodoxy, but rather to reject the accusation of idolatry brought by the Protestants (see Pierre Francastel, "La Contre-Réforme et les arts en Italie à la fin du XVIe siècle," in Francastel, *La Réalité figurative*, pp. 339–89).

22. Saint Ignatius of Loyola, *Exercices spirituels* (Paris, 1960), p. 44.

23. Ibid., p. 76, n. 1. "Where hell is concerned, the gaze of the imagination will see its immense fires, the ear will hear the shrieks, cries, and blasphemies, the sense of smell will take in the smoke, the sulphur, the stench, and the putrefaction, the sense of taste will absorb the bitterness, tears, and sorrow, and touch will feel the fire that is burning those souls." Ibid., pp. 53–54.

24. So called by the German Jesuit Jacob Masen in his *Ars nova argutiarum*

(Cologne, 1649), and the *Speculum imaginum veritatis occultae, exhibens symbola, emblemata, hieroglyphica, oenigmata* (Cologne, 1650), cited by Mario Praz, *Studies in Seventeenth-Century Imagery*, 2nd ed. (Rome, 1964), pp. 173–74.

25. The epithets and citation are from Father Richeome, *Tableaux sacrés des figures mystiques du très auguste sacrement de l'Eucharistie* (Paris, 1601), cited by Praz, *Studies in Seventeenth-Century Imagery*, p. 21.

26. See François Courel, introduction to Saint Ignatius, *Exercices spirituels*, p. 8.

27. Apoc. 1:7. Bible quotations in English are taken from the King James Version.

28. "The chief figure, Christ, is foreshortened in a truly frog-like manner." Burckhardt, *The Cicerone: An Art Guide to Painting in Italy* (London, n.d.), pp. 181–82.

29. Luke 24:51: "While he blessed them he was parted from them and carried up to heaven." And from Acts 1:9:

> And when he had spoken these things, while they beheld, he was taken up; and a cloud received him out of their sight. And while they looked steadfastly toward heaven as he went up, behold, two men stood by them in white apparel, which also said, "Ye men of Galilee, why stand ye gazing up into heaven? This same Jesus which is taken up from you into heaven shall so come in like manner as ye have seen him go into heaven."

30. Matt. 26:64: "Hereafter shall ye see the Son of Man sitting on the right hand of power and coming in the clouds of heaven." Matt. 24:30: "And then shall appear the sign of the Son of Man in heaven; and then shall all the tribes of the earth mourn and they shall see the Son of Man coming in the clouds of heaven with power and great glory."

31. The *Toldos Jeschu*, a Jewish work probably composed in the second century and published in 1681 by J. Wagenseil in his *Tela ignea satanae*, attributed to Christ a levitation of a magical nature. In this account, Jesus appears as a magician who owes his power to the possession of the secret name of Yahweh, which is hidden in the Temple. Thanks to this charm, he performs a number of miracles and, in particular, rises into the clouds in the presence of the queen and the wise men of Jerusalem. But one of the wise men, whose name is Judas and who also holds the key to the tetragrammaton, also rises up to the clouds with Jesus, and hurls him back down to earth. See Gustave Brunet, *Les Evangiles apocryphes* (Paris, 1863), pp. 390–91.

32. 1 Thess. 4:16–17: "For the Lord himself shall descend from heaven with a shout, with the voice of the archangel and with the trump of God, and the dead in Christ shall rise first. Then we, which are alive and remain, shall be caught up together with them in the clouds, to meet the Lord in the air."

33. Meyer Schapiro has shown that there was a similar conflict in the literature and art of the Middle Ages, and even among the Church Fathers. As early as the

sixth century, Gregory the Great distinguished between the Ascension of Christ and earlier ascents: whereas Enoch (*ante legem*) had needed the help of angels, and Elijah (*sub legem*) needed a chariot, Christ, who has all things at his disposal, rose of his own accord (*nimirum super omnia sua virtute ferebatur*). *Homiliae in Evangelia*, bk. 2, hom. 29, Migne, *Pat. lat.*, chap. 76, cols. 1216–17. See Meyer Schapiro, "The Image of the Disappearing Christ: The Ascension in English Art Around the Year 1000," *Gazette des Beaux-Arts* 23 (1943): 135–52, reprinted in Schapiro, *Late Antique, Early Christian, and Mediaeval Art: Selected Papers*, vol. 3, pp. 266–87.

34. Giovanni Battista Passeri, *Vite de' pittori, scultori, et architetti* (Rome, 1772), cited by Francis Haskell, *Patrons and Painters: A Study in the Relations Between Italian Art and Society in the Age of Baroque* (New York, 1963), p. 112.

35. Apoc. 10:1: "And I saw another mighty angel come down from heaven, clothed with a cloud; and a rainbow was upon his head and his face was as it were the sun, and his feet as pillars of fire."

36. When Moses begged Yahweh, "Show me thy Glory," he received the following reply:

> I will make all my greatness pass before thee, and I will proclaim the name of the lord before thee. . . . Thou canst not see my face; . . . Behold there is a place by me, and thou shalt stand upon a rock. And it shall come to pass, while my glory passeth by, that I will put thee in a cleft of the rock and will cover thee with my hand while I pass by. And I will take away mine hand and thou shalt see my back parts; but my face shall not be seen. (Exod. 33:18–21)

37. "And Moses went up into the mount and a cloud covered the mount. And the glory of the Lord abode upon Mount Sinai and the cloud covered it six days, and the seventh day he called unto Moses out of the midst of the cloud. And the sight of the glory of the Lord was like devouring fire on the top of the mount in the eyes of the children of Israel" (Exod. 24:15–18). Cf. Exod. 16:10: "And it came to pass, as Aaron spake unto the whole congregation of the children of Israel, that they looked toward the wilderness and, behold, the glory of the Lord appeared in the cloud." Hebrew legend stresses the role imparted to cloud in the Revelation. But it also accommodates the cloud that kept itself constantly at Moses's disposal, to carry him up to God and then bring him down again among men. When Yahweh asked Moses to join him on Mount Sinai, a cloud appeared and lay at Moses's feet. But Moses did not know whether he should climb on to it or simply hang on to it. Then the mouth of the cloud opened and Moses entered it and traveled through the sky as easily as a man walks upon the earth. But when he saw the guardian of the divine throne, the angel Sandalfon, he was so terrified that he almost tumbled off his vehicle. See Louis Ginzberg, *The Legends of the Jews*, 3rd ed. (Philadelphia, 1947), vol. 3, pp. 85, 109, 111.

The gift of the Tablets has been the subject of many representations, right from the start of Christian art. The Gebhardt Bible (or Admont Bible, Salzburg, twelfth century) produced a particularly interesting version, on two panels. In the left panel (fol. 68v), Moses has his feet on the mountain and his knees enveloped by a cloud from which the upper part of his body emerges, while he holds the Tablets upon which God is writing. In the right panel (fol. 69r), Moses is preparing to redescend to join his companions and is receiving God's blessing. The cloud has disappeared, but the quarter of a circle in which the figure of God is depicted is bedecked with a lacy kind of collar into which the cloud seems to have been transformed. Vienna, National Library, Ser. Nov. 2701; see André Grabar and Karl Nordenfalk, *La Peinture romane* (Geneva, 1958), p. 167.

38. Similarly, but in a Christian context, it is worth noting the treatment reserved for subjects such as the *Triumph of the Name of Jesus* (see Baciccia's fresco in the Gesù church) and the *Glory of the Name of God* (see Goya's fresco in the cathedral of Saragossa), on great painted ceilings, amid dazzling light and clouds.

39. 1 Cor. 10:1–6.

40. See Ludwig Volkmann, *Bilderschriften der Renaissance, Hieroglyphik und Emblematik in ihren Beziehungen und Fortwirkungen* (Leipzig, 1923).

41. Jean Baudoin, *Iconologie; ou, Explication nouvelle de plusieurs images, emblèmes, et autres figures hiéroglyphiques . . . , tirée des recherches et des figures de Cesare Ripa* (Paris, 1644).

42. Cesare Ripa, *Iconologia; overo, Descrittione dell'imagini universali cavate dall'antichità et da altri luoghi . . . , opera non meno utile che necessaria a poeti, pittori, et scultori per rappresentare le vitii, virtù, affetti, et passioni humane* (Rome, 1593), proem.

43. "Diro solo di quella, che appartiene a' Dipintori, overo a quelli che per mezzo di colori or d'altra cosa visibile possono rappresentare qualche cosa differente da essa, ed ha conformità con l'altra [I will only speak of (images) that belong to painters or to those who by means of colors or other visible means can represent something different from it and yet that resembles it]"; ibid.

44. Paolo Giovo, *Dialogo dell'imprese militari et amorose* (Rome, 1555).

45. "Con questo poi si forma l'arte dell'altre Imagini, le quali appartengono al nostro discorso, per la confomità che hanno con le definitioni"; ibid.

46. For an analysis of the philosophical background to emblematic literature (which, however, strangely enough does not mention Ripa's text), see Robert Klein, "La Théorie de l'expression figurée dans les traités italiens sur les *Imprese*, 1555–1612," in *La Forme et l'intelligible* (Paris, 1969), pp. 125–49.

47. Praz, *Studies in Seventeenth-Century Imagery*.

48. See, for example, Charles-Etienne Gaucher, preface to Hubert-François Gravelot and Charles-Antoine Cochin, *L'Iconologie par figures* (Paris, 1791).

49. "Nel numero dell'altre cose da anvertire sono tutte le parti essenziali della cosa istessa; e di queste serà necessario guardar minutamente le dispositioni e le qualità [In the number of other things to be converted are all the essential parts of

the thing itself, and it will be necessary to examine closely the way that they are disposed and their quality]"; Ripa, *Iconologia*, proem.

50. Baudoin, *Iconologie*, p. 29.

51. Ripa, *Iconologia*, pp. 41–42.

52. Plato, *Phaedrus* 265e.

53. Cited by Michel Foucault, *The Order of Things* (London, 1970), pp. 63–64.

54. Ibid., p. 65.

55. Ibid., p. 120.

56. Ibid., p. 66.

57. Ibid., pp. 17–45.

58. Ibid., p. 51.

59. Extra proof of this is provided by the fact that, in it, the "images" are presented in the alphabetical order of concepts, whereas, according to Michel Foucault, "the use of the alphabet as an arbitrary but efficacious encyclopaedic order does not appear until the second half of the seventeenth century"; ibid., p. 38.

60. Galileo Galilei, *Opere* (Florence, 1890–1909), vol. 9, p. 63.

61. See Jurgis Baltrušaitis, *Anamorphoses ou perspectives curieuses* (Paris, 1955).

62. See Erwin Panofsky, *Galileo as Critic of the Arts* (The Hague, 1954).

63. Foucault, *The Order of Things*, p. 32.

64. "The object of representation can be nothing but a representation of which the first representation is the interpretant. But an endless series of representations, each representing the one behind it, may be conceived to have an absolute object at its limit. The meaning of a representation can be nothing but a representation. In fact, it is nothing but the representation itself conceived as stripped of irrelevant clothing. But this clothing can never be completely stripped off; it is only changed for something more diaphanous. So there is an infinite regression here. Finally, the interpretant is nothing but another representation to which the torch of truth is handed along; and as a representation, it has its interpretant again. Lo, another infinite series." Charles Sanders Peirce, *Principles of Philosophy*, in Peirce, *Collected Papers* (Cambridge, Mass., 1965), vol. 1, p. 171.

65. Heinrich Wölfflin, *Renaissance et baroque* (Paris, 1967), pp. 74–75; English trans.: *Renaissance and Baroque*, trans. Kathrin Simon (London, 1964).

66. Praz, *Studies in Seventeenth-Century Imagery*, pp. 169–76.

67. D'Alembert, *Eloge de Despréaux*, n. 12, cited by Littré in "Représentation."

68. Such a ploy was not solely literary: the Elizabethan stage—and likewise the unified Italian stage—was organized in such a way as to permit a duplication of places and make it possible for one theatrical space to be contained within another (for example, for the representation of indoor scenes).

69. See Pierre Francastel, *La Figure et le lieu* (Paris, 1967), chap. 2, "Les Éléments figuratifs et la réalité."

70. George R. Kernodle, *From Art to Theatre: Form and Convention in the Renaissance* (Chicago, 1944).

71. Walter Benjamin, "L'Oeuvre d'art au temps de ses techniques de reproduction," French trans., in Benjamin, *Oeuvres choisies* (Paris, 1959), p. 206.

72. Giorgio Vasari, "Vita di Raffaello," in Vasari, *Le vite de' più eccellenti architetti, pittori, et scultori italiani* (Florence, 1550; 2nd ed., 1568), vol. 4, p. 365; English trans.: "Life of Raphael," in *Lives of the Painters, Sculptors, and Architects*, trans. Gaston du C. de Vere (London, 1912–14), vol. 4, pp. 209–49.

73. Karl-Friedrich Ruhmor, *Italienische Forschungen* (Berlin, 1827–31).

74. Herman Grimm, "Das Rätsel der Sixtinischen Madonna," in *Zeitschrift für bildende Kunst* (1922), 41–49.

75. Eugène Müntz, *Raphael* (Paris, 1901), p. 303.

76. Wölfflin, *L'Art classique*, pp. 158–60.

77. When referred to the tomb of Julius II, whose wish had been to be buried in the Sistine Chapel of Saint Peter's, Raphael's painting appears to constitute a temporary substitute for a monument that the pope's heirs were eventually to commission from Michelangelo. As Grimm notes, this provides an interesting example of the rivalry (*Wettstreit*) between the two artists. The theme of the Virgin and Child figured in Michelangelo's first plans for the monument. See Erwin Panofsky, "The First Two Projects of Michangelo's Tomb of Julius II," *Art Bulletin* (1937): 561–79.

78. Müntz, *Raphael*. As for whether it was possible for a painter of the early sixteenth century "to provide a glimpse of the infinite," see Chapter 4, pp. 163 ff.

79. On the development of sixteenth-, seventeenth-, and eighteenth-century theatrical machinery and mechanisms, often of an extremely complex nature (e.g., the sets by Sabbatini, in Parma), designed to move clouds around by means of mobile frameworks and equipped with seats for the actors, see the proceedings of the Centre National de Recherches Scientifiques colloquium, *Le Lieu théâtral à la Renaissance* (Paris, 1964).

80. Francastel, *La Figure et le lieu*, p. 72.

81. Kernodle, *From Art to Theatre*, pp. 70–72.

82. Renato Cipriani, *Tutta la pittura del Mantegna* (Milan, 1956), pl. 165. It is interesting to compare this to a painting by Mantegna on the same theme, now in the Prado. The upper part has been cut off, but the natural complement to the lower part is the Christ positioned in a mandorla that Roberto Longhi discovered in a private collection in Ferrara (ibid., pl. 77–79).

83. Seville museum, Soria, catal. no. 41, p. 27.

84. A detailed description may be found in Kernodle, *From Art to Theatre*, p. 102.

85. Francastel, *La Figure et le lieu*, p. 82.

86. Pierre Francastel, "Imagination plastique, vision théâtrale, et signification humaine," in Francastel, *La Réalité figurative*, pp. 211–38.

87. Vasari, "Vita di Cecca," in *Vite*, vol. 3, pp. 199–201; "Life of Cecca," in *Lives*, vol. 3, p. 195.

88. *Recitare* means to perform a play. But, exploiting a no doubt calculated ambiguity, Vasari here uses the word to signify the narration that accompanied the representation, a recital that was repeated year after year at a fixed date.

89. Vasari, "Vita di Cecca," p. 198; "Life of Cecca," p. 194.

90. Vasari, "Vita di Brunelleschi," in *Vite*, vol. 2, pp. 375–78; "Life of Brunelleschi," in *Lives*, vol. 3, pp. 193–236.

91. See the *Livre de conduite du régisseur et le compte des dépenses pour le mystère de la Passion joué à Mons en 1501*, published by Gustave Cohen (Strasbourg, 1925), p. 473: "Another piece of canvas, also clouds."

92. Gustave Cohen, *Histoire de la mise en scène dans le théâtre religieux français au Moyen Age*, 2nd ed. (Paris, 1926).

93. *Ibid.*, p. 153: "Icy doit descendre une nuée ronde en forme de couronne où doivent estres plusieurs anges faincts tenant espées nues [Here a round cloud in the shape of a crown must descend, bearing several holy actor-angels carrying drawn swords]."

94. Bibliotheque Nationale de France, MS fr. 972, cited in ibid., pp. 153–54; attributed to a certain Jean Michel, this was no doubt a book on stage sets that belonged to some brotherhood.

95. Francisco Javier Sanchez Canton, *Las adquisiciones del museo del Prado en los años 1952 y 1953* (Madrid, 1954), p. 2. But this kind of representation was not a prerogative of northern countries (see Fra Angelico's *Pala di San Marco*, etc.).

96. See Emile Mâle, *L'Art religieux de la fin du Moyen Age en France: Étude sur l'iconographie du Moyen Age* (Paris, 1908), p. 53.

97. Schapiro, "The Image of the Disappearing Christ," pp. 135–52.

98. "On the whole Mount of Olives the highest point is the one from which our Lord is said to have ascended to heaven; here stands a great round church with thee concentric porticos all roofed over. The inner chamber of the round church, having no roof, is exposed to the sky, but in its eastern part there is an altar, protected by a narrow roof. The inner space has no roof above it, so that from the spot where the Lord left his holy footprints when he was carried up to heaven in a cloud, the way is always open, and those who pray there may look up and see the sky directly." Adamnanus, *De locis sanctis*, bk. 1. For the original text, here translated by Schapiro ("The Image of the Disappearing Christ," p. 273), see P. Geyer, *Itinera hierosolymitana, saeculi IIII–XIII*, Corpus scriptorum ecclesiasticorum latinorum, vol. 38 (Vienna, 1898), pp. 246–51.

99. "The cloud did not make its appearance there, because our Lord had no need of the cloud's aid at the Ascension; nor did the cloud raise him up, but he took the cloud before him, since he hath all creatures in his hand, and by his divine power and by his eternal wisdom, according to his will, he orders and disposes all things. And he, in the cloud, disappeared from their sight and ascended into heaven, as a sign that from thence in like manner he will on Doomsday again come to earth in a cloud, with hosts of angels" (trans. Schapiro, "The Image of the Dis-

appearing Christ," p. 270; see R. Morris, ed., *The Blickling Homilies of the Tenth Century* (London, 1880), pp. 120–21.

100. During the Renaissance, the British solution underwent a transformation similar to that of the traditional mandorla. While in Mantegna's work the latter took on the appearance of a theatrical machine, in the solution arbitrarily dubbed "Gothic" the entire stage took on a unitary dramatic structure in which the body of Christ was partly concealed by a cloud. Meyer Schapiro astutely observes that the way in which the figure of Christ escapes from the common space (the horizon of which is defined by the gaze of the Apostles) is by passing beyond the pictorial field and being cut off by the frame of the image ("The Image of the Disappearing Christ," p. 284).

101. Francastel, *La Figure et le lieu*, p. 99.

102. *Toiles peintes et tapisseries de la ville de Reims; ou, La Mise en scène du théâtre des Confrères de la Passion*, a study and historical explanation by Louis Paris (Paris, 1849). In the opinion of this scholar, these images were just stage sets for the dramas that the actors employed by Charles VI had included in their repertory and that had been staged in Rheims between 1450 and 1496 (ibid., pp. lxi–lxvi).

103. Karl Adolf Knappe, *Dürer, gravures: Oeuvre complète* (Paris, 1964), p. xxxiii. As for the age of this schema of representation and its theatrical and figurative origins, apart from the information provided by Meyer Schapiro, it is worth noting that the Ascension is represented in more or less the same fashion (except that the hill is missing) in a work stemming from the same area of cultural diffusion and dated 1181, namely the famous altarpiece by Nicholas of Verdun, preserved in the Abbey of Klosterneuburg. But the panel bearing the Ascension seems to have been repainted or restored at some later date. See Louis Réau, "L'Iconographie du retable typologique de Nicolas de Verdun à Klosterneuburg," in *L'Art mosan*, ed. Pierre Francastel (Paris, 1953), p. 812.

104. Ferdinand de Saussure, *Cours de linguistique générale* (Paris, 1949), p. 43; English trans.: *Course in General Linguistics*, trans. Peter Owen (London, 1974), p. 23.

105. See Raymond Lebègue, "Quelques survivances de la mise en scène médiévale," in Lebègue, *Mélanges d'histoire du Moyen Age et de la Renaissance offerts à Gustave Cohen* (Paris, 1950), pp. 219–28.

106. It is thus that the Baroque *Theatrum sacrum* of Austria, which may appear to prolong the traditions of paraliturgical representations, in fact borrowed some of its procedures from the repertory of contemporary profane theater. See in Alpheus Hyatt Mayor, *Tempi e aspetti della scenografia* (Turin, 1954), p. 52, the description of a *Theatrum sacrum* presented in Vienna in 1670, probably by the Italian Ludovico Burnacini: "The Holy Sepulcher had been imagined by night, with the two guards sleeping near by. . . . Then a light shone in the sky and one beheld the dazzling apparition of the actor playing the eternal Father. A cloud opened, whence emerged

the Cross, held by two actors representing angels. The eternal Father disappeared in his Glory and the cloud that had held the Cross closed again."

CHAPTER 3 *Syntactical Space*

1. Reproduced in Erwin Panofsky, *Early Netherlandish Painting*, vol. 2, pp. 197–99.

2. Erwin Panofsky, *Studies in Iconology* (New York, 1972), p. 10.

3. See ibid., fig. 2, for the interpretation of an Ottonian miniature representing *The Resurrection of the Young Man of Naïm*. In this miniature, the miracle takes place on the "ground" where the figures are placed, and involves no communication between heaven and earth. But Panofsky claims to have produced his interpretation solely on the basis of his strictly iconological understanding, without resorting at all to an analysis of figurative functions.

4. See Panofsky, "Die Perspektive als 'Symbolische Form,'" in Panofsky, *Aufsätze zu Grundfragen der Kunstwissenschaft* (Berlin, 1964), p. 127, n. 3.

5. Panofsky, *Early Netherlandish Painting*, vol. 1, pp. 140–41.

6. See ibid., pp. 276–78, for an interpretation of Rogier van der Weyden's triptych in which Panofsky detects a term-for-term illustration of a passage from the *Golden Legend*.

7. See Jean-Louis Schefer, "Lecture et système du tableau," in Schefer, *Scénographie d'un tableau classique* (Paris, 1969).

8. See Noam Chomsky, "De quelques constantes de la théorie linguistique," *Diogène* 51 (July–September 1965): 14; and Chomsky, *Cartesian Linguistics* (New York, 1966).

9. Ferdinand de Saussure, *Cours de linguistique générale* (Paris, 1949), pp. 107–8; English trans.: *Course in General Linguistics*, trans. Wade Baskin (London, 1974).

10. Jacques Gernet, "La Chine: Aspects et fonctions psychologiques de l'écriture," in Gernet, *L'Ecriture et la Psychologie des peuples* (Paris, 1963), p. 29.

11. Roman Jakobson, *Essais de linguistique générale*, French trans. (Paris, 1963), p. 33; English trans.: *Fundamentals of Language* (The Hague, 1980).

12. See introduction to Panofsky, *Early Netherlandish Painting*.

13. Translator's note: "Cradles" is *assises* in French; the pun is unavoidable in French, perhaps, but untranslatable into English.

14. Leonetto Tintori and Millard Meiss, *The Painting of the Life of St. Francis of Assisi*, with notes on the Arena Chapel (New York, 1962). In an earlier work, Meiss concluded that the Assisi cycle's traditional attribution to Giotto should be rejected (Meiss, *Giotto in Assisi* [New York, 1960]). On the problem of Giotto's "name," see my article "Giotto," in *Encyclopaedia universalis*, vol. 7 (Paris, 1968), pp. 742–44.

15. See the remarks of Tintori and Meiss (*Painting of the Life of St. Francis*, pp. 43–58), and in particular the diagram (p. 195) of the distribution of light: it comes

from the right in scenes painted in the first three bays of the nave, and from the left in the bay closest to the choir.

16. "The impetus of the narrative," as Meiss puts it (*Giotto in Assisi*, p. 43).

17. "Il quale Giotto rimuto l'arte del dipingere di greco in latino, e ridusse al moderno." Cennino Cennini, *Il libro del arte* (Florence, 1437; reprint, 1859), p. 4.

18. "The cycle as a whole." Tintori and Meiss, *Painting of the Life of St. Francis*, p. 2.

19. See Roman Jakobson, "A la recherche de l'essence du langage," *Diogène* 51 (July–September 1965): 37.

20. P. Bonaventura has suggested that these should be replaced in their original positions. See *L'opera completa di Giotto* (Milan, 1966), pp. 91 ff.

21. Where dreams were concerned, Freud set up an opposition between an analysis *en masse* and an analysis *en détail* (in French in the text), designed to study a dream as a compound or conglomerate. Sigmund Freud, *L'Interprétation des rêves*, French trans. (Paris, 1967), p. 67.

22. Ibid., p. 432.

23. See John White, *The Birth and Rebirth of Pictorial Space*, 2nd ed. (London, 1967), p. 33.

24. In the church of San Fortunato in Montefalco, Benozzo Gozzoli painted a representation (1453–59) directly inspired by Giotto's, but in which the real space and the space of the dream are integrated within the unity of a single scene by geometrical means.

25. White, *Birth and Rebirth of Pictorial Space*, p. 37.

26. Emilio Cecchi (*Giotto*, French trans. [Paris, 1937], pp. 103 ff.) has shown how Giottesque space is primarily an architecture of gestures and glances. Psychological "triangulation" clearly preceded the geometrical construction of space

27. See Chapter 4, pp. 148 ff.

28. "Lodasi la nave dipinta ad Roma in quale il nostro toscano dipintore Giotto pose undici discepoli, tutti commossi da paura vedendo uno de' suoi campagni passeggiare sopra l'acqua, che ivi espresse ciascuno con suo viso et gesto porgere suo certo inditio d'anima turbato, tale che in ciascuno ermo suoi diversi movimenti et stati [They praise the ship painted in Rome by our Tuscan painter Giotto. Eleven disciples (are portrayed) all moved by fear at seeing one of their companions passing over the water. Each one expresses with his face and gesture a clear indication of a disturbed soul in such a way that there are different movements and positions in each one]." Leon Battista Alberti, *Della pittura* (Florence, 1436), bk. 2, p. 95; English trans.: *On Painting*, trans. Cecil Grayson (Harmondsworth, Eng., 1991), bk. 2, p. 78.

29. Pierre Francastel, *La Figure et le lieu*, p. 17.

30. Hans Jantzen, "Giotto und der gotische Stil," in Jantzen, *Die Aufsätze* (Berlin, 1951), pp. 35–40.

31. It is worth repeating that the notion of "Giottesque art" is here given a

deliberately loose meaning. The frescoes in which the delineation of figures emphasizes their solid positioning are also those in which critics for the most part refuse to see the hand of Giotto himself.

32. See below, pp. 111 ff.

33. Freud, *L'Interprétation des rêves*, pp. 269–70.

34. According to the text of the legend of Saint Bonaventure, the *Apparition* (chaps. 4, 10) ought to have been positioned between the *Vision of the Burning Chariot* and that of the *Thrones* (White, *Birth and Rebirth of Pictorial Space*, p. 55, n. 42). This shift and the juxtaposition, significant in itself, that it makes possible for scenes obeying an identical organizational principle (the division into two levels in the case of the Visions; all the figures placed in a strictly unified and delimited space in the case of *The Sermon* and *The Apparition*) seem to have been prompted by a figurative syntax, a few of whose rules I am attempting to identify.

35. See the excellent pages by André Chastel on Giotto, in his *Art italien* (Paris, 1956), vol. 1, pp. 145–50.

36. Cesare Gnudi, *Giotto* (Milan, 1958), p. 66.

37. Emile Benveniste, "Remarques sur la fonction du langage dans la découverte freudienne," in Benveniste, *Problèmes de linguistique générale* (Paris, 1966), pp. 75–87; English trans.: *Problems in General Linguistics*, trans. M. E. Meek (Coral Gables, Fla., 1971).

38. Francastel, *La Figure et le lieu*, p. 189.

39. Eugenio Battisti, *Giotto* (Geneva, 1960), p. 14.

40. Julia Kristeva, "Le Geste, pratique ou communication?" *Langages* 10 (June 1968): 52–53.

41. Antonin Artaud, *Le Théâtre et son double*, in Artaud, *Oeuvres complètes* (Paris, 1964), vol. 4, p. 139; English trans.: *The Theatre and Its Double*, in *Collected Works*, trans. Victor Corti (London, 1974), vol. 4, pp. 1–6.

42. Antonin Artaud, unpublished text cited by Jacques Derrida, "Le Théâtre de la cruauté et la clôture de la représentation," in Derrida, *L'Ecriture et la différence* (Paris, 1967), p. 342; English trans.: *Writing and Difference*, trans. Alan Bass (London, 1978).

43. Yvan Gobry, *Saint François et l'esprit franciscain* (Paris, 1967), p. 59.

44. See the analysis that Francastel gives of the *Crib of Grecchio*, where, he says, Giotto painted "a situation that corresponds neither to the *primitive scene*, nor exactly to the annual scene in which the miracle is commemorated" (*La Figure et le lieu*, pp. 189–90; my emphasis).

45. Derrida, "Le Théâtre de la cruauté," p. 346.

46. See below the *titulus* of *The Ecstasy of Saint Francis*.

47. Emilio Cecchi deliberately chose not to reproduce it in his *Giotto*, on the grounds of not only its poor state of preservation but also the "impoverished imagination" to which, he says, it testifies.

48. "Come il beato Francesco, pregando un giorno fervidamente, fu scorto dai

frati levarsi da terra con tutto il corpo, con le mani protese; una fulgidissima nuvoletta risplendette intorno a lui."

49. According to Monsignor A. Fargues (*Les Phénomènes mystiques distingués de leurs contrefaçons humaines et diaboliques* [Paris, 1923], vol. 2, p. 272), Saint Francis was the first mystic whose levitation was officially recognized. However, it is not mentioned in either the first or the second *Life* by Tommaso di Celano, who was in a position to question the saint's companions and who pays particular attention to his mystical trances. Only in Saint Bonaventure's *Life*, composed after the first of Celano's *Life*s and before the second, is the nocturnal levitation mentioned (*Life*, 143; see Olivier Leroy, *La Lévitation: Contribution historique et critique à l'étude du merveilleux* [Paris, 1928], pp. 7 and 224).

50. A separation that is by no means the rule in phonetic writing. The Greek scribes did not use it, and their practice defined a scriptural space very different from that of Latin. See James Février, *Histoire de l'écriture* (Paris, 1959), p. 31.

51. Ibid., p. 11.

52. See Jacques Derrida, "Freud et la scène de l'écriture," in *L'Ecriture et la différence*, pp. 293–340.

53. Erwin Panofsky, *Die Perspektive als "symbolische Form": Aufsätze zu Grundfragen der Kunstwissenschaft* (Berlin, 1964), p. 127, n. 5; English trans.: *Perspective as Symbolic Form*, trans. C. Wood (New York, 1991). Using remarkably rigorous terminology, Panofsky's text indicates that the perspective regime implies a repression, in the analytical sense of the term, of the *substratum*. He, like Freud, plays upon the conceptual distinction that the German language makes possible between *Darstellung* (presentation, representation in the sense of a depiction, either visual or dramatic: that is, in the present context, the representation of objects and of the portion of space that they occupy) and *Vorstellun*, (re)presentation in the mode of a symbolic process, that is — in Freudian terms — the fantastical reproduction of the initial perception that is repressed: a lost sign that analysis strives to reestablish in its position as signifier in a literal chain (Lacan). It is thanks to the "absence in its place" of this "sign" that a perspective order is constituted, governed by a perspective "code." I shall return elsewhere to this point, which is of crucial importance for any theory of representation. This fundamental text by Panofsky was published for the first time in the *Vorträge der Bibliotek Warburg* (1924–25).

54. Ibid., p. 108.

55. André Grabar, *Le Haut Moyen Age* (Geneva, 1957), p. 35. The mosaics in question are reproduced on pp. 34, 37, and 38. See Carlo Cecchelli, *I mosaici della basilica di S. Maria Maggiore* (Turin, 1956).

56. The decor of the upper level of the Assisi basilica, whose attribution to Giotto is challenged, includes on the interior wall of the facade an *Ascension* in which Christ is carried aloft on a cloud toward a "paradise" represented by a series of concentric circles, and also a *Pentecost* in which a dove is descending amid a crown of clouds.

57. Panofsky, introduction to *Early Netherlandish Painting*, vol. 1, p. 12.

58. The "cloud of light," as André Grabar puts it, going on to stress the temporal as well as the spatial significance of the cloud: the apparition is presented as being exceptional and momentary (André Grabar, *Christian Iconography: A Study of Its Origins* [Princeton, N.J., 1968], p. 116). It is worth noting that representations of Christ moving upward on a carpet of clouds only date from the sixth century. In the apse of Santa Pudenzia (fourth century, restored in the sixteenth century), Christ is enthroned on earth among his Apostles, while the Cross and the symbols of the Evangelists are set in a sky filled with small clouds.

59. On the absence of any differential characterization for the first two figures in the Trinity, in the paleo-Christian period, see ibid., pp. 118 ff.

60. Doura-Europus synagogue, the upper panel of the Torah screen (M. de Damas); cf. ibid., p. 20. Ravenna, San Vitale (mosaics in the choir), *The Sacrifice of Abel and Melchizedek, The Sacrifice of Abraham, Moses Receiving the Tablets of the Law.*

61. The separation between the two locations of the clouds is emphasized by the gap between the little clouds strung out toward the horizon and the solid bank of cloud upon which the medallion rests.

62. "Poi che noi non come Plinio recitiamo storie ma di nuovo fabrichiamo una arte di pittura della questa in questa hetà, quale io veggo, nulla si truova scritto [But we are not telling stories like Pliny. We are, however, building anew an art of painting about which nothing, as I see it, has been written in this age]." Alberti, *Della pittura*, bk. 2, p. 78; *On Painting*, bk. 2, p. 65.

63. *Della pittura*, bk. 1, p. 55; *On Painting*, bk. 1, p. 43.

64. "Dico l'officio del pittore esser così: descrivere con linea et tigniere con colori, in qual sia datoli tavola o parete, simile vedute superficie di qualunque corpo che quelle, ad una certa distanzia et ad una certa positione di centro, paiano rilevata et molto simili avere i corpi." *Della pittura*, bk. 3, p. 103; *On Painting*, bk. 3, p. 89. It is a definition that agrees with that of Piero: "La pictura non e se non dimostrationi de superficie e de corpi degradati o accresciuti nel termine [Painting is nothing but demonstrations of surfaces and bodies that are either made smaller or larger within the frame]." Piero della Francesca, *De prospectiva pingendi* (c. 1480; reprint, Florence, 1942), bk. 3, p. 128.

65. "Seguita la compositione de corpi nella quale ogni lode et ingegnio del pittore consiste. . . . Conviensi che i corpi insieme si confacciano in istoria con grandezza e con adoperarsi. . . . Adunque tutti i corpi per grandezza et suo officio s'aconfaranno a quello che ivi nella storia si facci [Now follows the composition of bodies, in which all the skill and merit of the painter lies. . . . All the bodies in the 'historia' must conform in function and size. . . . All bodies should conform in size and function to the subject of the action]." *Della pittura*, bk. 2, p. 91; *On Painting*, bk. 2, p. 74.

66. "Quello che prima da voluptà nelle storie viene della copia e varietà delle

cose [The first thing that gives pleasure in a 'historia' is a plentiful variety]." *Della pittura*, bk. 2, p. 92; *On Painting*, bk. 2, p. 75.

67. "Biasimo io quelli pittori quali, dove vogliono parere copiosi nullo lassando vacuo, ivi non compositione ma dissoluta confusione disseminano; pertanto non pare la storia facci qualche cosa degnia ma sia un tumulto aviluppata [I disapprove of those painters who, in their desire to appear rich or to leave no space empty, follow no system of composition, but scatter everything about in random confusion with the result that their 'historia' does not appear to be doing anything but merely to be in turmoil]." *Della pittura*, bk. 2, p. 92; *On Painting*, bk. 2, p. 75.

68. As Braque famously said, "I paint not things but the spaces between things."

69. Alberti remarks that it is a mistake to set a character in a building so small that there is hardly room for him to sit there: "Et sarebbe vitio se . . . , quello che spesso veggo, ivi fusse huomo alcuno nello hedificio quasi come in un scrignio inchiuso, dove apena sedendo vi si assetti [Another thing I often see deserves to be censured, and that is men painted in a building as if they were shut up in a box in which they can hardly fit sitting down]." Alberti, *Della pittura*, bk. 2, p. 91; *On Painting*, bk. 2, p. 75.

70. In conformity with the principle laid down at the outset, the painter can only express the affections of the soul through what is revealed by the movements of the body, which amount to a movement in space: "Ma noi dipintori i quali vogliamo coi movimenti della membra mostrare i movimenti dell'animo, solo riferiamo di quel movimento si fa mutando el luogo [We painters, however, who wish to represent emotions through the movements of limbs . . . may speak only of the movement that occurs when there is a change of position]." *Della pittura*, bk. 2, p. 95; *On Painting*, bk. 2, p. 78.

71. "Ma perché talora in questi movimenti si truova chi passa ogni ragione, mi piace qui de posari et de movimenti raccontare alcune cose quali ho raccolte della natura onde bene intenderemo con che moderatione si debbano usare [Since, however, the bounds of reason are often exceeded in representing these movements, it will be of help here to say some things about the attitude and movements of limbs which I have gathered from Nature, and from which it will be clear what moderation should be used concerning them]." *Della pittura*, bk. 2, pp. 95–96; *On Painting*, bk. 2, p. 79.

72. If a painter wishes to represent a figure standing balanced on one foot, that foot must be in line with the head "quasi come base d'una colonna [like the base of a column]." *Della pittura*, bk. 2, p. 96; *On Painting*, bk. 2, p. 79.

73. "Adunque il pittore volendo exprimere nelle cose vita farà ogni sua parte in moti. Ma in ciascuno modo terrà venustà et gratia; sono gratissimi i movimenti et ben vivaci quelli e quali si muovano in alto verso l'aëre [So the painter who wishes his representations of bodies to appear alive should see to it that all their members

perform their appropriate movements. But in every movement beauty and grace should be sought after. Those movements are especially lively and pleasing that are directed upwards into the air]." *Della pittura*, bk. 2, p. 90; *On Painting*, bk. 2, pp. 73–74.

74. "Ma dove cosi vogliamo ad i panni suoi movimenti, sendo i panni di natura gravi et continuo cadendo a terra per questo sarà bene in la pittura porvi la faccia del vento zeffiro o austro che soffi fra le nuvole onde i panni ventoleggiono [Since by nature clothes are heavy and do not make curves at all, as they tend always to fall straight down to the ground, it will be a good idea, when we wish clothing to have movement, to have in the corner of the picture the face of the West or South wind blowing between the clouds and moving all the clothing before it]." *Della pittura*, bk. 2, p. 98; *On Painting*, bk. 2, p. 81. On the fact that the wind, itself invisible, can only be represented by the things that it blows along, see Chapter 4 of the present book, pp. 138–41ff. Curiously enough, on this point Alberti adopts an altogether allegorical and literary position (see above). On loosened hair, blowing in the wind, which became as it were the distinctive trademark of the *maniera antica* in the Quattrocento, see Erwin Panofsky, "Albrecht Dürer and Classical Antiquity," in Panofsky, *Meaning in the Visual Arts* (Garden City, N.J., 1955), pp. 243–44, n. 23.

75. See Robert Klein, "Pomponius Gauricus on Perspective," *Art Bulletin* 43 (1961): 211–30.

76. "Constat enim tota hec in universum perspectiva dispositione, ut intelligamus. . . . Quot necessariae sint ad illam rem significandum personae ne aut numero confundatur, aut rarite deficiat intellectio [All of this generally depends upon the perspective disposition, as we may understand. . . . However many figures as may be necessary for signifying that thing, the intellect should not confuse them in number, nor underestimate them because of their rarity]." Pomponius Gauricus, *De sculptura* (reprint, Leipzig, 1889), cited by Klein, "Pomponius Gauricus on Perspective," p. 244; trans. David Money (Darwin College, Cambridge, Eng., 1999).

77. "Questo ragione di dividere il pavimento s'appartiene ad quella parte quale al suo luogo chiameremo compositione [This method of dividing up the pavement pertains especially to that part of the painting which, when we come to it, we shall call composition]." Alberti, *Della pittura*, bk. 1, p. 244; *On Painting*, bk. 1, p. 58.

78. "Principio dove io debbo dipigniere, scrivo uno quadrangolo di retti angoli quanto grande io voglio, el quale reputo essere una finestra aperta per donde io miro quello che quivi sarà dipinto [First of all, on the surface on which I am going to paint, I draw a rectangle of whatever size I want, which I regard as an open window through which the subject to be painted is seen]." *Della pittura*, bk. 1, p. 70; *On Painting*, bk. 1, p. 54.

79. "Adunque chi mira una pictura vede certa intersegatione d'una pirramide. Sarà adunque pictura non altro che intersegatione della pirramide visiva secondo

data distantia, posto il centro e constituti i lumi in una certa superficie con linee et colori artificiose rappresentata [The viewers of a painted surface appear to be looking at a particular intersection of the pyramid. Therefore a painting will be the intersection of a visual pyramid at a given distance, with a fixed centre and certain position of lights, represented by art with lines and colours on a given surface]." *Della pittura*, bk. 1, p. 65; *On Painting*, bk. 1, p. 48.

80. According to Panofsky, Dürer's journey to Bologna in 1506 to become initiated into the secret art of perspective (*Kunst in heimlicher Perspectiva*) testifies to this painter's desire to understand the theoretical basis of procedures that he had already mastered perfectly on a technical level. See Erwin Panofsky, *The Life and Works of Albrecht Dürer*, 3rd ed. (Princeton, N.J., 1948), vol. 1, p. 248.

81. See Alexandre Koyré, *Études galiléennes* (Paris, 1966), p. 13; English trans.: *Galileo Studies*, trans. John Mepham (Hassocks, Eng., 1978).

82. See Alessandro Parronchi, "Le due tavole del Brunelleschi," *Paragone* 107 (November 1958): 3–32, and *Paragone* 109 (January 1959): 3–31; reprinted in *Studi su la "dolce" prospettiva* (Milan, 1964), pp. 226–95.

83. "Et saratti ad conoscere buono giudice lo specchio nè se come le cose ben dipinte molto abbino nello specchio gratio; cosa maravigliosa come ogni vitio della pittura si manifesti diforme nello specchio [A mirror will be an excellent guide to knowing this. I do not know how it is that paintings that are without fault look beautiful in a mirror; and it is remarkable how every defect in a picture appears more unsightly in a mirror]." Alberti, *Della pittura*, bk. 2, p. 100; *On Painting*, bk. 2, p. 83.

84. From a structural point of view, the job of a painter, as conceived by Alberti, is not unlike that attributed by Georges Dumézil firstly to the priests of the Vedic hymns, whose function was to show the king the *perspectives* (my emphasis) of assured sovereignty; and secondly to the Roman *fetiales*, who were supposed, by means of their ritual operations, to open up the space into which the army would then advance, under the protection of the gods. Both cases involve a *repetitio rerum* that a priori defines the field of whatever is happening and that *underpins*—as the hymns felicitously put it—the scene (see Georges Dumézil, "Ius fetiale," in Dumézil, *Idées romaines* [Paris, 1969], pp. 63–78).

85. "La circunscrittione, cioè il modo del disegnare [Circumscription is simply the recording of the outlines]." Alberti, *Della pittura*, bk. 2, p. 87; *On Painting*, bk. 2, p. 65.

86. "Principio, vedendo qual cosa, diciamo questa essere cosa quale occupa un luogo [In the first place, when we look at a thing, we see it as an object which occupies a space]." *Della pittura*, bk. 2, p. 81; *On Painting*, bk. 2, p. 64.

87. As late as the seventeenth century, the disagreement between André Bosse (inspired by the mathematician Desargues) and the Académie Française hinged on Bosse's claim that the goal of perspective was to represent things, "not as the eye sees them or thinks it sees them, but as the laws of perspective impose them

upon our reason." See *La Méthode universelle de mettre en perspective les objets donnés réellement ou en devis* (Paris, 1636).

88. "Voglio che i giovanni quali ora nuovi si danno a dipignere, cosi facciano quanto veggo di chi inpara a scrivere. Questi in prima sepato insegniano tutte le forme delle lettere, quali li antiqui chiamano helementi, poi insegniano le silabe, poi appresso insegniano componere tutte le dizzioni. Con questa ragione ancora seguitino i nostri a dipignere. Inprima inparino ben disegniare li orli delle superficie et qui si exercitino quasi come ne' primi helementi della pittura." Alberti, *Della pittura*, bk. 2, p. 46; *On Painting*, bk. 2, p. 92.

89. "Language and writing are two distinct systems of signs; the second exists for the sole purpose of representing the first." Jacques Derrida, *Of Grammatology*, trans. G. C. Spivak (Baltimore, 1976), p. 30.

90. Ibid.

91. See for example the *De archtectura libri decem*, bk. 8, chap. 4: "The Egyptians employed symbols in the following manner: they engraved an Eye, by which they understood God; a Vulture for Nature; a Bee for a King; a Circle for Time; an Ox for Peace; and so on. And the reason why they expressed the meaning by such symbols was that, since words are only understood by the peoples who speak the language in question, inscriptions using ordinary characters would soon be lost, as has indeed happened in the case of our Etruscan characters." Based on the translation by Giacomo Leoni (London, 1955), p. 169.

92. "Segnio qui appello qualunque cosa stia alla superficie per modo che l'occhio possa vederla." Alberti, *Della pittura*, bk. 1, pp. 55–56; *On Painting*, bk. 1, p. 37.

93. "Il punto essere segnio quale non si possa dividere in parte. . . . I punti, se innordine costati l'uno al altro crescono una linea et apresso di noi sarà linea segnio la cui longitudine si puo dividere ma di larghezza tanto sarà sottile che non si potrà fendere [The first thing to know is that a point is a sign which one might say is not divisible into parts. . . . Points joined together continuously in a row constitute a line. So for us a line will be a sign whose length can be divided into parts, but it will so slender in width that it cannot be split]." *Della pittura*, bk. 1, pp. 55–56; *On Painting*, bk. 1, p. 37.

94. "Più linee, quasi come nella tela più fili accostati, fanno superficie et è superficie certa parte estrema del corpo quale si conosce non per sua alcuna profondità ma solo per sua longitudina e per sue ancora qualità." *Della pittura*, bk. 1, p. 56; *On Painting*, bk. 1, p. 38.

95. As Giulio-Carlo Argan seems to think, in "The Architecture of Brunelleschi and the Origins of Perspective Theory in the XVth Century," *Journal of the Warburg and Courtauld Institute* 9 (1946): 102.

96. "Quello ultimo orlo quale chiude la superficie [The outermost boundary which encloses the surface]." Alberti, *Della pittura*, bk. 1, p. 56; *On Painting*, bk. 1, p. 44.

97. "Et dove la circonscriptione non altro sia che certa ragione di segniare l'orlo delle superficie [Circumscription is the process of delineating the external outlines on the painting]." *Della pittura*, bk. 2, p. 85; *On Painting*, bk. 2, p. 65.

98. "Adunque l'orlo e dorso danno suoi nomi alle superficie [The properties inherent in the periphery and conformation of bodies have determined the names given to surfaces]." *Della pittura*, bk. 1, p. 57; *On Painting*, bk. 1, p. 39.

99. "Non credo io dal pittore si richiegga infinita faticha ma bene s'aspetti pittura quale modo paie rivelata et simigliata a chi si ritrae: qual cosa non intendo senza aiuto del velo alcuno mai possa [If I am not mistaken, we do not ask for infinite labour from the painter, but we do expect a painting that appears markedly in relief and similar to the objects presented. I do not understand how anyone could ever even moderately achieve this without the help of the veil]." *Della pittura*, bk. 2, p. 84; *On Painting*, bk. 2, p. 67.

100. Giorgio Vasari, "Vita di Paolo Uccello," in *Le vite de' più eccellenti architetti, pittori, et scultori italiani* (Florence, 1550), vol. 2, pp. 203–17; English trans.: "Life of Paolo Uccello," in *Lives of the Painters, Sculptors, and Architects*, trans. Gaston du C. de Vere (London, 1912–14), vol. 2, pp. 131–40.

101. "Onde Donatello scultore, suo amicissimo, gli disse molte volte, mostrandoli Paolo mazzochi a punte a quadri tirati in prospettiva per diverse vedute, e palle a settentadue facce a punte di diamanti, e in ogni faccia bruciloli avvolti sur per e bastoni, e altre bizzarrie, in che spendeva e consumava il tempo: Eh! Paolo, questa tua prospettiva ti fa lasciare il certo per l'incerto: queste sono cose che non servono se non a questi che fanno le tarsie [The sculptor Donatello, who was very much his friend, said to him very often — when Paolo showed him *mazzocchi* (ducal caps) with pointed ornaments and squares drawn in perspective from diverse aspects, spheres with seventy-two diamond-shaped facets, with woodshavings wound round sticks on each facet, and other fantastic devices on which he spent and wasted his time — 'Ah, Paolo, this perspective of thine makes thee abandon the subject for the shadow; these are things that are only useful to men who work at the inlaying of wood']." Vasari, "Vita di Uccello," pp. 205–6; "Life of Uccello," pp. 132–33.

We need not discuss the interpretation of Uccello's art implied by Vasari's text (on this point, see my preface to the volume devoted to this painter in the "Classiques de l'art" series [Paris, 1972]), but should note the comparison that he draws between the research of the painter and the practice of marquetry. By an irony of fate (and/or as an invention of Vasari), an unkind remark on the part of Donatello prompted Uccello to retire and return to the studies on perspective that then occupied him right up to his death. Vasari, "Life of Uccello," p. 140.

102. See Adolfo Venturi, "Intarsi marmorei di Leon Battista Alberti," *L'Arte* 72 (1990): 34–36.

103. On the vogue for marquetry in Florence from the first half of the fifteenth century on, and the initiatory role played in this respect by Brunelleschi,

see André Chastel, "Marqueterie et perspective au quinzième siècle," *La Revue des arts* (September 1953): 141–54. As for Piero, from whom several generations of marquetry experts drew their inspiration, his influence countered "Donatellism" and its tendency toward expressionism. See, Chastel, *Le Grand atelier d'Italie, 1460–1500* (Paris, 1965), p. 91.

104. "The basis for a system of signs is not the relation between a signifier and the signified (that relation may be the basis of a symbol, but not necessarily of a sign); it is the relations between signifiers among themselves. The *depth* of a sign adds nothing to its definition; it is its extension that matters, the role that it plays in relation to other signs, the way in which it resembles them or differs from them: the being of every sign depends on what surrounds it, not on its roots." Roland Barthes, *The Fashion System* (New York, 1983).

105. "Principio, vedendo qual cosa, diciamo questo essere cosa quale occupa uno luogo. Qui il pictore, descrivendo questo spatio, dirà questo, suo guidare uno orlo con linea essere circonsciptione [In the first place, when we look at a thing, we see it as an object which occupies a space. The painter will draw around this space, and he will call this process of setting down the outline, appropriately, circumscription]." Alberti, *Della pittura*, bk. 2, p. 81; *On Painting*, bk. 2, p. 64.

106. See Julia Kristeva, "Le Sens et la mode," pp. 66–67.

107. "Il lume et l'ombra fanno parere le cose rilevate [The incidence of light and shade makes it apparent where surfaces become convex or concave]." Alberti, *Della pittura*, bk. 2, p. 99; *On Painting*, bk. 2, p. 82.

108. "Che già, ove sia la pictura fiore d'ogni arte, ivi tutta la storia di Narcisso viene a proposito. Che dirai tu essere dipigniere, altra cosa che simile abracciare con arte quella ivi superficie del fonte? [For as painting is the flower of all the arts, so the tale of Narcissus fits our purpose perfectly. What is painting but the act of embracing by means of art the surface of the pool?]." *Della pittura*, bk. 2, pp. 78–79; *On Painting*, bk. 2, p. 61.

109. "Diceva Quintiliano che pittori antichi soleano circonscrivere l'ombra al sole et cosi indi poi si truovo quest'arte cresciuta [Quintilian said that the ancient painters used to circumscribe shadows cast by the sun, and from this our art has grown]." *Della pittura*, bk. 2, pp. 78–79; *On Painting*, bk. 2, p. 64. See Quintilian, *Institutio oratoria* 10.2, trans. H. E. Butler (Cambridge, Mass., 1953), vol. 4, pp. 76–79, in which delineation is explicitly assimilated to *imitation*; see also Pliny, *Historia naturalis* 35.15.

110. Vasari, "Vita di Filippo Brunelleschi," in *Vite*, vol. 2, p. 332; "Life of Filippo Brunelleschi," in *Lives*, vol. 2, p. 198.

111. "Nè resto ancora di mostrare a quelli che lavorano le tarsie, che è un arte di commettere legni di colori; e tanto gli stimolo, che fu cagione di buono uso e molte case utili che si fece di quel magisterio, ed allora e poi di molte cose eccelenti che hanno recato e forma utile a Firenze per molti anni [Nor did he refrain from teaching it, even to those who worked in tarsia (marquetry), which is the art

of inlaying coloured woods; and he stimulated them so greatly that he was the source of a good style and of many useful changes that were made in that craft, and of many excellent works wrought both then and afterwards, which have brought fame and profit to Florence for many years]." "Vita di Brunelleschi," p. 333; "Life of Brunelleschi," p. 198.

112. Chastel, *Le Grand atelier d'Italie*, p. 142. Confirmation of the antiquity of marquetry perspectives and of the role played by Brunelleschi and Uccello is provided by the "Life of Benedetto da Maiano," in which Vasari recounts that this sculptor had started out by working with wood and excelled "as a craftsman in that form of work, which . . . was introduced at the time of Filippo Brunelleschi and Paolo Uccello — namely, the inlaying of pieces of wood tinted with various colours, in order to make views in perspective, foliage, and many other diverse things of fancy" (*Lives*, vol. 3, p. 257).

113. Chastel, *Le Grand atelier d'Italie*, p. 145 (translator's note: but is not the emphasis placed on "amusing" in danger of obscuring the *theoretical* significance of such a paradox?). It is interesting to note that critics are divided over the identity of the person who answered to the name Manetti and who figures alongside Giotto, Donatello, Brunelleschi, and Uccello, among the "inventors" of the Renaissance on a little panel preserved in the Louvre and often attributed to Uccello himself. In the opinion of Vasari and that of his publisher, Milanesi, the man concerned was Antonio di Tuccio Manetti (born in 1423), a mathematician and the biographer of Brunelleschi, and also a friend of Uccello, with whom, according to Vasari, he used to like to discuss Euclid ("suo amico con quale conferiva assai e ragionava delle cose di Euclide"; Vasari, "Vita di Uccello," p. 214, "Life of Uccello," p. 139). In the opinion of Boeck, it was, on the contrary, Antonio di Ciaccheri, known as Manetti (born in 1405), a specialist in *intarsia* and cited among the "masters of perspective" in the *Memorie storiche di Benedetto Dei*, published by Semper in the *Quellenschriften für Kunstgeschichte* 9 (1875): 263. In any event, the very idea that a specialist of marquetry might be substituted for a mathematician among the "inventors of the Renaissance" is significant in itself. See John Pope-Hennessy, *The Complete Work of Paolo Uccello* (London, 1950), p. 155.

114. "Et per quanto s'aveva a dimostrare di cielo, cioé le muraglie del dipinto stampassono nell'aria, messo d'argento brunito, accio che l'aria e cieli naturali vi si specchiassono dentro; e cosi e nugoli, che si veggono in quello argento esser menati dal vento, quand'e'trae [As he had to show the sky on which the walls shown in perspective were stamped, he put darkened silver so that the natural air and sky would be mirrored there, and also the clouds to be seen in the air, pushed along by wind when it blew]." Cited by Parronchi in "Le due tavole del Brunelleschi." See Antonio Manetti, *Vita di Filippo Brunelleschi*, ed. Domenico de Robertis (Milan, 1976).

115. "The visible bodies are of two kinds only, of which the first is without shape or any distinct or definite extremities, and these though present are imperceptible

and consequently their colour is difficult to determine. The second kind of visible bodies is that of which the surface defines and distinguishes the shape.

"The first kind, which is without surface, is that of the bodies which are thin or rather liquid, and which readily melt into and mingle with other thin bodies, as mud with water, mist or smoke with air, or the element of air with fire, and other similar things, the extremities of which are mingled with the bodies near to them, whence by this intermingling their boundaries become confused and imperceptible, for which reason they find themselves without a surface, because they enter into each other's bodies, and consequently such bodies are said to be without surface." Leonardo da Vinci, C.A., 132rb; *The Notebooks of Leonardo da Vinci*, ed. Edward MacCurdy (London, 1938), vol. 2, pp. 363–64.

116. It is worth remembering that Aristotelian tradition regularly assimilated a cloud to a mirror. On the basis of that assimilation, medieval optics produced an explanation for the formation of rainbows. The *Quaestiones perspectivae*, attributed to Biaggio Pelicani, explains the "miraculous" apparition of angels carrying swords and trumpets in the sky above Milan by a reflection in the clouds of the golden angel that topped the belltower of the church of San Gottardo. Alessandro Parronchi ("Le fonti di Paolo Uccello," pp. 491–93) uses this argument to suggest that the *reversed* figure that appears in a cloud above the sacrificer, in Uccello's *The Sacrifice of Noah*, is not supposed to be God, as Vasari insisted, but was a reflection in the cloud of Noah himself.

CHAPTER 4 *The Powers of the Continuum*

1. The *complete* theory, as will be shown elsewhere. See Hubert Damisch, *L'Origine de la perspective* (Paris, 1987); English trans.: *The Origin of Perspective*, trans. John Goodman (Cambridge, Mass., 1994).

2. Emile Benveniste, *Problèmes de linguistique générale* (Paris, 1966), p. 131.

3. A famous passage from Vitruvius, skillfully analyzed by Panofsky (*Die Perspektive als "symbolische Form": Aufsätze zu Grundfragen der Kunstwissenschaft* [Berlin, 1964], pp. 137–39, n. 18), shows that antiquity was familiar with a form of perspective. But the theoretical question concerns (a) the respective principles and methods of ancient perspective and Renaissance perspective, and (b) the status of the perspective code, its position, and its function in the organization of representation.

4. Erwin Panofsky, "*Nebulae in pariete*: Notes on Erasmus' Eulogy on Dürer," *Journal of the Warburg and Courtauld Institutes* 14 (1951): 34–41.

5. "Et unquam vidisti nebulam pictam in pariete? Vidisti utique et memnisti [And have you ever seen a cloud painted on a wall? You have certainly seen one and remembered it]." *Decimi magni a ausonii burdigalensis opuscula*, ed. Peiper (Leipzig, 1886), cited by ibid., p. 222. In a letter to Aurelius Symmachus, the same Ausonius writes of golden leaves and painted clouds that give pleasure only for as long

as one is looking at them ("Hoc velut aerius bratteae fucus aut picta non longius, quandam videtur, oblectat"; ibid.).

6. Aloïs Riegl, "Zur Kunsthistorischen Stellung der Becher von Vafio," in Riegl, *Gesammelte Aufsätze* (Vienna, 1929), pp. 77–79; see Friedrich Matz, *La Crète et la Grèce primitive* (Paris, 1962), pp. 121–25.

7. John Ruskin, *Modern Painters* (London, 1855), vol. 3, pt. 4, chap. 16, "On Modern Landscape."

8. See the two versions of *Venus Arming Aeneas*, by Poussin (the museums of Toronto and Rouen); *Catalogue de l'exposition Nicolas Poussin* (Paris, 1960), nos. 44 and 59.

9. Jacques Derrida, *Of Grammatology*, trans. G. C. Spivak (Baltimore, 1976).

10. Giulio-Carlo Argan ("The Architecture of Brunelleschi and the Origins of Perspective Theory in the XVth Century," *Journal of the Warburg and Courtauld Institute* 9 [1946]: 105) detects a satirical allusion to the gold backgrounds of medieval painting.

11. See Pierre Francastel, *Peinture et société* (Lyons, 1951), pt. 1.

12. "Onde mi sono meravigliato assai che un tanto accurato e diligente facesse un errore cosi notabile [I marvel greatly that a man so accurate and diligent could make an error so notable]." Giorgio Vasari, "Vita di Paolo Uccello," in *Le vite de' più eccellenti architetti, pittori, et scultori italiani* (Florence, 1550), vol. 2, p. 210; English trans.: "Life of Paolo Uccello," in *Lives of the Painters, Sculptors, and Architects*, trans. Gaston du C. de Vere (London, 1912–14), vol. 2, p. 136.

13. Bibliotheque Nationale de France MS 2038, 132r. See Jean-Paul Richter, ed., *The Literary Works of Leonardo da Vinci*, 2nd ed. (London, 1939), vol. 1, p. 127. For Leonardo, as for Alberti before starting to compose his story, a painter needed to have thoroughly learned about perspective. It should be noted that Cennino Cennini already considered the study of nature to be both the rudder and the triumphal arch (*timone e porta trionfale*) of a *disegno* (Leonardo, *Treatise on Painting*, trans. Philippe MacMahon [Princeton, N.J., 1956], chap. 28), which he regarded as being dependent upon a power—defined as the power of sight—which, in its turn, took as its rudder and its light (a) the light of the sun, (b) the light of the painter's eye, and (c) the painter's hand: according to him, without an interchange between pictorial practice and the basic facts of optical theory, nothing could be done in a reasonable manner ("E'l timone e la guida di questo potere veder, si è la luce del sole, la luce dell'occhio tuo, e la man tua: che senza queste tre cose nulla non si puo fare con ragione"). *Treatise*, chap. 8, p. 6.

14. "Durerus quanquam et alias admirandus, in monochromatis, hoc est, nigris lineis, quid non exprimit? Umbras, lumen, splendorem, eminentias, depressiones: ad haec, ex situ rrei unius, non unam speciem sese oculis intuitium offerentem. Observat exacte symmetrias et harmonias. Quin ille pinget et quae pingere non possunt, ignem, radios, tonitrua, fulgetra, fugura, vel nebulas, ut aiunt, in pariete, sensus, affectus omnes denique totum hominis animum in habitu corporis relu-

centem, ac pene vocem ipsam. Haec felicissimus lineis iisque nigris sic ponit ob oculos, ut si colorem illinas, iniuriam facias operi. An non hoc mirabilius, absque colorum lenocinio praestare, quod Apelles praestitit colorum praesidio?" Erasmus, *Dialogus de recta Latini Graecique sermonis pronuntiatione,* cited by Panofksy from the Leiden edition (1643), in *"Nebulae in pariete,"* pp. 35–36; trans. David Money (Darwin College, Cambridge, Eng., 1999).

15. Goethe, *Conversations avec Eckermann* (Paris, 1949), p. 75.

16. "Ut autem qui musici periti sunt, rectius pronuntiant etiam non cantantes: ita qui ducendis in omnem formam lineis digitos habet exercitatos mollius ac felicius pinget literas [Just as those who are skilled in music pronounce better, even when not singing, so those who have fingers skilled at drawing lines in all shapes make letters more smoothly and happily]." Erasmus, *Dialogus,* cited in Panofsky, *"Nebulae in pariete,"* p. 35; trans. David Money.

17. The expression is borrowed from Pliny: "Ambire enim se ipsam debet extremitas et sic desinere, ut promittat alia et post se, ostendatique etiam quae occultat." Pliny, *Historia naturalis* 35.36.

18. Referring, in his *Adagia,* to the poem by Ausonius cited above, Erasmus concludes, contrary to the Latin poet, that it is not possible to paint a cloud: "Nam nebula res inanis quam ut coloribus exprimi queat [For a cloud is too empty a thing to be expressed in colors]." Erasmus, *Dialogus,* cited in Panofsky, *"Nebulae in pariete,"* p. 39, n. 4; trans. David Money.

19. Wölfflin, *Principes,* p. 13. It should be noted that Wölfflin's text deals with precisely the domain of culture that the present study seeks to investigate, a field in which the names of Pliny, Erasmus, and Leonardo arise time and again:

> The art of the Renaissance . . . was capable of representing all that it *wished* to, and one gets a fair idea of its power when one remembers that in the last analysis it managed to find a linear expression for even the least plastic of objects, bushes, hair, water and clouds, smoke, and flames. Is it certain that it is true to say that such things are more difficult to apprehend using lines than bodies with determinate shapes? Just as one may detect in the sound of bells all possible words, one can render what one sees in various different ways, and nobody is justified in saying that one expression is truer than another. (ibid., p. 33)

20. If painting were a language in the strict sense, it ought to be possible to translate any text, in any language, into its terms, and vice versa. See Louis Hjelmslev, "La Structure fondamentale du langage," in *Prolégomènes à une théorie du langage,* French trans. (Paris, 1968), p. 179.

21. Pontormo, *Lettre à Benedetto Varchi* (see above, Chapter 1, pp. 32 ff.).

22. See above, Chapter 1, pp. 30 ff.

23. The expression is taken from Giuseppe Fiocco, *Mantegna* (Paris, 1938), p. 121.

24. In truth, Mantegna and the frescoes of Padua and Mantua were not the first to produce perspective models that lent themselves to variations that appear to contradict the literal doctrine expounded by Alberti in *Della pittura*, which left the painter entirely free to establish the central point, provided it was *within the limits that he had assigned to his composition*. Alessandro Parronchi has shown how two works only slightly later than Brunelleschi's experiments had already been constructed according to a perspective *dal sotto in su* that was perfectly justifiable in terms of the theory. They were the tondo bearing *The Ascension of Saint John*, by Donatello, in the old sacristy of San Lorenzo, and Massaccio's *Trinity*, in Santa Maria Novella (see Chapter 4, pp. 156 ff.; also Parronchi, "Le due tavole del Brunelleschi," in *Studi su la "dolce" prospettiva* [Milan, 1964], pp. 226–95). One reservation needs to be made where the *Trinity* is concerned, however, for it constituted but one part of a vaster composition, treated in trompe l'oeil, in which the perspective effect was designed to reinforce an illusion (see U. Schlegel, "Observations on Massaccio's Trinity Fresco in Santa Maria Novella," *Art Bulletin* 55 (1963): 19–33).

25. Maurice Merleau-Ponty, *Signes* (Paris, 1960), p. 63.

26. Jurgis Baltrušaitis, *Anamorphoses et perspectives curieuses* (Paris, 1955).

27. Baltrušaitis interprets the letter cited in Chapter 3 from Dürer to Pirkheimer as expressing a desire to be initiated into what were then considered to be secret practices. As we have seen, Panofsky regarded the letter merely as proof of Dürer's interest in the *theoretical* bases of perspective. However, the two interpretations are not mutually exclusive, given that initiation into the mysteries of anamorphosis implied being shown an example of specular coincidence between the viewing point and the vanishing point, something that is not mentioned in Alberti's treatise but that is nevertheless a basic principle of Renaissance perspective, as is proved by Brunelleschi's experiment.

28. Baltrušaitis, *Anamorphoses et perspectives curieuses*, p. 5.

29. "Una di quelle pitture, le quali, perché (benché) riguardate in scorcio da un luogo determinato, mostrino una figura umana, sono con tal regola di prospettiva delineate che, vedute in faccie e come naturalmente e communemente si guardano le altre pitture, altro non rappresentano che una confusa e inordinata mescolenza di linee e di colori, dalle quali anco si potriano malamente raccapezzare imagini di fiumi o sentier tortuosi, ignude spaggie, nugoli, o stranissime chimere [One of those paintings which, when (although) seen obliquely from a predetermined spot, show a human figure, are delineated according to such a rule of perspective that, seen from in front and naturally in the usual fashion as other paintings are, represent nothing but a confused and disorganized jumble of lines and colours in which one may with difficulty make out images of rivers or winding paths, empty beaches, clouds, or the strangest of chimeras]." Galileo Galilei, *Considerazioni al Tasso*, in *Opere*, vol. 9, p. 129, cited by Panofsky, *Galileo as Critic*, pp. 13–14, n. 4.

30. As is suggested by Lomazzo in bk. 6, chap. 19 ("Manière de faire la perspec-

tive inverse qui paraisse vraie étant vue par un seul trou") in his *Traité de la peinture* (Milan, 1584); see Baltrušaitis, *Anamorphoses et perspectives curieuses*, pp. 18–21.

31. "Subito che l'aria fia alluminata s'empirerà d'infinite spetie, le quali son causate da vari corpi e colori, che infra essa sono collocati, delle quali spetie l'occhio si fa bersaglio e calamita [The instant the atmosphere is illuminated it will be filled with an infinite number of images which are produced by the various bodies and colours assembled in it. And the eye becomes the target and the loadstone of these images]." C.A., 27r; see Richter, ed., *Literary Works of Leonardo da Vinci*, vol. 1, p. 135.

32. "The air is full of an infinite number of images of the things which are distributed through it, and all of these are represented in all, all in one and all in each" (C.A., 138rb; *The Notebooks of Leonardo da Vinci*, ed. Edward MacCurdy [London, 1938], vol. 2, p. 364). "Show how nothing can be seen except through a tiny fissure through which the atmosphere passes full of the images of objects that cross over one another, in between the thick and opaque sides of those fissures" (C.A., 345r).

33. *Notebooks*, vol. 1, p. 300.

34. Richter, ed., *Literary Works of Leonardo da Vinci*, vol. 1, pp. 159–60.

35. C.A., 182r.

36. C.A., 68r; *Notebooks*, vol. 1, p. 393; Richter, ed., *Literary Works of Leonardo da Vinci*, vol. 1, p. 126. "La superficie è termine del corpo. E'l termine d'un corpo non è parte d'esso corpo, e'l termine d'un corpo è principio d'un altro quell è niente che non è parte d'alcuna cosa. Quella è niente che niente occupa [The surface is a limitation of the body and the limitation of the body is no part of that body, and the limitation of one body is that which begins another. That which is no part of any body is a thing of naught. A thing of naught is that which fills no space]."

37. On the three kinds of perspective used in painting, see *Notebooks*, vol. 2, pp. 263, 372, 374. It should be noted that Alberti himself declared that he thought nothing of a painter who did not understand the power that shade and light exerted upon surfaces, or whose figures had no qualities apart from outline (Leon Battista Alberti, *Della pittura* [Florence, 1436], bk. 2, p. 99; English trans.: *On Painting*, trans. Cecil Grayson [Harmondsworth, Eng., 1991], bk. 2, p. 82). His *Della pittura* does make room for atmospheric perspective, which it describes in purely pictorial terms; charged with light and color, rays pass through the damp, thick air (*l'aere quale humido di certa grassezza*) and grow weaker (*indeboliscono*) the more distant they are. Hence the rule: "Quanto maggiore sarà la distantia tanto la veduta superficie parrà più fusca [As the distance becomes greater, so the surface seen appears more hazy]." *Della pittura*, bk. 1, pp. 60–61; *On Painting*, bk. 1, p. 48.

38. "The true knowledge of the form of an object becomes gradually lost in proportion as distance decreases its size." *Notebooks*, vol. 2, p. 365.

39. "Painting is concerned with all the ten attributes [*ofiti*] of sight, which are: Darkness, Light, Solidity, Colour, Form and Position, Motion and Rest. This little work of mine will be a tissue [of the studies] of these attributes, reminding the

painter of the rules and methods by which he should use his art to imitate all these things, the works of Nature which adorn the world." Bibliotheque Nationale de France MS 2038, 22b; Richter, ed., *Literary Works of Leonardo da Vinci*, vol. 2, p. 120.

40. "Ogni forma corporea, in quanto allo ofitio dell'occhio, si divide in tre parti cioè corpo, figura e colore; la similitudine corporea s'astende più lontana dalla sua origine che non fa il colore o la figura, di poi il colore s'astende più che la figura [Every visible body, in so far as it affects the eye, includes three attributes; that is to say: mass, form, and colour; and the mass is recognizable at a greater distance from the place of its actual existence than either colour or form. Again, colour is discernible at a greater distance than form]." Bibliotheque Nationale de France MS 2038, 12v; Richter, ed., *Literary Works of Leonardo da Vinci*, vol. 1, pp. 210–11.

41. "If you see a man near at hand you will be able to recognize the character of the substance, of the shape and even of the colour, but if he goes some distance away from you you will no longer be able to recognize who he is because his shape will lack character, and if he goes still further away from you, you will not be able to distinguish his colour but he will merely seem a dark body, and further away still he will seem a very small round, dark body." Bibliotheque Nationale de France MS 2038, 12v; *Notebooks*, vol. 2, p. 238.

42. See the proceedings of the Centre National de Recherches Scientifiques colloquium, *Léonard de Vinci et l'expérience scientifique au XVIe siècle* (Paris, 1953).

43. *Notebooks*, pp. 248 and 262.

44. Paris, Institut de France MS G, 37a; Richter, ed., *Literary Works of Leonardo da Vinci*, vol. 1, p. 129.

45. "That dimness [*il mezzo confuso*] which occurs by reason of distance, or at night, or when mist comes between the eye and the object, causes the boundaries of this object to become almost indistinguishable from the atmosphere." C.A. 316v; *Notebooks*, vol. 2, p. 368.

46. "How to Paint the Wind," *E*, 6v; Richter, ed., *Literary Works of Leonardo da Vinci*, vol. 1, p. 298.

47. "On a Deluge and Its Depiction in Painting"; Windsor MS 12665v; Richter, ed., *Literary Works of Leonardo da Vinci*, vol. 1, p. 608.

48. "On a Deluge and Its Depiction in Painting"; Windsor MS 12665r; Richter, ed., *Literary Works of Leonardo da Vinci*, vol. 1, p. 606.

49. "How to Represent a Storm"; Bibliotheque Nationale de France MS 2038, 21r; Richter, ed., *Literary Works of Leonardo da Vinci*, vol. 1, p. 351.

50. "E ti parra forse potermi riprendere dell'avere io figurato le vie fatte per l'aria dal moto del vento, che'l vento per se non si vede infra l'aria. A questa parte si risponde che non il moto del vento, ma il moto delle cose da lui portate è sol quel che per l'aria si vede." "On a Deluge and Its Depiction in Painting"; Windsor MS 12665; Richter, ed., *Literary Works of Leonardo da Vinci*, vol. 1, p. 345.

51. See Chapter 2, pp. 51 ff.

52. For Peirce, a sign or *representamen* is a term established in relation to another (its object) in a relationship that is such that it makes a third (its *interpretant*) assume in relation to the second a similar triadic relationship, and so on ad infinitum. Even if a *representamen* does not function as such until the moment when it actually determines an *interpretant*, nevertheless its representational quality does not depend upon the effective determination of an *interpretant* nor even upon the fact that it in effect has an object. See Charles Sanders Peirce, *Elements of Logic*, vol. 2, chap. 3; *Collected Papers* (Cambridge, Mass., 1965), vol. 2, pp. 156–57.

53. It is in this triple sense that it also has a part to play in the program designed for "the way of representing a battle." "And if you introduce horses galloping outside the crowd, make the little clouds of dust distant from each other in proportion to the strides made by the horses; and the clouds which are furthest removed from the horses should be the least visible; make them high and spreading and thin, and the nearer ones will be more conspicuous and smaller and denser." Richter, ed., *Literary Works of Leonardo da Vinci*, vol. 1, p. 348.

54. Max Dvořàk, "Uber Greco und den Manierismus," in Dvořàk, *Kunstgeschichte als Geistesgeschichte* (Vienna, 1928), pp. 261–76.

55. Walter Friedländer, *Mannerism and Anti-Mannerism in Italian Painting* (New York, 1965), p. 17.

56. Dvořàk, "Uber Greco und den Manierismus," p. 270.

57. Cited in ibid., p. 273.

58. See Karl Mannheim, "Beiträge zur Theorie der Weltanschauungs-Interpretation," *Jahrbuch für Kunstgeschichte* I, no. 15 (1921–22); an English translation may be found in *Essays on the Sociology of Knowledge* (New York, 1952), pp. 33–83.

59. "Et piacemi sia nella storia chi admonisca et insegni ad noi quello che ivi si facci: o chiami con la mano a vedere o, con viso cruccioso e con li occhi turbati, minacci che niuno verso loro vada; o dimostri qualche pericolo o cosa ivi maravigliosa o te inviti ad piagniere con loro insieme o a ridere [I like there to be someone in the 'historia' who tells the spectators what is going on, and either beckons them with his hand to look, or with a ferocious expression and forbidding glance challenges them not to come near . . . or points to some danger or remarkable thing in the picture, or by his gestures, invites you to laugh or weep with them]." Alberti, *Della pittura*, bk. 2, p. 94; *On Painting*, bk. 2, pp. 77–78.

60. See Max Dvořàk, "Dürers Apokalypse," in *Kunstgeschichte als Geistesgeschichte*, pp. 193–202. As Wölfflin perceived (*Die Kunst Albrecht Dürers*, 6th ed. [Munich, 1943]), an echo of this crisis is detectable even in Dürer's work, divided between the two tendencies of the *Apocalypse*, in which the measurements of human space are undermined, and the *Life of the Virgin*, which abides by the dimensions of daily experience.

61. See M. R. James, *The Apocalypse in Art* (London, 1931).

62. Dvořàk, "Uber Greco und den Manierismus," p. 195.

63. Panofsky, *The Life and Works of Albrecht Dürer*, 3rd ed. (Princeton, N.J., 1948), pp. 56–58.

64. See Karl Mannheim, "Towards the Sociology of the Mind," in Mannheim, *Essays on the Sociology of Culture* (London, 1956), p. 33.

65. See Chapter 3, p. 84.

66. Ferdinand de Saussure, *Cours de linguistique générale* (Paris, 1949), p. 111.

67. Juri Tynianov and Roman Jakobson, "Les Problèmes des études littéraires et linguistiques," *Novyj Lef*, no. 12 (1928); French trans. in *Théorie de la littérature* (the texts of the Russian Formalists), trans. Tzvetan Todorov (Paris, 1965), pp. 138–40.

68. Representational painting, which sets out to institutionalize a trompe l'oeil space, in this "perspective" subordinates color to drawing. Nonfigurative art, on the contrary, strives to liberate color, to produce it as *a nonfigure*.

69. See E. T. De Wald, *The Illustrations of the Utrecht Psalter* (Princeton, N.J., 1933), pl. 2, 5, 7, 10, 11, 18, 22, 29, and others.

70. See Jean Porcher, *Le Sacramentaire de Saint-Etienne de Limoges* (Paris, [1953]), pl. 11.

71. Paris, Bibliotheque Nationale de France, lat. 10525; see Abbé V. Leroquais, *Les Psautiers manuscrits des bibliothèques publiques de France* (Paris, 1940–41), vol. 2, pp. lxxxiii–lxxxv.

72. See Meyer Schapiro, "From Mozarabic to Romanesque in Silos," *Art Bulletin* 22 (1939): 313–74; reprinted in Schapiro, *Romanesque Art: Selected Papers* (New York, 1977), vol. 1, pp. 28–101.

73. Carlo Cecchelli, *San Clemente* (Rome, n.d.), p. 146 and pl. 25.

74. Bayer Staatsbibliothek, Munich, *Livre des péricopes d'Henri II*, CLM 4452, fol. 152v. See André Grabar and Karl Nordenfalk, *La Peinture du haut Moyen Age* (Geneva, 1957), reprod. p. 204.

75. Bibliothèque de l'Arsenal, MS 1186, fol. 1v; see Bibliotheque Nationale de France, *Catalogue de l'exposition des manuscrits à peinture en France du XIIIe au XVIe siècle* (Paris, 1955), pl. 1.

76. When at the turn of the seventeenth century, painting, especially so-called decorative painting, tried to widen the perspective order and tip it over in order to open onto the sky, it was no longer flights up into the sky and overflowing clouds that seemed unsuitable, but rather the obstinacy of a painter such as Caravaggio who persisted in keeping his figures riveted to the ground, where they dirtied their feet with dust that was in no sense metaphysical or—as Poussin was to remark—not at all "ancient."

77. Tynianov and Jakobson, "Les Problèmes des études littéraires et linguistiques," p. 139.

78. Giulio-Carlo Argan, *L'Europe des capitales* (Geneva, 1964), p. 28.

79. See George R. Kernodle, *From Art to Theatre: Form and Convention in the Renaissance* (Chicago, 1944).

80. Similarly, within a single series: it makes no sense to speak of "survivals" in connection with the machines of Baroque theater. The elements "borrowed" from medieval theater are integrated with new, original structures. They were treated in a "realistic" way by the medieval theater directors, and continued to be used thus by Mahelot at the Hôtel de Bourgogne, but were used for the stage scenery of Spanish theater in purely symbolic and conventional ways.

81. See the famous *Tragic and Comic Scenes* by the school of Piero della Francesca, preserved in Urbino, Baltimore, Berlin, and elsewhere.

82. See *The Dispute of Saint Catherine*, by Masolino, in the church of San Clemente in Rome, where the side walls, the end wall, and the ceiling of the hall containing the picture are decorated by a regular checkerboard pattern, while the floor remains uniformly neutral.

83. John White, *The Birth and Rebirth of Pictorial Space*, 2nd ed. (London, 1967), p. 139, pl. 30b. Masaccio's design is all the more remarkable in that it amounts to installing a geometrically constructed, blind, coffered vault, in place of the luminous cloud that the texts regularly associated with the presence of the Trinity's Holy Spirit, both on the occasion of the Baptism and on that of the Transfiguration. "Spiritus fuit nubes lucida in die transfigurationis": Saint Augustine, letter to Evodius, epistle 169, *Patrologia latina*, vol. 33, col. 745. See Saint Thomas Aquinas, *Summa theologiae* q. 43, a. 7, ad. 6, and the Roman Breviary (6 August, resp. 2): "In splendente nube Spiritus Sanctus visus est."

84. "And when the cloud is created it also generates wind, since every movement is created from excess or scarcity; therefore in the creation of the cloud it attracts to itself the surrounding air and so becomes condensed, because the damp air was drawn from the warm into the cold region which lies above the clouds; consequently as it has to make water from air which was at first swollen by it, it is necessary for a great quantity of air to rush together in order to create the cloud; and since it cannot make a vacuum, the air rushes in to fill up with itself the space that has been left by the [former] air, which was first condensed and then transformed into a dense cloud. In this circumstance the wind rushes through the air, and does not touch the earth, except on the summits of the high mountains; it cannot draw air from the earth, because there would then be a vacuum between the earth and the cloud; and it draws but little air through the traverse and draws it more abundantly through every line." Leonardo da Vinci, *Leic.*, 28r; *Notebooks*, vol. 2, pp. 122–23.

85. "When the air is still and a full company of clouds have risen to a height, and there above as has been said press themselves together, they squeeze out so much air from themselves, which through the violence exerted creates such movement in the air, that as you may see it communicates its movement to the other lesser clouds. And as they also drive the air forwards in the same way they even furnish themselves with a reason for greater flight; for when a cloud either finds itself in the midst of others or apart from them, if it produces the wind behind

itself that air which is between it and its neighbour following comes to multiply, and by multiplying acts in the same way as the powder does in the mortar, for this expels from the position near to it the less heavy body and the lighter weight. And this being the case it follows that the cloud in driving the wind towards the others which offer resistance is the cause of putting these themselves to flight. And by sending this vanguard of the winds before itself it also adds volume to the rest." C.A., 212va; *Notebooks*, vol. 1, pp. 400–401.

86. *Notebooks*, vol. 2, p. 112.

87. "The elements are changed one into another, and when the air is changed into water by the contact that it has with its cold region this then attracts to itself with fury all the surrounding air which moves furiously to fill up the place vacated by the air that has escaped; and so one mass moves in succession behind another, until they have in part equalized the space from which the air has been divided, and this is the wind" (C.A., 169ra; *Notebooks*, vol. 1, p. 397). "The elements, that is to say water, air and fire, for the clouds are composed of warmth and humidity and (in summer) dry vapours" (C.A., 212 va; *Notebooks*, vol. 1, p. 401). "But dust, which is of a terrestrial nature, may be dispersed in the air and affect the shape of a cloud" (C.A., 79r).

88. On the cloud in the form of a high mountain in flames that Leonardo saw over Lake Maggiore, see *Notebooks*, vol. 2, p. 123.

89. Richter, ed., *Literary Works of Leonardo da Vinci*, vol. 1, p. 38.

90. Ibid., p. 31.

91. "Nissuna humana investigatione si po dimandare vera scientia, se essa non passa per le matematiche dimostrationi [No human investigation can be called true science without passing through mathematical tests]." *Trattato*, entry 1; Richter, ed., *Literary Works of Leonardo da Vinci*, vol. 1, pp. 31–32.

92. "Il principio della scientia della pittura è il punto; il secondo è la linea; il terzo è la superficie [The science of painting begins with the point, then comes the line, the plane comes third]. *Trattato*, entry 3; Richter, ed., *Literary Works of Leonardo da Vinci*, vol. 1, p. 32.

93. "Il pittore è quello, che per necessità della sua arte ha partorito essa pro-spettiva." *Trattato*, entry 17; Richter, ed., *Literary Works of Leonardo da Vinci*, vol. 1, p. 37.

94. Ibid.

95. "E se'l geometra riduce ogni superficie circondata da linee alla figura del quadrato et ogni corpo alla figura del cubo, e l'Aritmeticha fa il simile con le sue radici, cube e quadrate, queste due scientie non s'estendono, se non alla notitia della quantità continua e discontinua, ma della qualità non si travaglia, la quale è bellezza delle opere di natura e ornamento del mondo [And as the geometrician reduces every area circumscribed by lines to the square and every body to the cube; and arithmetic does likewise with its cubic and square roots, these two sciences do not extend beyond the study of continuous and discontinuous quantities; but they

do not deal with the quality of things which constitutes the beauty of the works of nature and the ornament of the world]"; ibid.

96. "E veramente non senza caggione non l'hanno nobilitata, perchè per se medesima si nobilita, senza l'aiuto de l'altrui lingue, non altrimente, che si facciano l'eccellenti opere di natura. E se li pittori non hanno di lei descritto e ridotta in scientia, non è colpa della pittura [And truly it was not without reason that they did not confer honours upon her, since she proclaims her own glory without the help of tongues in the same way as do the excellent works of nature. And it is not the fault of painting if painters have not described their art and reduced it to a science]." *Trattato*, entry 17; *Richter*, ed., *Literary Works of Leonardo da Vinci*, vol. 1, pp. 37–38.

97. Richter, ed., *Literary Works of Leonardo da Vinci*, vol. 1, p. 32.

98. D, 10v; *Notebooks*, vol. 1, p. 212.

99. Alexandre Koyré, *Études galiléennes* (Paris, 196), p. 23, n. 3; English trans.: *Galileo Studies*, trans. John Mepham (Hassocks, Eng., 1966).

100. Alexandre Koyré, "Léonard de Vinci 500 ans après," in Koyré, *Études d'histoire de la pensée scientifique* (Paris, 1966), pp. 85–100.

101. Ibid., pp. 95–97.

102. See Michel Fichant, "L'Idée d'une histoire des sciences," in Michel Fichant and Michel Pécheux, *Sur l'histoire des sciences* (Paris, 1969), pp. 72–73.

103. Pierre Duhem, *Études sur Léonard de Vinci, ceux qu'il a lus, ceux qui l'ont lu*, 2nd ed. (Paris, 1955), vol. 2, p. 17.

104. See Chapter 3, pp. 116 ff.

105. B.M., 160r and 190v; *Notebooks*, vol. 1, p. 645.

106. See Alexandre Koyré, "Sens et portée de la synthèse newtonienne," in Koyré, *Études newtoniennes* (Paris, 1968), pp. 31–32; English trans.: *Newtonian Studies* (London, 1965).

107. Ibid.

108. B.M., 173v and 190v; *Notebooks*, vol. 1, p. 82; Richter, ed., *Literary Works of Leonardo da Vinci*, vol. 2, pp. 137–38.

109. "Scrivi la qualità del tempo, separata dalla geometrica [Describe the nature of time as distinguished from the geometrical definitions]." B.M., 176r; Richter, ed., *Literary Works of Leonardo da Vinci*, vol. 2, p. 138; *Notebooks*, vol. 1, p. 82.

110. Koyré, "Léonard de Vinci 500 ans après," pp. 98–100.

111. Ibid.

112. Ibid., p. 94.

113. *Trattato*, entry 1: "Geometry is infinite because every continuous quantity is divisible to infinity in one direction or the other. But the discontinuous quantity commences in unity and increases to infinity and, as it has been said, the continuous quantity increases to infinity and decreases to infinity. And if you allow yourself to say that you give me a line of twenty *braccia* [armlengths], I will tell you how to make one of twenty-one." B.M., 18r; *Notebooks*, vol. 1, p. 641.

114. See Pierre Duhem, "Leonardo da Vinci et les deux infinis," in *Études sur Leonardo da Vinci*, vol. 2, pp. 4–53.

115. "Qual è quella cosa che non si dà, e s'ella si dessi non sarebbe? Egli è l'infinito, il quale se si potesse dare, e'sarebbe terminato e finito, perchè cio che si puo dare, a termine colla cosa che la circuisce ne'suoi stremi, et cio che non si puo dare a quella cosa che non ha termini [Which is the proposition which cannot be granted? and, if granted, would have no sense? It is the infinite, which thereby would become circumscribed and limited. For what the proposition assumes is of necessity bounded by something that surrounds it, and that which cannot be granted is that which has no limits]." C.A., 131r; Richter, ed., *Literary Works of Leonardo da Vinci*, vol. 1, p. 257.

116. Not only Leonardo, but already the contemporaries and even immediate predecessors of Alberti, in particular Biagio Pelicani, who studied under Buridan in Paris and then taught at Pavia, Bologna, Florence (where he was living in 1389), and Padua (where he died in 1416). According to Robert Klein ("Pomponius Gauricus on Perspective," *Art Bulletin* 43 [1961]: 238), he was the author of the anonymous *Perspectiva* given to Alberti by Bonnucci (*op. volg.*, vol. 4). In his *Quaestiones perspectivae* of 1390, a copy of which Paolo Toscanelli brought with him when he returned to Florence in 1424, Pelicani set out the rule governing all perspective constructions, in the optical sense of the expression — the rule according to which a limit must be assigned to infinitely vanishing lines by the apparent introduction of parallels: "Cum lineae equidistantes et oculus fuerint in eadem superficie, extimatur lineas illas debere concurrere et nullomodo esse equidistantes [When equidistant lines and the eye are on the same surface it is thought that these lines must converge and cannot in any way be equidistant]." *Quaestiones perspectivae* 1.1, *quaest.* 7, art. 1, corol. 2; cited in Parronchi, "Le due tavole del Brunelleschi," p. 241. Leonardo's formula reiterates Vitellion's assertion (*Perspectiva* 1.4, th. 21), according to which "lineae . . . videbuntur quasi concurrere, non tamen videbuntur unquam concurrentes, quia semper sub angulo quodam videbuntur [lines will seem as it were to converge, but they will never be seen to converge because they are always seen under a particular angle]." Panofsky (*Die Perspektive als "symbolische Form,"* pp. 140–41, n. 22) has shown that, at the time when it was stated (thirteenth century), Vitellion's theorem could not have consituted a critique of the theory of the vanishing point, which, mathematically speaking, is linked to the "finitist" concept of a *limit*, that is to say to the possibility of imagining that, given that parallels extend ad infinitum, their relative distance and hence the angle at which their most distant points are seen become equal to zero. It was Desargues who produced the first really pertinent definition of the concept of the vanishing point. Nevertheless, the Brunelleschian model did lead to the posing, in pictorial terms, of the question of the infinite, which had already been described metaphorically as a point (*quasi persino in infinito*, as Alberti was to put it). But far from being founded upon an explicit infinitization of space, the perspective construction, through the *limit* that

it assigns itself, on the contrary implies a closure of the representational scene. In truth, as we shall see, the question of the infinite was far from being a solely mathematical question, and it did not cease, in various forms, some of them pictorial, to be at work within the entire culture of the Renaissance. So it is not possible to follow Panofsky when he declares categorically that the concept of space conveyed by the perspective construction with a central vanishing point is the very same that was later to be rationalized by Descartes and formalized by Kantian theory (*Die Perspektive als "symbolische Form,"* pp. 121–22). The *ideological* censure directed at the progress of mathematics, and the remarkable *advances* made by art and philosophy (to which we shall be returning), manifest the complexity of cultural work that cannot be classified under the traditional rubrics of the history of art, science, or ideas, and also the need for a critical redistribution of the domains and roles that are assigned, in an overall cultural structure, to "art," "science," and "ideology."

117. *Notebooks*, vol. 2, p. 363.

118. Francastel (*Peinture et société*, pp. 36–37 and 68) believes he can trace to Masolino the origin of speculation on open space and of an attempt to organize unitary, cubic space without resorting to a back cloth. Similarly, the most important invention that he attributes to Uccello was his creation of an imaginary opening in the background of the painting in his *Story of Noah*. This work (which Mantegna was to carry on in different ways) stood in contrast to the elaboration of a closed scenographic cube, of which Andrea del Castagno's *Last Supper* provides a good example (Florence, Cenacolo di Sant'Apollonia), and which Poussin was again to construct materially, two centuries later, in the course of his experiments on the lighting of figures. We should recognize, however, along with Francastel, that in this case representation still required simultaneous notation of both the container and what was contained, the container (the perspective cube open on one side) being itself contained in an abstract space. Francastel rightly regards this paradox as an "aesthetic" version of the problem of the continuum. This is a problem that Maurice Blanchot (*L'Entretient infini* [Paris, 1969], p. 11, n. 1) has accurately described:

> When one assumes (usually implicitly) that "reality" is continuous and that only our awareness or expression of it introduces discontinuity, we forget in the first place that "the continuum" is only a model, a theoretical form which, in forgetting, we take to be pure experience, a pure empirical affirmation. However, "the continuum" is merely an ideology that is ashamed of itself, just as empiricism is merely knowledge that repudiates itself. Allow me to recall what set theory has made it possible to posit: contrary to what has long been maintained, there is a *power* of the infinite that raises the infinite above the continuum or, as J. Vuillemin puts it: "the infinite is a genus of which the continuum is a species." (*La Philosophie de l'algèbre*)

119. "Quali segniate linee amme dimostrino in che modo, quasi per sino in infinito, ciascuna traversa quantità sequa alterandosi [This inscribed line indicates to me in what way, as if looking into infinity, each transverse quantity is altered visually]." Alberti, *Della pittura*, bk. 1, p. 71; *On Painting*, bk. 1, p. 56.

120. Alexandre Koyré, "Les Étapes de la cosmologie scientifique," in Koyré, *Études d'histoire de la pensée scientifique*, p. 83.

121. Alexandre Koyré, "Galileo and Plato," *Journal of the History of Ideas* 4, no. 4 (October 1943).

122. Koyré, "L'Apport scientifique de la Renaissance," *Journal of the History of Ideas* 4, no. 4 (October 1943): 45.

123. See Alexandre Koyré, *From the Closed World to the Infinite Universe* (Baltimore, 1957), pp. 42–57.

124. Ibid., pp. 6–24.

125. Marcellus Stellatus Palingenius, *Le Zodiaque de la vie* (The Hague, 1732), cited in Koyré, *From the Closed World to the Infinite Universe*, pp. 25–27.

126. "Therefore the reigne and position of the world consists in three,/ Celestiall, Subcelestiall, which with limits compass be;/ The Rest no boundes may comprehend which bright above the Skye/ Doth shine with light most wonderful." Cited in Koyré, *From the Closed World to the Infinite Universe*, pp. 26–27.

127. Ibid., fig. 1; as we shall see, the schema corresponds, on all points, to that used by Botticelli to illustrate the second canto of the *Divina commedia*. See Schuyler Camman, "The Symbolism of the Cloud Collar Motif," *Art Bulletin* 33 (1951): 1–9; and Marcel Granet, *La Pensée chinoise*, 2nd ed. (Paris, 1950), pp. 358 ff.

128. Johannes Sambucus, *Emblemata et aliquot Nummi antiqui operis* (Antwerp, 1564). See Mario Praz, *Studies in Seventeenth-Century Imagery*, 2nd ed. (Rome, 1964), pp. 34 ff.

129. Koyré, *From the Closed World to the Infinite Universe*, pp. 98–99.

130. Kepler, *Opera omnia*, cited in ibid., p. 84.

131. Ibid., p. 69 (my emphasis). It is all the more remarkable that Kepler, before Desargues (see above), came to consider a point at infinity as a particular case of a finite point. See Kepler, *Ad vitelloniem paralipomena* (Frankfurt, 1604), p. 93 (the letters refer to a figure given in the text): "Focus igitur in circulo unus est *A*, isque idem qui et centrum: in Ellipsi foci duo sunt *B C*, aequaliter a centro figurae remoti et acutiore. In Parabole unus *D* ist intra sectionem, alter vel extra vel intra sectionem in axe fingendus est infinito intervallo a priore remotus, adeo ut educta *H G* vel *I G* ex illo caeco foco in quodcunque punctum sectionis *G*, sit axi *D-K* parallelos [Therefore there is one focus in a circle (*A*) which is the same as the centre; in an ellipse there are two focal points (*B* and *C*), equally distant from the centre and the end-point of the figure. In a parabola one (*D*) is inside the section, another must be imagined either outside or inside the section on the axis, separated from the former by an undefined interval, such that a line drawn from *H* to

G, or *I* to *G* from this blind focus into any point of the section *G* should be parallel to the axis *D-K.*]" Cited in H. F. Baker, *Principles of Geometry* (Cambridge, Eng., 1929), vol. 1, p. 178; trans. David Money.

132. Koyré, *From the Closed World to the Infinite Universe*, p. 81.

133. Ibid., p. 84.

134. Ibid., pp. 86–87. Newton was to be the first to affirm the infinity of the universe, for reasons as much theological as scientific. See Koyré, *Études d'histoire de la pensée scientifique*, p. 84.

135. "Since Ptolemy was once mistaken over his basic tenets, would it not be foolish to trust what moderns are saying now? Is it not more likely that this huge body which we call the Universe is very different from what we think?" "The Apology for Raimond Sebond," *The Essays of Michel de Montaigne*, trans. and ed. M. A. Screech (Harmondsworth, Eng., 1991). The diameter of Copernicus's world is at least two thousand times greater than that of the world of Aristotle and Ptolemy. See Koyré, *From the Closed World to the Infinite Universe*, pp. 34–36.

136. Koyré, *From the Closed World to the Infinite Universe*, p. 31.

137. Koyré, *Études newtoniennes*, p. 30.

138. See the marquetry pictures executed in about 1480 for the study of Federico da Montefeltro in Gubbio (now in the Metropolitan Museum, New York); see E. Winternitz, "Quattrocento Science in the Gubbio Study," *Metropolitan Museum Bulletin* (October 1942): 104–16, and Baltrušaitis, *Anamorphoses et perspectives curieuses*, p. 59.

139. On the enlargement and restructuring of the musical space of the Renaissance, and its relation to the enlargement and restructuring of physical and cosmic space, see Edward E. Lowinsky, "The Concept of Physical and Musical Space in the Renaissance," in the papers for 1940–41 for the American Musicological Society, pp. 57–84.

140. Alberti, *Della pittura*, bk. 1, pp. 71–72; *On Painting*, bk. 1, pp. 56–57. The method criticized by Alberti consisted in arbitrarily drawing a first line parallel to the base line, then a second whose distance from the first was two-thirds of the latter's distance from the base line, and so on. According to Alberti, the defect of this method was a *theoretical* one: it is not governed by the rule of the central point.

141. See Argan, "The Architecture of Brunelleschi and the Origins of Perspective Theory"; and Francastel, *Peinture et société*, pp. 17–19.

142. See Karl Lehmann, "The Dome of Heaven," *Art Bulletin* 27 (1945): 1–27.

143. Andrea Palladio, *I quattro libri dell'architettura* (Venice, 1570), bk. 1, vol. 4, preface.

144. See the catalogue of the "Symbolisme cosmique et monuments religieux" exhibition, Paris, Guimet Museum, July 1953, catal. no. 167. Campanella's "city of the sun" exalted the cosmic value of utopia: "In the firmament of the cupola are all the great stars of the sky, with their earthly names and properties. . . . There you

can see the poles and the meridians, incomplete because the lower part is missing, but completed ideally by the spheres of the altar. Seven lamps named after the seven planets are always kept alight there"; ibid., catal. no. 169.

145. André Chastel, *Art et humanisme à Florence au temps de Laurent le Magnifique* (Paris, 1959), p. 213. In Botticelli's series of drawings for the *Divina commedia*, a plan of the cosmos illustrates the second canto: it consists of eleven concentric circles, the central one of which represents the earth (*terra*); next come the zones of the air (*aria*) and fire (*fuoco*); then the various heavens. It is the same schema as that later used by Peter Appianus in his *Cosmographia*. See Yvonne Batard, *Les Dessins de Sandro Botticelli pour la Divine Comédie* (Paris, 1952), p. 80, reprod. p. 82.

146. See Camman, "The Symbolism of the Cloud Collar Motif," p. 8, n. 52. The English language expresses this duality by the opposition between "sky" and "heaven," which repeats the distinction that the Romans made between *coelus* and *polus*. It is worth remembering that the church built on the Mount of Olives to commemorate the Ascension of Christ, a round church (which would normally have been covered by a cupola), was left open to the skies. Pilgrims were invited to position themselves on the very spot from which the Apostles watched the miracle, and to raise their eyes to the infinite depth of the sky, just as they had.

147. Koyré, *From the Closed World to the Infinite Universe*, pp. 48–49.

148. Riegl, *Die Entstehung der Barockkunst*, pp. 85–86.

149. "*The Madonna of Foligno*. . . . The theme is that of the Madonna in glory, an old motive, but to some extent novel since it was not often treated in the Quattrocento. That century, unembarrassed by ecclesiastical conceptions, preferred the Madonna to be seated on a solid throne rather than floating in the air, but the changed emotional climate of the sixteenth and seventeenth centuries, which tended towards a sharper separation between earthly and heavenly things, preferred this idealized scheme. . . . The halo around the Madonna is softened and dissolved in a painterly way, but not completely so, for the old-fashioned, rigidly designed disc is retained as part of the background, but the clouds are made to flow around it while the accompanying *putti*, to whom the Quattrocento conceded at most a little scrap of cloud on which to rest, can now tumble about in their element like fish in water." Heinrich Wölfflin, *Classic Art*, trans. Linda Murray and Peter Murray (London, 1994), pp. 128–30.

150. The same applies for a construction with a single vanishing point as applies to one with a series of layers spread out along the same horizon (see the *sinopia* of the *nativity* painted by Uccello for the church of San Martino alla Scala, in Florence; see the catalogue for the exhibition "The Great Age of Fresco," New York Metropolitan Museum [1968], catal. no. 36, reprod. p. 149). And it also applies for the constructions that Panofsky calls "fishbones" (*Fischgräte*), in which vanishing lines run two by two to points regularly disposed along the same vertical axis (or *line*), which, according to Panofsky, was the schema characteristic of ancient perspective. Panofsky, *Die Perspektive als "symbolische Form,"* pp. 106 ff., fig. 5.

151. See Philippe Sollers, "Un Pas sur la lune," *Tel Quel* 39 (Autumn 1969): 3–12.

152. Giordano Bruno, *De l'infinito universo e mondi* (1584), dedicatory epistle, cited in Koyré, *From the Closed World to the Infinite Universe*, p. 42. The allusion to Pluto and Jupiter probably refers to the above cited text in which Palingenius drew a distinction between the shadowy realms the empire of which extends beneath the clouds, and the realm of the gods, situated above the clouds.

153. Koyré, *Études galiléennes*, p. 19.

154. Although Alberti appropriates the project for the geometrization of pictorial space that is associated with the name of Brunelleschi, he does not confuse physics and geometry. The concept of movement that affects an *istoria* is still strictly Aristotelian: movement assimilated to change, and changes of *place*, the only changes that a painter can portray, are no different in nature from that of other movements of the body, such as generation and corruption, growth and wasting, and the change from health to sickness. It is also in Aristotelian terms that Alberti distinguishes between seven kinds of changes of place: upward, downward, to the left, to the right, forward, backward, circular (an allusion to the circular movement so important to traditional cosmology). The space of a representation is not the empty, qualityless, uncentered void of geometry. Governed as it is by the rule of the vanishing point, it, like Kepler's space, only exists thanks to the bodies that occupy it, bodies whose movements, interplay, spacing, superpositioning, and mutual overlapping define the perspective of the *istoria*. See Alberti, *Della pittura*, bk. 2, pp. 92–98; *On Painting*, bk. 2, pp. 75–81.

155. Roberto Longhi does not wish to do Correggio the injustice of believing that, since painting was *cosa mentale* (a thing of the mind), he was content with purely imaginary light. But he does admit that it is necessary to make up for the inadequate light in order to study the painting thoroughly. The paradox is that it was not until the advent of electric light that Correggio's work could produce all its potential effects. See A. G. Quintavalle, *Gli affreschi del Correggio*, presentation.

156. "Jede isolierende Schranke zwischen Götlichen und Weltlichem ist gefallen [Every frontier between the divine and the earthly is abolished]." Riegl, *Die Entstehung der Barockkunst*, p. 53.

157. Koyré, *Études d'histoire de la pensée scientifique*, pp. 176–77.

158. Galileo Galilei, *Sidereus nuncius*, in *Opere*, vol. 3, p. 80; French trans., *Le Messager céleste* (Paris, 1964) (my emphasis).

159. Aristotle, *Meteorologica* 1.9.

160. See Galileo's letter to Cigoli (1624), in *Opere*, vol. 6.

161. "Come la terra è una stella [How the earth is a star]." Richter, ed., *Literary Works of Leonardo da Vinci*, vol. 2, p. 111.

162. Galileo, letter to Benedetto Castelli, 21 December 1613; French trans. in Galileo, *Dialogues et lettres choisies* (Paris, 1966), p. 81.

163. Dante, *Purgatorio* 28.81; Galileo, "Discours des comètes" (1619), in *Dialogues et lettres choisies*, p. 81.

164. Paul Henri Michel, preface to Galileo, *Dialogues et lettres choisies.*

165. Cited by Stefano Bottari, *Correggio* (Milan, 1961), pp. 36–37.

166. Giovanni-Battista Armerini, *Veri precetti della pittura* (Ravenna, 1587; reprint, Milan, 1820), p. 231.

167. See Panofsky, *Galileo as a Critic*, p. 5, n. 2.

168. See Koyré, *La Révolution astronomique: Copernic, Kepler, Borelli* (Paris, 1961).

169. Francastel opportunely notes that it was only in the seventeenth century that Euclid's *Elements* were arranged in their present order, more or less consciously so as to justify a vision of the world that the plastic arts had begun to elaborate as early as the fifteenth century (*Peinture et société*, p. 109).

170. Giordano Bruno, cited in Koyré, *From the Closed World to the Infinite Universe*, p. 45.

171. See Marie-Christine Gloton, *Trompe-l'oeil et décor plafonnant dans les églises romaines de l'âge baroque* (Rome, 1965).

172. This is the point at which to stress that, in Rome at least, a considerable time lag separated so-called Baroque architecture and so-called Baroque painting. Most of the major decorative works designated as such date from the late seventeenth century, that is to say from a considerably later period than that of "high Baroque" as defined by art historians. The Gesù church remained with no decor of this kind for over a century (see Andrea Sacchi's painting representing *The Visit of Urban VIII to the Gesù in 1639*, Rome, Museo Nazionale di Roma). In truth, the architecture of Bernini, Borromini, Pietro di Cortona, and Rainaldi seems to exclude any kind of painted decor: it produces architectural entities that are complete in themselves and do not lend themselves to illusory openings. Ceiling painting only became widely accepted at a quite late date, in buildings of a classical structure that had not been designed for it. Wölfflin's thesis according to which Baroque art replaced the quest for "solid" effects that characterized the linear perspective of the early Renaissance by immaterial effects and atmospheric perspective is, at least where Baroque decor is concerned, not acceptable: just as the painters of stage scenery took over from the architects who had defined the structure of the stage upon which Baroque spectacles were performed (see Kernodle, *From Art to Theatre*, p. 186), similarly so-called Baroque decoration developed as it were in counterpoint on architectonic structures that were in many cases Baroque in name only but with which it was integrally connected in its very principles.

173. For another example of an eventual rejection of perspective *dal sotto in su*, see "Memoir on the Life and Works of Mr. Mengs," by the Chevalier d'Azara, in Anton Raphael Mengs, *Works* (London, 1796), vol. 1, p. 16:

> Returning to Rome, he undertook to paint the ceiling of the gallery of the Villa of Cardinal Alexander Albani, in which he represented Apollo with the Goddess of Memory and the Muses, their offspring. In this work he

availed himself much of that which he had observed in the paintings of Herculaneum, to be seen in the Portico Museum: he attached a piece in continuation to this painting, to avoid the great error of works of this kind, which is to terminate of itself, as is the modern custom; for by this method he avoided the disagreeable breakings off or curtailing, which always destroy the beauty of the figures. Nevertheless, not to controvert entirely that mode, he painted the two pieces laterally, where entered only one figure to each, terminating according to the modern taste.

174. See the letter from Father Pozzo to the prince of Liechtenstein (1694), in *Significati delle pitture fatte nella volta della chiesa di Sant' Ignazio a Roma* (Rome, 1828). It is known that Father Pozzo was later to decorate the Liechtenstein palace in Vienna with a famous fresco with a celestial perspective.

175. Cesare Ripa, *Iconologia; overo, Descrittione dell'imagini universali cavate dall'antichità et da altri luoghi . . . , opera non meno utile che necessaria a poeti, pittori, et scultori per rappresentare le vitti, virtù, affetti, et passioni humane* (Rome, 1593), proem; see Gloton, *Trompe-l'oeil*, p. 158.

176. It was even to reveal itself to be remarkably loquacious; the prolixity of theologians is somewhat surprising when compared with the reserve of many mathematicians and scientists. One such was Buffon who, in the eighteenth century still, in the preface to his translation of Newton's *Method of Fluxions* (Paris, 1740), was still defending a strictly privative concept of the infinite:

> Most of our errors in metaphysics stem from the reality that we ascribe to ideas of privation. We are familiar with the finite, we see real properties in it, we remove them from it and when we consider it following that removal we no longer recognize it and believe we have created something new, when all we have done is destroy part of something that we used to know.
>
> So we should only consider the infinite, whether it be the infinitely small or the infinitely great, as a privation, a denial of the idea of the finite, which can be used as an assumption that in some cases can serve to simplify ideas, and must generalise their results in the practice of the sciences. All the skill thus lies in making the most of this assumption and trying to apply it to the subjects under consideration. All the merit thus lies in the application or, in a word, in the use that one makes of it. (pp. x–xi)

177. A formula first used by Nicholas of Cusa but frequently ascribed to Hermes Trismegistus.

178. Pascal, *Pensées*, ed. Brunschvicg, sec. 2, no. 72; English trans.: *The Essential Pascal*, ed. Robert W. Gleason, trans. G. F. Pullen (London, 1966).

CHAPTER 5 *Our Sheet's White Care*

1. As I myself attempted to do in a much earlier draft of the present chapter, "Un 'Outil' plastique: Le nuage," *Revue d'esthéthique* (January–June 1958): 104–48.

2. Alexander Cozens, *A New Method of Assisting Invention in Drawing Original Compositions of Landscape*, 1st ed. (London, 1785); see Paul Oppé, *Alexander and John Robert Cozens* (London, 1952).

3. Cozens, preface to *A New Method*, cited in Ernst Gombrich, *Art and Illusion*, 2nd ed. (London, 1962), pp. 156–57. On Cozens, see H. W. Janson, "The 'Image made by Chance' in Renaissance Thought," in *De Artibus Opuscula XL: Essays in Honour of Erwin Panofsky*, ed. Millard Meiss (New York, 1961), pp. 254–66; and a forthcoming work by Jean-Claude Lebensztjen.

4. Gombrich, *Art and Illusion*, pp. 157–58.

5. John Ruskin, *Modern Painters* (London, 1855), pt. 4, chap. 16; citing here from the 1856 ed., vol. 3, p. 255.

6. "We turn our eyes as boldly and as quickly as may be from the serene fields and skies of medieval art to the most characteristic example of modern landscape. And, I believe, the first thing that will strike us or that ought to strike us is their cloudiness" (ibid., p. 254).

7. Ibid., vol. 1, p. 214.

8. Ibid., pt. 2, sec. 3, chaps. 1 and 2. In the last volume of *Modern Painters* (pt. 7, "Of Cloud Beauty"), Ruskin compares a cloud to a leaf that comes between man and the earth, just as a cloud comes between man and the sky.

9. Ibid., vol. 1, p. 218.

10. Ibid., chap. 4.

11. Ibid., vol. 3, pt. 4, chap. 15. See Leonardo da Vinci, "Buildings Appear Larger When There Is Mist," in *The Notebooks of Leonardo da Vinci*, ed. Edward Mac-Curdy (London, 1938), vol. 2, p. 324.

12. Ruskin, *Modern Painters*, vol. 3, p. 257.

13. Ibid., pt. 4, chap. 16.

14. Ibid., vol. 1, p. 213.

15. "There is not a moment of any day of our lives where nature is not producing scene after scene, picture after picture, glory after glory, and working still upon such exquisite and constant principles of the most perfect beauty that it is quite certain it is all done for us, and intended for our perpetual pleasure"; ibid., p. 201.

16. In the first volume (p. 36) of *Modern Painters*, Ruskin classifies mystery among the six qualities (truth, simplicity, mystery, inadequation, decision, rapidity) of execution.

17. "There are landscapes that are volatized, dawns that fill the sky; celestial and fluvial festivals of a sublime, stripped nature, rendered completely fluid by a great poet." Joris-Karl Huysmans, *Certains* (Paris, 1889), p. 202.

18. Ruskin, *Modern Painters*, vol. 4, pt. 5, chap. 4, p. 56 ("On Turnerian Mystery").

19. For a particularly virulent attitude in this respect, see William Blake, "Annotations to Reynolds," in Blake, *Complete Writings*, ed. Geoffrey Keynes (London, 1957), pp. 445–79.

20. Ruskin, *Modern Painters*, vol. 4, p. 57.

21. Ibid., p. 62.

22. Ibid., p. 58.

23. Ibid., vol. 5, p. 115, note.

24. Ibid., p. 122.

25. Ibid., pt. 1, sec. 1, chap. 5 ("Of Ideas of Truth").

26. Ibid., vol. 5, p. 111.

27. "Where ride the captains of their armies? Where are set the measures of their march?" Ibid., vol. 4, p. 106.

28. Ibid., p. 111.

29. Ibid., vol. 1, p. 204.

30. Ibid., vol. 5, p. 121.

31. It is remarkable that the paintings of Turner in which the dissolution (but not the deconstruction) of the perspective cube is the most extreme are not landscapes, seascapes, or studies of the sky, but interior scenes (see the series of *Interiors at Petworth*, c. 1830, now in the Tate Gallery), in which the cube, without however ceasing secretly to structure the composition, loses all oversharp shapes. In one famous picture (*Rain, Steam, Speed*, 1844, National Gallery), Turner had no hesitation in associating "service of the clouds" with the emblem of modernity par excellence, the railway, as Monet was to in his Gare Saint-Lazare series, in which the linear structure corresponding to the metal roof is *repulsed*, significantly, by the steam emitted by the engines.

32. "Accidental shadows can also be used successfully, that is to say shadows caused by something outside the picture, provided they look real, as in a landscape, in which clouds naturally produce effects which make certain terrains disappear into the distance." Alexandre-François Desportes, cited in André Fontaine, *Conférences inédites de l'Académie de peinture* (Paris, 1903), p. 66.

33. "It is one of the most discouraging consequences of this work of mine that I am wholly unable to take notice of the advance of modern science. What has conclusively been discovered or observed about clouds, I know not; but by the chance inquiry possible to me I find no book which fairly states the difficulties of accounting for even the ordinary aspects of the sky." Ruskin, *Modern Painters*, vol. 5, p. 107.

34. Luke Howard, *Essay on the Modifications of Clouds* (London, 1803), reprinted in *The Climate of London, Deduced from Meteorological Observations* (London, 1818–20). See Goethe's two studies on Howard (*Naturwissenschaftlichen*

Schriften) and his series of poems entitled "Atmosphäre," "Howards Ehrengedächt-
nis," "Stratus," "Cumulus," and so on:

> The world which is so great and spreading,
> The sky so high and distant,
> All this my eyes can take in
> But not my thoughts.
>
> To find your way in infinity,
> You must first distinguish, then gather things together.
> That is why my winged song gives thanks
> To the man who distinguished between the clouds.
>   (from "Atmosphäre")

35. Kurt Badt, *Wolkenbilder und Wolkengedichte der Romantik* (Berlin, 1960).

36. See Louis Hawes, "Constable's Sky Sketches," *Journal of the Warburg and Courtauld Institutes* 32 (1969): 344–65.

37. Example: "5 Sept. 1822: 10 o'clock morning, looking south-east. Brisk wind at west. Very bright and fresh grey clouds running fast over a yellow bed, about half way in the sky."

38. In his *Lectures on Landscape Painting*, published by Charles Robert Leslie; see Badt, *Wolkenbilder und Wolkengedichte der Romantik*, pp. 74 and 77.

39. "And indeed, shapeless forms leave no memory except that of a possibility, . . . just as a series of notes struck at random is no melody, so a puddle, a rock, a cloud, a fragment of coastline are not reducible forms." Paul Valéry, *Dégas, danse, dessin*, in Valéry, *Oeuvres*, vol. 2, p. 1194.

40. Goethe, letter to Zelter, 24 July 1823, cited in Badt, *Wolkenbilder und Wolkengedichte der Romantik*, p. 19 (my emphasis).

41. See the letter from Goethe published by Carl Gustav Carus as an introduction to his *Neuen Briefe über die Landschaftmalerei* (1831), in Badt, *Wolkenbilder und Wolkengedichte der Romantik*, p. 36.

42. Goethe, "Wolkengestalt nach Howard" (1820), cited in Badt, *Wolkenbilder und Wolkengedichte der Romantik*, p. 24.

43. See Ruskin, *The Storm-Cloud of the Nineteenth Century: Two Lectures Delivered at the London Institution* (Orpington, Eng., 1884), pp. 34–35. This is the very subject addressed by Goethe's *Farbenlehre*.

44. Epicurus, *Letter to Pythocles*, in Diogenes Laertius, para. 100, trans. R. D. Hicks, Loeb Classical Library (London and Cambridge, Mass., 1958).

45. Lucretius, *De natura rerum* 6.160–61, trans. W. H. D. Rouse, Loeb Classical Library (Cambridge, Mass., and London, 1953).

46. Ibid., ll. 246–47.

47. Epicurus, *Letter to Pythocles*, trans. Hicks, p. 84.

48. See Goethe, "Ganymede":

Higher, higher I aspire
The clouds float
Toward me, bowing, the clouds
Toward my nostalgic love
Help me! Help me!
In your bosom? Raise me up!
Intertwining, intertwined,
Up there, on your chest; O all loving Father!

49. Lucretius, *De rerum natura* 6.451–54.

50. Ibid., ll. 462–66.

51. Ruskin, *Modern Painters*, vol. 5, pp. 108–9.

52. Friedrich Nietzsche, "Le Cas Wagner," in *Le Crépuscule des idoles* (Paris, 1952), p. 35.

53. Friedrich Nietzsche, *Le Gai Savoir*, in Nietzsche, *Oeuvres complètes* (Paris, 1967), p. 18 (my emphasis).

54. Friedrich Nietzsche, *Considérations*, p. 63.

55. Richard Wagner, "Le Théâtre des festivals scéniques de Bayreuth" (1873), in Wagner, *Oeuvres en prose* (Paris), vol. 11, pp. 170–74.

56. Richard Wagner, "Lettre sur la musique" (1860), *Oeuvres en prose*, vol. 6, p. 226.

57. Wagner, "Théâtre des festivals scéniques de Bayreuth," pp. 174–77.

58. See René Leibowitz, introduction to *Schoenberg et son école* (Paris, 1947).

59. "Only where sight and hearing combine mutually is the true artist satisfied." Wagner, "L'Oeuvre d'art de l'avenir," in *Oeuvres*, vol. 3, p. 137.

60. Richard Wagner, "Opéra et drame," in *Oeuvres*, vol. 4, p. 220.

61. Wagner, "L'Oeuvre d'art de lavenir," p. 212.

62. Richard Wagner, "Sur la représentation du *Tannhaüser*" (1852), in *Oeuvres*, vol. 7, pp. 215–16.

63. Leibowitz, introduction to *Schoenberg et son école*, p. 59.

64. Richard Wagner, "Une communication à mes amis" (1851), in *Oeuvres*, vol. 6, pp. 57–58.

65. Debussy's *Nocturnes* (in particular the first of them, entitled *Clouds*) occupy a decisive historical position in this respect. In them, breaking away, despite appearances, from impressionism and the attractions of vagueness, Debussy strove to liberate the melodic line from thematic constraints, so that its instability and regulated energy would contribute to define the new musical space as an analytical field, founded on symbiosis, displacement, and transformation.

66. Wagner continually criticized the mechanisms that conferred a *tactile* reality upon the image of the stage, and insisted on the need to isolate that image and get it to emerge in all its essential ideality. See "Un Coup d'oeil sur l'opéra allemand contemporain," in *Oeuvres*, vol. 11, p. 86: "What disgusted me most about

it was the crudeness in laying bare all the mysteries of the scenery to the eyes of those spectators. Some believe that things that can only be effective from a carefully calculated distance must on the contrary be displayed as much as possible in the harsh footlights right at the front of the stage."

67. The very same year, 1888, in which Nietzsche wrote the letter from Turin that marked his break with Wagner, also saw the beginning of the publication, in monthly installments, of *Japon artistique,* edited by Samuel Bing, the future founder of the Salon de l'Art Nouveau.

68. Henri Focillon, *La Peinture au XIXe et XXe siècle, du réalisme à nos jours* (Paris, 1928), pp. 206–8.

69. "Never has painting been so mysteriously discreet, you would think it a piece of silk imbued with deep and pale colors. . . . Whistler's nocturnes are unique in the history of painting and represent in its highest form the art of suggestion or, if you like, the lyrical intimacy of Western landscape"; ibid., pp. 198–99.

70. "The examination and contemplation of paintings [that represent] the Buddha and Buddhist scenes has an edificatory value that places such paintings in the highest rank. Next come landscapes, [the sources] of inexhaustible delights; views of mists and clouds are particularly beautiful. Bamboos and rocks come in the next class, then flowers and plants. As for pretty women and subjects 'with feathers or fur,' they are simply for the amusement of worldly people. They cannot be included among the pure jewels of art." Mi Fu, *Huashi,* chap. 3, par. 158; see Nicole Vandier-Nicolas, *Le Houa-che de Mi Fou; ou, Le Carnet d'un connaisseur de l'époque des Song du Nord* (Paris, 1964), p. 147.

71. Mi Fu, *Huashi,* chap. 3, pars. 19, 21, and 45; also pp. 35–36 and 49.

72. *Jie zi yuan huazhuan,* vol. 3, chap. 21; French trans. Raphaël Petrucci, *Encyclopédie de la peinture chinoise* (Paris, 1918), p. 173.

73. See Pierre Ryckmans, ed., *Les "Propos sur la peinture" de Shitao: Traduction et commentaire pour servir de contribution à l'étude terminologique et esthétique des théories chinoises de la peinture* (Brussels, 1970), pp. 63–64.

74. See Schuyler Camman, "The Symbolism of the Cloud-Collar Motif," *Art Bulletin* (1951): 1–9.

75. *Qing lü*: aerial perspective obtained by colors merging from blue into green, with gold and brilliant colors also used, along with very fine lines.

76. Petrucci, *Encyclopédie,* p. 173.

77. *Jin bi*: a technique similar to *qing lü,* but with a more pronounced use of gold.

78. Petrucci, *Encyclopédie,* pp. 173–74.

79. Zhang Yanyuan, *Lidai ming hua ji* (annals of celebrated painters in the course of successive dynasties; a work completed in 847), cited by Nicole Vandier-Nicolas, *Art et sagesse en Chine: Mi Fou (1051–1107), peintre et connaisseur d'art dans la perspective de l'esthéthique des lettrés* (Paris, 1963), p. 63.

80. On the aesthetics of the scholars, see Vandier-Nicolas, *Art et sagesse en Chine.*

81. See Peter Swann, *Chinese Painting* (London, 1958). It is in this context that the text of Song Di (in the volume published by H. Giles as *An Introduction to the History of Chinese Pictorial Art* [Shanghai, 1905]), cited at note 75 of Chapter 1, should be judged.

82. *Lidai ming hua ji,* cited in Vandier-Nicolas, *Art et sagesse en Chine,* p. 62. The same rejection of heterodox techniques is to be found in the eighteenth century on the part of Zou Yigui. See Ryckmans, ed., *Les "Propos sur la peinture" de Shitao,* pp. 46–47.

83. Ibid., pp. 132–34. See, for the Ming dynasty, the text of Dong Qichang cited in Ryckmans, ed., *Les "Propos sur la peinture" de Shitao,* pp. 34–35, n. 6: "Some people claim that every painter should found his own school. But that is nonsense. To paint willow trees, use the manner of Zhao Qianli; for pines, follow Ma Hezhi; for dead trees, follow Li Cheng; for those various models are eternal and could never be changed; and even if you reinterpret them in your own way, in the essentials you cannot diverge from this fundamental source. It would indeed be inconceivable to presume to creation of your own, scorning the classical rules."

84. "The paintings of Huang Cuan do not deserve to be collected together, they are easy to copy; the paintings of Su Hi cannot be copied." Mi Fu, *Huashi,* chap. 3, par. 108; Vandier-Nicolas, *Art et sagesse en Chine,* p. 97.

85. "Antiquity is the instrument of knowledge; to transform consists in knowing that instrument but without becoming its servant. Knowledge that is closely linked to imitating is bound to lack breadth; so a fine man only borrows from Antiquity in order to found something in the present. . . . As for me, I exist through myself and for myself. . . . And if it so happens that my work resembles that of some other master, it is he who is following me and not I who have sought him out." Ryckmans, ed., *Les "Propos sur la peinture" de Shitao,* chap. 3 ("La Transformation"), pp. 31–32. In a similar vein, see *Jie zi yuan huazhuan,* vol. 1, chap. 10 ("The Talent of Transformation"), and the texts cited in ibid., pp. 35–36.

86. See Mi Fu: "With a single movement, I have swept away the dreadful style of the two Wang [the two most universally venerated calligraphers] and have illuminated the Song dynasty for ten thousand ages." Cited by Tang Hou; cf. Ryckmans, "Les Propos sur la peinture de Shitao," *Arts asiatiques* 14 (1966): 95, n. 4.

87. See Ryckmans, ed., *Les "Propos sur la peinture" de Shitao,* chap. 1, p. 11.

88. *Jie zi yuan huazhuan,* vol. 1, chap. 12, Petrucci, *Encyclopédie de la peinture chinoise,* p. 33.

89. Ryckmans, ed., *Les "Propos sur la peinture" de Shitao,* p. 16.

90. See James Cahill, *La Peinture chinoise* (Geneva, 1960), p. 11.

91. *Jie zi yuan huazhuan,* vol. 1, chap. 13, "Ink and Brush," Petrucci, *Encyclopédie de la peinture chinoise,* p. 35.

92. Dong Qichang, cited in Ryckmans, ed., *Les "Propos sur la peinture" de Shitao*, p. 46, n. 1.

93. Han Zhuo (twelfth century), *Shanshui chuan quanzi* (Complete treatise on landscape), cited in Ryckmans, ed., *Les "Propos sur la peinture" de Shitao*.

94. Henri Focillon, *Hokusai* (Paris, 1925), pp. 19 ff.

95. See Ryckmans, ed., *Les "Propos sur la peinture" de Shitao*, p. 80, n. 2; Vandier-Nicolas, *Art et sagesse en Chine*, pp. 131, 230, etc.

96. Ryckmans, ed., *Les "Propos sur la peinture" de Shitao*, p. 47, n. 1.

97. See the texts cited in ibid., p. 73, n. 1.

98. Ibid., chap. 15, p. 115.

99. Ibid., chap. 9, p. 71.

100. *Jie zi yuan huazhuan*, chap. 12. In the thirteenth century, Li Song described meticulous architectural drawings as the "painting of limits," *jiai hua*; see Cahill, *La Peinture chinoise*, p. 54.

101. Ryckmans, ed., *Les "Propos sur la peinture" de Shitao*, p. 73, n. 1 (my emphasis).

102. Ibid., chap. 6, p. 51.

103. "To have the brush but not the ink does not mean that a painting is literally without ink, but that the wrinkles and washes are reduced to very little; the outlines of rocks are starkly drawn, the trunks and branches of trees stand out harshly, giving the impression of a lack of ink or, as they say, of 'bones predominating over flesh.' To have the ink but no brush does not mean that the painting was truly produced without a brush; it means that, in the tracing of rocks and the painting of tree-trunks, the brush made only light touches while washes were used to excess so that they came to hide the brush strokes and obliterate the presence of the brush. This produces an impression of a painting devoid of a brush; it is what is known as 'flesh predominating over bones.'" Tang Dei, cited in ibid., pp. 46–47.

104. Leon Battista Alberti, *On Painting*, trans. Cecil Grayson (Harmondsworth, Eng., 1991), bk. 1, pp. 55–56; see Chapter 3, p. 111.

105. "Schauplatz ist die Fläche genauer die begrenzte Ebene." Paul Klee, *Das bildnerische Denken* (Basel, 1956), p. 39. It would be interesting to study how far medieval painting, if not the painting of antiquity, proceeded in the West, as did the later painting of the Renaissance, from a preliminary delimitation of the field or, as Klee called it, the pictorial "scene," for such an enquiry would be an indispensable preliminary to the elaboration of any theory of Western painting as such.

106. See Viviane Alleton, *L'Ecriture chinoise* (Paris, 1970), pp. 28–30.

107. Ibid.

108. In contrast, the base line of the quadrangle (*la linea di sotto qual giace nel quadrangolo*), as Alberti puts it, in Western painting takes on the importance of a *foundation*, since it is on its basis and thanks to its division into equal segments

that it becomes possible to plot out the checkerboard paving that constitutes the "floor" of the perspective construction.

109. See Chang Tung-Sun, "La Logique chinoise," *Yenching Journal of Social Studies* I, no. 2 (1939); French trans. in *Tel Quel* 38 (Summer 1969): 3–22.

110. *Jie zi yuan huazhuan*, vol. I, chap. 16, Petrucci, *Encyclopédie de la peinture chinoise*, p. 44.

111. "The division into two sections consists in placing the scene below, the mountain above, and, conventionally, one then adds clouds in the middle to mark out more clearly the separation between the two sections." Ryckmans, ed., *Les "Propos sur la peinture" de Shitao*, chap. 10, p. 79.

112. "When divisions are made according to the method of three successive planes or of two sections, they seem to ensure that the painting is spoilt. . . . If, in each landscape, one embarks upon a kind of clearing operation or cutting it up into pieces, the result will not be at all alive, for the eye can immediately detect how it has been fabricated. . . . If one paints according to this method of three planes, how can the result be any different from an engraved plaque?" Ibid.

113. Chang, "La Logique chinoise," p. 12.

114. *Gu fa yong bi shi ye* 骨法用筆是也, "The law of bones, using a brush" (Petrucci, *Encyclopédie de la peinture chinoise*, p. 7), or, as James Cahill translates it, "The Six Laws and How to Read Them" (*Ars Orientalis* 4 [1961]: 372–81): "using the brush [according to] the method for bones." We need not go into the problems posed by the translation of Xie He's "Principles." But it is worth pointing out that the main difficulty encountered by Western specialists who are attempting to produce an equivalent translation in a European language is connected with the recurrence, in many statements, of the formula *shi ye*, which, according to the traditional interpretation, serves simply to identify the principle with its numerical order (see Cahill, ibid.; Alexander Soper, "The First Two Laws of Hsieh Ho," *Far Eastern Quarterly* 8 [1948]: 412–23), while others assimilate it to a copula that gives articulation to the two pairs of characters that are in play in the manner of a definition ("which means," "that is to say"). Thus R. W. Acker (*Some T'ang and Pre-T'ang Texts on Chinese Painting* [Leiden, 1954], p. 4) translates the second principle as follows: "Bare method which is [a way of] using the brush." Quite apart from the fact that Chinese possesses no verb "to be," the difficulty stems from the inability of Western logic to cope with a form of thought that is not familiar with the problem of equating two terms. As Chang Tung-Sun ("La Logique chinoise," p. 10) declares, "a formula such as '*shi* . . . *ye* . . .' does not signify that anything is identical to anything else and consequently does not constitute a logical proposition such as appears in the Western structure." The point is made simply to underline, yet again, that, quite apart from the problem of the context in which these notions appear, it is impossible to transfer some notions from a Far Eastern text to a Western one.

115. "To draw the forms in conformity with the things" (Petrucci); "responding to things, images depict their form" (Cahill); "Correspondence to the object, which means the depicting of forms" (Acker).

116. "Apply the colour in accordance with the similarity of the objects" (Petrucci); "According to kind, set forth (describe) colours (appearances)" (Cahill); "Suitability to type, which has to do with the laying on of colours" (Acker).

117. "Distribute the lines and give them their hierarchical place" (Petrucci); "Dividing and planning, positioning and arranging" (Cahill); "Division and planning, i.e. placing and arrangement" (Acker).

118. Dai Xi, cited in Ryckmans, ed., *Les "Propos sur la peinture" de Shitao*, p. 47.

119. *Meng yang*, "technical training," as Ryckmans translates it (ibid., p. 48). The origin of this concept is to be found in *I Ching*: "The task of the saint is *to distinguish rectitude from chaos.*" From this the expression, in everyday language, comes to mean the basic instruction given to a child beginning to learn to read Chinese characters.

120. See Ryckmans, ed., *Les "Propos sur la peinture" de Shitao*, chap. 2, p. 27: "Ink comes from Nature, thick or fluid, dry or unctuous, however one wants it. The brush is controlled by man in order to express outlines, wrinkles, and different kinds of washes, as he wishes to."

121. Ibid., chap. 6, p. 51.

122. "Painting results from the reception of ink; ink from the reception of a brush; a brush from reception by a hand; a hand from reception from a mind." Ibid., chap. 4, p. 41.

123. Ibid., chap. 8, p. 63.

124. Chang, "La Logique chinoise," p. 12.

125. When referring to *the* pictorial practice or *the* pictorial theory of China, I am, clearly, not unaware of the fact that that practice and—to a lesser degree—that theory have a *history*, and that it would not be right to set on the same level a painter from the Han period and a Ming landscape painter. Quite apart from the fact that Chinese art is governed by a kind of historicity very different from the Western kind (a subject to which I hope to return), the reason why I here refer to Chinese painting and theory as a whole is simply in order to clear the way for a *general* theory in which Western art would lose its centralizing preeminence.

126. Jing Hao (?), *Bi fa ji* (Notes on the method of the brush; tenth century), cited in Vandier-Nicolas, *Art et sagesse en Chine*, p. 188.

127. Ibid., p. 187.

128. "When tackling the mountain, painting finds its soul; when tackling water, it finds its movement." Ryckmans, ed., *Les "Propos sur la peinture" de Shitao*, chap. 7, p. 57.

129. "A girl, perceiving the shadow of her lover on a wall, will draw the outline of that shadow." Chateaubriand, *La Génie du christianisme*, bk. 1, chap. 3 (a fable taken over from Pliny, *Historia naturalis* 35).

130. Guo Xi, *Linquan gao zhi*, cited in Vandier-Nicolas, *Art et sagesse en Chine*, p. 205.

131. Mi Fu, *Huashi*, chap. 3, par. 24; Vandier-Nicolas, *Art et sagesse en Chine*, p. 36.

132. Marcel Granet, *La Pensée chinoise*, pp. 115–48.

133. Ryckmans, ed., *Les "Propos sur la peinture" de Shitao*, chap. 17, p. 115.

134. Ibid., chap. 7, p. 57.

135. Marcel Granet, *La Pensée chinoise*, pp. 123–24.

136. Chang, "La Logique chinoise," p. 12.

137. Ryckmans, ed., *Les "Propos sur la peinture" de Shitao*, chap. 13, p. 89.

138. Ibid., pp. 49 and 58.

139. "As men of the past have said, a poem is a painting without shapes, a painting is a poem that takes a [visible] shape. Philosophers have often expounded on this theme and I, for myself, have regarded this as a guiding principle. Thus in my days of leisure I have often meditated upon the poems of the Jin and the Tang, both those of the past and those of the present. Some of these fine lines give perfect expression to the deep activity of man (*fu zhong zhi shi*: visceral activity), others, in ornate fashion, describe the spectacle presented to their eyes. . . . What is the best way to discern the idea by which painting is guided? When everything around me becomes familiar, and my heart and my hand are at one, I can at last freely conform to the rules and discover in all that surrounds me a way to return to the source." Guo Xi, *Linquan gao zhi*, cited in Vandier-Nicolas, *Art et sagesse en Chine*, pp. 195–96.

140. Granet, *La Pensée chinoise*, p. 118, n. 5. See E. J. Eitel, "Feng-shuei ou principe de sciences naturelles en Chine," *Annales du musée Guimet* 1 (1880): 203–53.

141. The journeys that painters were traditionally expected to make may be compared to the epic theme of the forays made by emperors. The training of a painter is not limited to literary culture and a knowledge of the works of the ancients: "It is also necessary for an artist either in a carriage or on horseback, to make tracks, in his journeys, across a good half of the universe. Only then will he be able to wield the brush." Guo Xi, cited in Ryckmans, "Les Propos sur la peinture de Shitao," p. 112, n. 6. The journeys of Zong Bing (375–443) took him first to the east (the starting point for the activity of Yang, which is linked to the spring), next to the south. Only then, forced by ill health to return home, did he paint the sites that had delighted him on the [four] walls of his room.

142. Ryckmans, ed., *Les "Propos sur la peinture" de Shitao*, chap. 17, p. 115. The eight directions correspond to the eight trigrams in an octagon in the *I Ching*. As for the number five (the five mountains, but also the five elements, the five sounds, the five colors, etc.) situated in the middle of the nine prime numbers (the nine provinces, the nine rivers, the nine heavens, the nine rubrics of the *Hong Fang*, the nine halls of the *Ming Tang*, etc.), it is regarded as the symbol of the Center (see

Granet, *La Pensée chinoise*, chap. 3, "Le Système du monde"). In other words, the four pillars and the four cardinal mountains that in nature play a role analogous to that of the leaders of society now seem to be complemented by a fifth, corresponding to the central pillar and assimilated to the axis of the universe. The four (barbarian) seas correspond to the inorganic space that circumscribes the saints. Ibid., p. 359.

143. Granet, *La Pensée chinoise*, p. 318.

144. Ibid., p. 125.

145. "To enable the sovereign to exercise his central action, it was necessary, between the sixth month, which marked the end of the summer, and the seventh, which was the first month of autumn, to institute a kind of time of rest, counted as one month although not attributed any definite duration. It only possessed an intellectual duration, and this in no way encroached upon the twelve months or the seasons; yet it was far from being of no account: it was the equivalent of a whole year, for within it seemed to lie the motor of the year"; ibid., p. 103.

146. Ryckmans, ed., *Les "Propos sur la peinture" de Shitao*, chap. 8, p. 64.

147. "Lu Chaishi said: Su Wen Chang, speaking of painting, values surprising mountain peaks, steep cliffs, wide rivers, waterfalls, weird rocks, old pine trees, and Daoist hermits and priests. In general, he values a painting onto which the ink has dripped drop by drop, filled with cloud and mist, which is empty as if one could not see the sky, yet full as if one could not see the earth. In those circumstances, the picture is a superior one. These words do not seem to tally with what has been said above (concerning opening up the Sky and the Earth); but Wenchang is a scholar with a free soul. Amid extreme fullness, he had ideas about extreme emptiness. He says 'empty' and 'full': those two words reveal his character." *Jie zi yuan huazhuan*, vol. 1, chap. 16, Petrucci, *Encyclopédie de la peinture chinoise*, p. 44.

148. According to Granet (*La Pensée chinoise*, p. 125, n. 3), "the word *jie* means 'articulation,' and evokes the image of a length of bamboo. It designates the instrument used to beat out the rhythm (*the king gets the Yin and the Yang to act in concert by beating the rhythm of the four seasons*) and the divisions of time that serve to space out and regulate the passing of the seasons."

149. "As for the immensity of the landscape: with its land stretching for a thousand leagues, its series of peaks, its ranks of cliffs, even an immortal who, in his flight wished only to take a superficial glance could not take it all in." Ryckmans, ed., *Les "Propos sur la peinture" de Shitao*, chap. 8, p. 64.

150. Ryckmans rightly emphasizes the ambivalence of the sign *yi* 一 in the concept of *yi hua* 一畫; *yi* means not only "one" but also the absolute One of the *I Ching* (the bar, the original fundamental emblem that, through successive divisions and combinations, expresses the totality of phenomena), the One that, upon being divided, produced the Sky and the Earth. The supposed etymology of the character *tian* 天, "sky," in typical fashion associates the notion of one with the notion of the absolute: 天 signifies extreme height that nothing can surpass; this

character comes from the association of *yi* 一, "one," and *da* 大, "large." Ibid., p. 17.

151. "I have often seen beginners grab a brush and fill a picture with blots and clumsy lines. The very sight was painful to the eyes; one immediately felt disgusted. How could such a picture please connoisseurs?" *Jie zi yuan huazhuan*, vol. 1, chap. 16, Petrucci, *Encyclopédie de la peinture chinoise*, p. 44.

152. Mi Fu, *Huashi*, chap. 3, par. 76, cited in Vandier-Nicolas, *Art et sagesse en Chine*, p. 73. That is something that admirers of Japanism, starting with Whistler, persisted in not perceiving ("We have come to speak of painting that is elevating, of a painter's duty, and of particular paintings that are full of thought"; *Ten O'clock*), thereby clearing the way for an idealist interpretation of Far Eastern art and for the development, in the West, of pictorial practices that claim to be inspired by it (in particular by calligraphy), practices whose function of ideological conceal-ment — by borrowing the outward aspects of Far Eastern practices but obliterating the theory, the "thought" behind them — has by now been fully demonstrated.

153. See the technique known as *yun suo*, "enclosed in cloud," that is applied to the mists that float before waterfalls, half concealing them: "When one paints those clouds, one must leave no trace of the brush and ink. One simply makes an outline with a faint colour, thereby showing the skill of one's hand." Petrucci, *Encyclopédie de la peinture chinoise*, p. 161, fig. 77.

154. Dong Qichang, *Bi mo pingmiao*, according to Vandier-Nicolas, *Art et sa-gesse en Chine*, p. 240.

155. Bertolt Brecht, "Sur la peinture chinoise," in Brecht, *Écrits sur la littérature et l'art* (Paris, 1970), vol. 2, p. 69 (my emphasis).

156. Zhang Yanyuan, *Lidai ming hua ji*, chap. 6, cited in Shio Sakanishi, *The Spirit of the Brush* (London, 1939), p. 40.

157. Mi Fu, *Huashi*, chap. 2, par. 160, according to Vandier-Nicolas, *Art et sagesse en Chine*, p. 149. Li Cheng, *zi* (alias) Xianxi, one of the greatest landscape painters of the period of the Five Dynasties, who died in 967. So many works were attributed to him in the Song period that Mi Fu declared that he would like to write "an essay on his non-existence." *Huashi*, chap. 3, par. 18, cited in Vandier-Nicolas, *Art et sagesse en Chine*, p. 34.

158. "The wind and the rain, obscurity and clarity constitute an atmospheric mood; dispersion and grouping, depth and distance constitute the schematic orga-nization; verticals and horizontals, hollows and relief create the rhythm; shade and light, thickness and fluidity create spiritual tension; rivers and clouds, clustered together or scattered, create the link; the contrast between crevices and outcrops create an alternation of action and withdrawal." Ryckmans, ed., *Les "Propos sur la peinture" de Shitao*, chap. 8, p. 63.

159. Ibid., p. 90, n. 2.

160. *Jie zi yuan huazhuan*, vol. 3, chap. 18, Petrucci, *Encyclopédie de la peinture chinoise*, pp. 165–66.

161. *Jie zi yuan huazhuan*, vol. 3, chap. 1, Petrucci, *Encyclopédie de la peinture chinoise*, pp. 121 ff.

162. *Jie zi yuan huazhuan*, vol. 3, chap. 21, Petrucci, *Encyclopédie de la peinture chinoise*, pp. 172–73.

163. "The Mountain, with its peaks rising one above the other, its succession of cliffs, its secret valleys and deep precipices, its high sharply pointed crags, its vapors, mists and dews, its wisps of smoke and clouds, puts one in mind of the unfurling flow and ebb of the Sea; but all that is not the soul manifested by the Sea itself; those are simply qualities of the Sea that the Mountain appropriates." Ryckmans, ed., *Les "Propos sur la peinture" de Shitao*, chap. 13, p. 89.

164. *Jie zi yuan huazhuan*, vol. 3, chap. 21, Petrucci, *Encyclopédie de la peinture chinoise*, p. 173.

165. Ryckmans, ed., *Les "Propos sur la peinture" de Shitao*, chap. 8, p. 64.

166. Zhang Yanyuan, *Lidai*, cited by Acker, *Some T'ang and Pre-T'ang Texts*, pp. 154–59. "Once one has understood that there are two styles of painting, one abbreviated (*shu*), the other detailed (*mi*), then one can engage in discussion about it." Gu Kaizhi, *zi* (alias) Chang Kang, was a calligrapher and portraitist of the Jin period (fourth century). Lu Tanwei worked in the fifth century, Zhang Sengyou in the sixth century. Wu Daozi (eighth century), the great ancestor of the "scholars," was famous for the speed with which he worked: he is supposed to have painted from memory, in a single day, a panorama of the River Jialing.

167. Ryckmans, ed., *Les "Propos sur la peinture" de Shitao*, p. 16.

168. Vandier-Nicolas, *Art et sagesse en Chine*, p. 16.

169. The text is attributed to the calligrapher Cai Dong (132–92), the inventor of the cursive (*bafen*) style; see Vandier-Nicolas, *Art et sagesse en Chine*, p. 55.

170. See *Xuanhe Shu Pu* (a catalogue of the autographs of the emperor Hui Zong, twelfth century), cited in Vandier-Nicolas, *Art et sagesse en Chine*, p. 55.

171. Dong Qichang, *Hua chansi suibi*, cited in Vandier-Nicolas, *Art et sagesse en Chine*, pp. 122–23.

172. Dong Qichang, *Hua yan*, cited in Vandier-Nicolas, *Art et sagesse en Chine*, p. 141.

173. Vandier-Nicolas, *Art et sagesse en Chine*, p. 64, according to a text borrowed from Shen Zongqian, a calligrapher and painter active in the eighteenth century.

174. "As for spattered ink, that is using ink delicately and subtly without allowing the brush strokes to appear, as if [the image] simply spurted forth." Li Ri-hua, a painter of the seventeenth century, cited in Vandier-Nicolas, *Art et sagesse en Chine*, p. 64.

175. Tang Zhiqi, the author of the *Huishi weiyan*, cited in Vandier-Nicolas, *Art et sagesse en Chine*, pp. 241–42.

176. Translator's note: The French word *trait* can mean either "trait" or "feature" and also "line."

177. Brecht, "Sur la peinture chinoise."

178. See Vandier-Nicolas, *Art et sagesse en Chine*, p. 219.

179. Huysmans, "Whistler," in *Certains*, p. 66.

180. See Meyer Schapiro, *Paul Cézanne* (London, 1988), p. 4.

181. Paul Cézanne, letter to Emile Bernard, 23 October 1905, in *Correspondance* (Paris, 1937), p. 276.

182. "The reading of the model and its realization is sometimes long in coming for the artist." Cézanne, letter to Charles Camoin, 9 December 1904, *Correspondance*, p. 267.

183. See Clement Greenberg, "Cézanne," in Greenberg, *Art and Culture* (Boston, 1961), p. 55.

184. See Cézanne, letter to Charles Camoin, *Correspondance*, p. 268; to Emile Bernard, 23 December 1904, *Correspondance*, p. 269; also to Emile Bernard, 23 October 1905, *Correspondance*, p. 276.

185. Schapiro, *Paul Cézanne*, p. 17.

186. Cézanne, letter to Emile Bernard, 23 October 1905, *Correspondance*, p. 277 (my emphasis).

187. "I regret my advanced age, on account of my coloring sensations." Cézanne, letter to his son, 3 August 1906, *Correspondance*, p. 281.

188. Letter to Emile Bernard, *Correspondance*, p. 277.

189. Letter to Ambroise Vollard, 9 January 1903, *Correspondance*, p. 252.

190. Mallarmé, *Oeuvres complètes* (Paris, Bibliothèque de la Pléiade), p. 455.

191. Lactantius, *Divinae institutiones* 3.17, cited by Maurice Solovine, in *Epicure: Doctrines et maximes* (Paris, 1937), pp. 118–19.

192. Joachim Gasquet, *Cézanne* (Paris, 1926), pp. 117–18.

193. See Meyer Schapiro, "On Some Problems in the Semiotics of Visual Art: Field and Vehicle in Image-Signs," *Semiotica* 1, no. 3 (1969): 223–24; reprinted in A. Kroeber, ed., *Theory and Philosophy of Art: Style, Artist, and Society: Selected Papers* (New York, 1994), pp. 1–3.

194. El Lissitzky, "New Russian Art: A Lecture" (1922); see Sophie Lissitzky-Küppers, *El Lissitzky* (London, 1968), p. 334.

Cultural Memory | *in the Present*

Sarah Winter, *Freud and the Institution of Psychoanalytic Knowledge*

Samuel Weber, *The Legend of Freud: Expanded Edition*

Aris Fioretos, ed., *The Solid Letter: Reading of Friedrich Hölderlin*

J. Hillis Miller / Manuel Asensi, *Black Holes / J. Hillis Miller; or, Boustrophedonic Reading*

Miryam Sas, *Fault Lines: Cultural Memory and Japanese Surrealism*

Peter Schwenger, *Fantasm and Fiction: On Textual Envisioning*

Didier Maleuvre, *Museum Memories: History, Technology, Art*

Jacques Derrida, *Monolingualism of the Other; or, The Prosthesis of Origin*

Andrew Baruch Wachtel, *Making a Nation, Breaking a Nation: Literature and Cultural Politics in Yugoslavia*

Niklas Luhmann, *Love as Passion: The Codification of Intimacy*

Mieke Bal, ed., *The Practice of Cultural Analysis: Exposing Interdisciplinary Interpretation*

Jacques Derrida and Gianni Vattimo, eds., *Religion*

The authorized representative in the EU for product safety and compliance is:
Mare Nostrum Group
B.V Doelen 72
4831 GR Breda
The Netherlands